Heat Stress in Sport and Exercise

Julien D. Périard • Sébastien Racinais
Editors

Heat Stress in Sport and Exercise

Thermophysiology of Health and Performance

Springer

Editors
Julien D. Périard
Research Institute for Sport & Exercise
University of Canberra
Canberra
ACT
Australia

Sébastien Racinais
Athlete Health and Performance
Aspetar
Doha
Qatar

ISBN 978-3-319-93514-0 ISBN 978-3-319-93515-7 (eBook)
https://doi.org/10.1007/978-3-319-93515-7

Library of Congress Control Number: 2019933902

This Springer imprint is published by the registered company Springer Nature Switzerland AG
The registered company address is: Gewerbestrasse 11, 6330 Cham, Switzerland

Foreword

I was lucky enough to be exposed to practical and academic experiences of the effects of heat stress on performance early on in my career and almost at the same time. The practical experience came in 2008 when I qualified for the Beijing Olympics as an outsider. Heat was rightly identified as a major factor for the race in Beijing, so I spent the few months before trying out various strategies and preparing for it. This mostly involved training in hot places, but we were also experimenting with sodium loading, pre-cooling and strategies to keep cool in the race, as well as tracking core temperature with an ingestible pill. At university in Leeds, I chose an environmental physiology module which taught me the basics of how the body responds to exercise in the heat and the adaptations that occur when training in hot environments. It was a subject that fascinated me, not only for its relevance but because I felt understanding it required a really good grasp of general physiology.

A few years on, at a race in London, I had my first run-in with heat stress. The race had gone well. With 400 m to go, I was running in second place on the shoulder of big rival, Javier Gomez. That's about the last thing I remember until I woke up on an emergency medical bed, covered in ice with wires and drip lines coming out of me. I was discharged relatively quickly and didn't feel right for months afterwards, especially in hot conditions. Later on that year, I travelled to Prof. Mike Tipton's lab for some testing to learn more about this mysterious illness and learnt that the recovery process is a long one.

Armed with the knowledge of how to acclimatise, prepare, get the nutrition right and moderate my effort, I haven't had another episode. I've definitely been close. My next experience of this dangerous condition came after a race in Australia. I was having a bad day and Jonny was racing for the win. With less than a kilometre to run, I saw him start to wobble. By the time I got to the finish (5 min later), Jonny was semi-unconscious on a stretcher in the sun. This was very concerning and I urged the medical staff to act quickly, aware of the dangers. The medical provision was poor, no ice, no drips and no cool area, and we had to supply shop-bought ice to cool him down. This concerned us both, and we appealed to the International Federation afterwards to implement minimum standards for medical provision.

Later the same year, after the Olympics in Rio, the now infamous event in Mexico happened. I was unlucky enough to see it play out in front of my eyes. Running along in ridiculously hot conditions (28 °C wet-bulb globe temperature), Jonny sped up to win the race. That spike in intensity was enough to push him over the edge, and a few kilometres later, I saw him stumbling around with the tell-tale signs of heat illness. Thankfully this time, the medical provision he received was exemplary. The effects of that day, probably compounded with the episode earlier in the year, took Jonny at least a year to get over.

After our personal experiences, Jonny and I consider heat illness to be a very dangerous condition and one that doesn't get the attention it deserves. Any further study and attention it garners is welcomed enthusiastically by both of us, and so it is a great honour to write the foreword for this book. Whether you are a researcher, student, coach or interested athlete, I hope you find it a valuable and enjoyable read.

Bramhope, UK

Alistair Brownlee
2× Olympic Gold Medallist (2012, 2016)
2× ITU Triathlon World Champion (2009, 2011)
3× European Champion (2010, 2011, 2014)
Commonwealth Games Champion (2014)

Contents

Part I

The Physiology of Heat Stress

Human Thermoregulation

1

Andreas D. Flouris

1.1 Introduction

Thermoregulation has always been an integral component of exercise physiology and it is becoming increasingly pertinent to sport performance due to the occurring climate change [1, 2]. This is because heat and temperature are two of the most fundamental properties of biological organisms, with a vast area of research and topics exploring fascinating connections with, and consequences to cells, tissues, and systems [3]. This all-encompassing nature of thermoregulation represents a major challenge. Most areas of physiology, as wide-ranging as they may be, have an inherently limited scope. But what does not fall under the influence of thermoregulation? What cell, tissue, system, or process can be understood or studied as devoid of it or unaffected by its change? As a result, body temperature is monitored in all hospitalized individuals because a deviation from its typical range (36.5–37.5 °C) is a clear indication of a pathological condition, while the clinical thermometer is one of the most essential instruments for the practice of medicine [4].

Those who wish to understand human thermoregulation and its impact on sport and exercise must acquire knowledge of nearly all bodily systems. This is exemplified by the topics covered in the first two sections of the present book, which range from systems theory and biophysics to neural, cardiovascular, behavioural, and pathophysiological mechanisms, as well as cellular fluid regulation and physiological adaptation to extreme environments. Not to mention, of course, the applied knowledge provided in the third section covering how heat exposure impacts performance in various sports and environments, the technical and biophysical countermeasures that can be used to mitigate its effects, as well as the relevant implications

A. D. Flouris (✉)
FAME Laboratory, Department of Exercise Science, University of Thessaly, Trikala, Greece

J. D. Périard, S. Racinais (eds.), *Heat Stress in Sport and Exercise*,
https://doi.org/10.1007/978-3-319-93515-7_1

for health policy. So, a book about heat stress in sports and exercise is about as fundamental, and hence as important, as it gets in human physiology.

In order to allow readers without a solid background in human thermoregulation to appreciate some of the finer points of the book provided in the following chapters, the present chapter includes an overview of the general principles of human thermoregulation at the systems level (i.e. the level of architecture and organization) as well as the relevant afferent and efferent thermoeffector pathways in the central and peripheral nervous systems. The focus will be primarily on systems and pathways related to warm/hot stimuli, as they are more pertinent to sports and exercise under heat stress.

1.2 Autonomic and Behavioural Thermoregulation

Like all endothermic organisms (those that regulate their body temperature through controlled responses in heat production and heat loss), humans achieve thermoregulation via autonomic (i.e. involuntary) and behavioural (i.e. voluntary) means. These are two entirely different branches of the thermoregulatory system, sharing the same sensors for detecting changes in thermal homeostasis but having discrete pathways for afferent and efferent information relay, as well as for central (i.e. brain) information processing (Fig. 1.1).

1.2.1 Autonomic Thermoregulation

Autonomic thermoregulation is based on sensing changes in temperature by thermosensors which are spread throughout the body and transmit afferent information

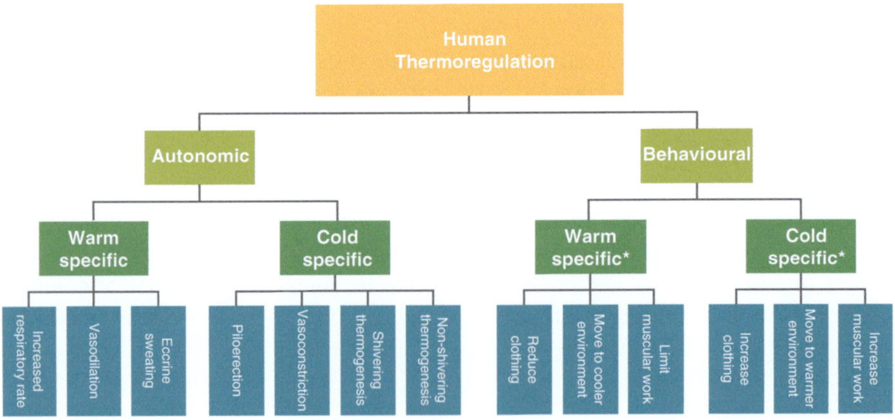

Fig. 1.1 The autonomic (involuntary) and behavioural (voluntary) branches of human thermoregulation together with warm- and cold-specific thermoeffector responses. The asterisks denote that the provided responses are some examples of the multitude of behavioural response options

to the central nervous system. These thermal signals are integrated and, subsequently, transmitted downstream via the autonomic nervous system to the thermoeffectors (e.g. cutaneous vasculature, sweat glands, skeletal muscle, white or brown adipose tissue) to initiate heat dissipating or heat conserving/generating responses [5]. Specifically, increases in peripheral and/or core temperature, which are typically due to exposure to a warmer environment and/or exercise/work-induced increased metabolic heat production, stimulate heat dissipation mechanisms; that is, eccrine sweating, skin vasodilation, and tachypnea [6–8]. On the other hand, reductions in peripheral and/or core temperature, which are typically due to exposure to a colder environment and/or removal of clothing insulation, stimulate heat conservation and generation mechanisms [5]. Heat conservation is achieved via peripheral/skin vasoconstriction, while heat generation occurs through shivering (which takes place in the muscle) [6–8] and non-shivering (which takes place in the muscle and the brown adipose tissue) [9–11] thermogenesis.

The thermoregulatory mechanisms which are most pertinent for sports and exercise under heat stress include eccrine sweat secretion and skin vasodilation. Eccrine sweat glands are cholinergically innervated tubular glands with a twisted coil and an undulating or coiled duct which leads to a sweat pore at the skin surface [12, 13]. There are 1.6–4 million glands in adults and are located in the dermis or hypodermis across almost the entire body [14, 15] (100–250 glands/cm^2 assuming equal distribution across the body). Variability in sweat gland distribution across body segments, secretion rates, thermal sensitivity, and sweating onset lead to regional differences in sweat output during both rest and exercise [16–20]. Typically, a greater number of eccrine sweat glands are found at the forehead, hand, foot, thigh, and leg [19]. Yet, sweat output varies highly even within regions—as has been previously demonstrated at the foot [21], head [22], and torso [22, 23]—with increasing variation appearing at higher levels of hyperthermia [22]. In total, the liquid secreted by eccrine glands can reach 4 L h^{-1} or 14 L day^{-1} (15 g min^{-1} m^{-2}) [24] and it is made of 99% water [25]. The evaporation of high energy water molecules in sweat from the skin surface leads to energy release from the body to the surrounding environment. This attenuates the rise in skin temperature and, subsequently, blood redistribution to cutaneous vessels, resulting in a cooling effect of 2426 J/g of sweat [26]. As such, evaporation of eccrine sweat is the most effective autonomic thermoregulatory mechanism during sports and exercise under heat stress.

Changes in cutaneous circulation can significantly augment heat loss—or minimize dry heat gain—during sports and exercise in hot environments. While resting in a thermoneutral environment, about 0.5 L min^{-1} of blood (5–10% of cardiac output) is supplying the cutaneous circulation, but during heat stress the cutaneous circulation receives up to 8 L min^{-1} (50–70% of cardiac output) [27] through an increase in and redistribution of cardiac output (e.g. a significant reduction in the splanchnic and renal blood flow during heat stress) [28]. To achieve such drastic changes in perfusion, the cutaneous vasculature is under dual autonomic nervous control [29, 30] containing both vasoconstrictor and vasodilator nerves [31, 32]. The afferent and efferent pathways involved in this process are described in detail in a subsequent section.

1.2.2 Behavioural Thermoregulation

Regulation of thermal homeostasis via behavioural means comprises of conscious decisions—controlled by the nervous system—aiming to defend or restore heat balance. In humans, these conscious decisions can involve a wide array of complex somatomotor activities [e.g. changing environment, posture, or microclimate (clothing), altering physical activity and/or food consumption] and, importantly, voluntary adaptation of work rate [33–35]. Behavioural thermoregulation reduces the requirement for autonomic thermoregulation, and it has been proposed as the first line of defence of heat balance [36]. This is based on the notion that autonomic thermoregulatory mechanisms are characterized by an inherent finite capacity for heat conservation/loss/production. In contrast, behavioural thermoregulation mechanisms have a near-infinite capacity to defend/restore heat balance.

Behavioural thermoregulation is a major element of exercise performance in the heat. This is because probably the most notable thermoregulatory response observed during exercise in a hot environment is a voluntary reduction in exercise work rate [33–35]. This has been confirmed in both laboratory [37–40] and real-life competition [41–45] studies where participants were in control of their exercise work rate [46–56] even at minimal levels of hyperthermia [48, 51–56]. The main aim of these voluntary reductions in work rate during exercise or work in the heat is to attenuate the rate of metabolic heat production [57, 53] and, thus, mitigate hyperthermia [35]. In a recent conceptual model (Fig. 1.2; [35]), we proposed that the behavioural drive to reduce exercise work rate in hot environments is driven primarily by increases in skin temperature which, in turn, generate: (1) changes in thermal perception [34], and (2) reductions in peak oxygen uptake due to increased cardiovascular strain [50, 58–61]. With regard to the first point, increased skin temperature is known to impact thermal perception (i.e. generates the feeling of being hot) via stimulation of peripheral thermosensors [35]. In turn, changes in thermal perception may in some cases lead to increased perceived exertion (i.e. the subjective feeling of fatigue) [62], leading to a voluntary reduction in exercise work rate [55, 63] by attenuating motivation to continue exercising in a hot environment [64]. Regarding the second point, increased skin temperature leads to an autonomic redistribution of blood towards the cutaneous vasculature (analysed in the previous section). This leads to reduced peak cardiac output and blood flow to the working muscles due to an inability to maintain blood pressure [65, 66]. In turn, peak oxygen uptake is attenuated [67–69] and the requirements of the exercise elicit a greater relative work intensity (i.e. percentage of peak oxygen uptake) and increased perceived exertion [50, 70, 71].

It is important to note that not all studies support the notion that increases in skin temperature generate changes in thermal perception (first point above) [72, 73]. Thus, it is likely that the relative contribution of thermal perception and cardiovascular strain in eliciting behavioural thermoregulatory responses in work output during exercise in the heat is dependent on the level of hyperthermia [35]. According to this notion, increased thermal discomfort and sensations of warmth generate reductions in self-selected exercise work rate during the early stages of self-paced exercise in the heat. As exercise progresses, leading to significant blood redistribution

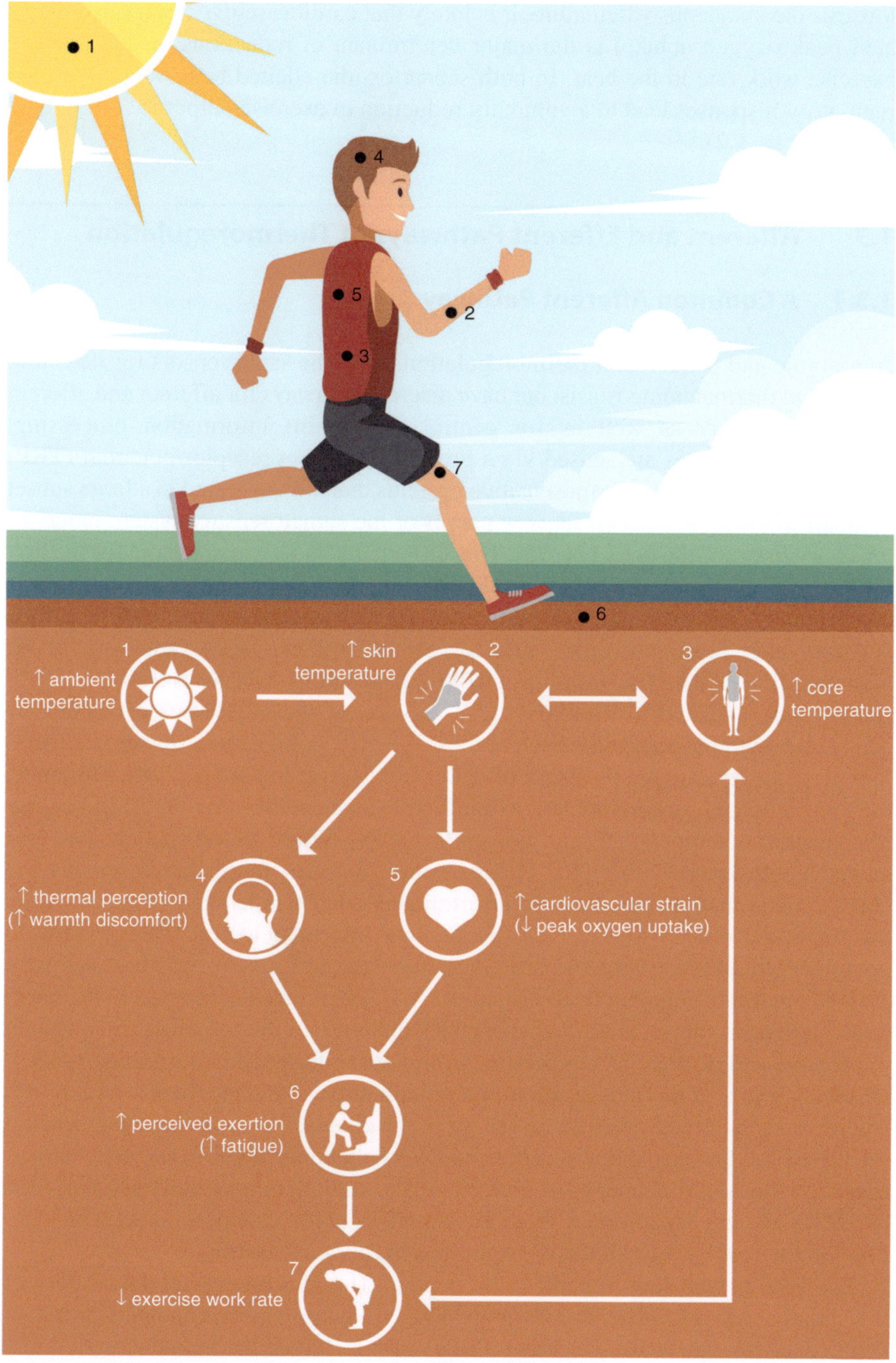

Fig. 1.2 A conceptual model of behavioural thermoregulation during exercise in hot environments illustrating how thermal perception and/or cardiovascular strain mediate reductions in exercise work rate in the heat through their impact on perceived exertion. The model has been previously proposed by Flouris and Schlader [35] and appears here redrawn and slightly modified

towards the cutaneous vasculature, it is likely that cardiovascular strain (i.e. attenuated peak oxygen uptake) is the major determinant of reductions in self-selected exercise work rate in the heat. In both scenarios, the elicited behavioural thermoregulatory responses lead to a voluntary reduction in exercise output to restore heat balance (Fig. 1.2).

1.3 Afferent and Efferent Pathways of Thermoregulation

1.3.1 A Common Afferent Pathway

Autonomic and behavioural thermoregulation share the same sensors for detecting changes in thermal homeostasis, but have discrete pathways for afferent and efferent information relay as well as for central (i.e. brain) information processing. Temperature changes are sensed via a group of transient receptor potential (TRP) ion channels, which are thermosensitive proteins that are expressed in a large subset of small-diameter peripheral sensory nerves of the somatosensory system innervating every tissue of the body [74, 75]. The TRP channels were discovered for the first time in 1997 [76] and, since then, our knowledge on their properties and function is continuously increasing. To date, a total of 28 TRP channels have been described and they have been categorized into seven sub-families: TRPA (ankyrin), TRPC (canonical), TRPML (mucolipin), TRPM (melastatin), TRPN (NOMPC), TRPP (polycystin), and TRPV (vanilloid) [77]. Several of these 28 types of nerves are activated by temperature, with each of these types responding to a specific temperature range generating the sensation of cold, cool, warm, and hot [77, 78]. Moreover, they are activated by certain plant extracts such as menthol, eucalyptol, and capsaicin generating temperature sensations [10, 11, 79]. While all ion channels are sensitive to temperature, the TRP channels are 25 times more thermosensitive, a characteristic that enables them to discriminate small temperature changes within the range of 0–60 °C [80–82]. Nevertheless, it is important to note that responses to noxious cold and heat remain partially active even when cold- and heat-sensitive TRP channels are completely abolished [83, 84]. This comes in contrast to the once-hoped universal role of thermosensitive TRP channels in the somatosensory system to perceive and distinguish temperatures, suggesting the existence of other molecular mechanisms of thermosensation that remain elusive to date. To this end, recent studies have identified several new groups of ion channels—including the mechanosensitive PIEZOs [85], the store-operated ORAI channels [86, 87], and the channels related to the ANO1 channel that is expressed in smooth muscle and epithelial cells [88–90]—which, in addition to other novel classes of ion channels that will be identified in the future, may fill the missing links in thermosensation.

The TRP nerves send signals to the lamina I (i.e. the superficial dorsal horn of the spinal cord) [91, 92] either through slow-conducting unmyelinated C fibres of small diameter (in warm-sensitive TRPs) or via larger, fast-conducting, thinly myelinated Aδ fibres (in cold-sensitive TRPs) [76, 77]. In turn, the output neurons of the lamina I relay their thermoafferent signals to the brainstem [91, 93]. Once the temperature-specific signals reach the brainstem, the information is diffused to a widely distributed network of brain loci that probably participate to the multifaceted processes (i.e. autonomic and behavioural) related to thermoregulation. Differences in methodology and outcomes across human studies conducted to date do not allow for conclusive statements regarding a putative thermoregulatory network in the brain. Despite the existing challenges, it is important to note that the functional processes of most brain regions proposed to date as part of the brain thermoregulatory network correlate with the properties and roles ascribed to them within thermoregulation: (1) the somatosensory, insular, and cingulate cortices for the integration of thermoafferent signals; (2) the insular and cingulate cortices for the autonomic thermoregulatory control; (3) the premotor and prefrontal cortices for the behavioural thermoregulatory control [94]. Importantly, neuroimaging studies suggest that these brain regions share functional connections when it comes to the coding of temperature-specific signals. The following sections describe in detail the thermoefferent pathways of autonomic and behavioural thermoregulation related to warm/hot stimuli, as they are more pertinent to sports and exercise under heat stress.

1.3.2 Efferent Pathways for Autonomic Thermoregulation to Warm Stimuli

1.3.2.1 Eccrine Sweating

The hypothesis that the preoptic hypothalamic area plays a central role in thermoregulation has received wide support for the past 50 years [36, 95, 96]. Yet, it is important to note that the majority of studies demonstrating the involvement of this region in thermoregulatory control have been conducted in animals [97]. The first neuroimaging study to implicate the human preoptic anterior area in thermoregulation was published in 2014 and provided concrete data that this region is a likely source of the thermal drive to sweat [97]. This finding was confirmed in a subsequent study published during 2015 [98]. Interestingly, activation of the preoptic hypothalamic area during thermal sweating is associated with a number of brain regions, suggesting functional connectivity with these areas when the body is heated [97]. These findings suggest that the preoptic anterior hypothalamic area is a central piece in a network of cortical loci that appear to contribute to thermoregulation. Due to the technical challenges involved in distinguishing neural tracks in the human brain, the precise neural pathways remain partly understood. Thus, our knowledge to date is based primarily on animal studies, the majority of which point towards the following paradigm: efferent information from the preoptic anterior area pass through the pontine tegmentum and the medullary raphe regions, they proceed to the ventral horn of the spinal cord, and then reach the sweat glands through sympathetic nonmyelinated C fibres [13, 99, 100]. A few of the sympathetic nerve terminals extend to the sweat duct, while the majority cluster around the sweat gland secretory coil [101]. They secrete acetylcholine which connects to muscarinic receptors on the sweat gland to induce sweating [102].

The requirement for increased sweat output is met by augmenting the population of activated sweat glands (near maximal recruitment can be achieved within 8 min [103]) and, if further increase is needed, the sweat output per gland [103]. The latter is modulated by the amount of acetylcholine released by the sympathetic nerve terminals as well as its breakdown rate by the enzyme acetylcholinesterase [104]. Under profuse sweating, the capacity of acetylcholinesterase to hydrolyze acetylcholine is attenuated due to its increased concentration in the synaptic cleft [104]. Finally, it is important to note that another factor affecting sweat rate is nitric oxide concentration, whereby increased nitric oxide release augments sweat gland output by influencing acetylcholine release from sympathetic sudomotor nerves [105].

1.3.2.2 Skin Vasodilation

Although the study of human efferent pathways to the skin vasculature has been increasing, they have not been completely characterized to date. This is in contrast to our knowledge about these pathways in the rat, where the majority of efferent pathways have been mapped, showing that the cutaneous vasculature is controlled by sympathetic neurons which receive input from the intermediolateral spinal column that, in turn, is mainly connected to the medullary raphe/peripyramidal region [106, 107]. In turn, these medullary sympathetic ganglia are controlled by pontine (i.e. locus coeruleus), midbrain (i.e. retrorubral field, ventral tegmental area, and periaqueductal gray matter), and hypothalamic (i.e. dorsomedial and paraventricular nuclei) neurons which are functionally connected to warm-sensitive neurons in the preoptic hypothalamic area [107–110].

As indicated above, the human cutaneous vasculature can rapidly achieve vast increases in blood flow as large as 16-fold between thermoneutral and hyperthermic conditions [27, 28]. To achieve such drastic changes in perfusion, the cutaneous vasculature is under dual autonomic nervous control [29, 30] containing both vasoconstrictor and vasodilator nerves [31, 32]. The initial increase in cutaneous blood flow during heat stress is achieved via sympathetic vasoconstrictor tone withdrawal, while further increase in skin perfusion is achieved by active cutaneous vasodilation [111]. The cutaneous vasoconstrictor system acts via the binding of norepinephrine to α1- and α2-adrenergic receptors in response to cold and for the brief attenuation in cutaneous blood flow observed at the start of intense exercise [112, 113]. On the other hand, active vasodilation depends on functional cholinergic fibres and requires a co-transmission of nitric oxide for maximum activation (i.e. 30–45% of maximum cutaneous active vasodilation is depended on nitric oxide) [29, 114–118]. Moreover, factors including prostaglandins [119], vasoactive intestinal peptide [120], substance P/NK-1 receptors [121], and H1 histamine receptor activation [122] are considered to play a role in this process. At this point, it is important to note that heat-induced cutaneous vasodilation through the sympathetic nervous system prevails over the need to supply blood to the working muscles. Therefore, the heat-induced peripheral vasodilation during exercise contributes to reduce peak cardiac output [66] and, as a consequence, blood flow to the working muscles [65]. As a consequence, peak oxygen uptake is attenuated during exercise in the heat [67–69].

1.3.3 Efferent Pathways for Behavioural Thermoregulation to Warm Stimuli

As indicated above, thermoafferent signals generated at the TRP ion channels are diffused to a number of brain regions that show functional interconnectivity/co-processing [94]. While the available data has yet to show a clear pattern of response [94], the brain regions that have been implicated to date in the regional brain responses generating behavioural adaptations to temperature changes are as follows:

- *Brainstem*: behavioural thermoregulation and other whole-body homeostatic behaviours are controlled by the mammalian brainstem [91, 123]. Thermoafferent signals reach the brainstem through output neurons of the lamina I [91, 93]. A functional human study showed that skin warming is linked with increased activity in the ventral midbrain, particularly in the ventral tegmental area [124] which, in animal studies, has been associated with hypothermic responses [125–127]. Another functional human study demonstrated increased activity throughout the rostral and caudal medulla during both warming and cooling [128]. These regions have also been associated with cold-induced vasoconstriction [129] in animals as well as thermal sweating in humans [130] and animals [131–133].
- *Primary and secondary somatosensory cortices*: thermal sensations and feelings are associated with activity in these cortical areas [134].
- *Anterior region of the mid cingulate cortex*: activity in this region has been implicated in coding the pleasantness or unpleasantness of thermal stimuli [128] as well as in thermal comfort/discomfort [135–137].
- *Orbitofrontal cortex*: increased activity in this region has been associated with sensations of pleasantness in response to warm stimuli applied to the hand [136].
- *Insula*: While this region is primarily linked with the processing of thermoafferent inputs and autonomic responses, it has also been implicated in monitoring interoceptive signals [94].

Given that some form of skeletal muscle activation plays a central role in the vast majority of behavioural thermoregulatory responses, it is logical to postulate that, once the above-mentioned extensive hemispheric activations occur leading to appropriate decision-making, behavioural responses materialize in the motor, premotor, and prefrontal cortices. This notion is supported by reports from functional human brain studies [138–141] demonstrating that whole body warming and cooling is associated with increased activity in the premotor and prefrontal cortices. Such responses have been interpreted as attempts to initiate movement in response to thermal sensations triggering actions to restore heat balance [94]. The descending axon potentials from the motor and premotor cortices reach the spinal cord either directly or indirectly through axons projecting to the brainstem which, then, conveys the information to the spinal cord [142]. Within the spinal cord, signals travel in the lateral column with axons projecting throughout to innervate proximal and distal muscles across the body [143, 142]. Once the axon potentials reach the skeletal muscle fibres—the effector organ of behavioural thermoregulation—acetylcholine is released to induce muscle fibre contractions which are the final step in the thermoefferent process of behavioural thermoregulation. The resulting locomotion materializes the conscious decision(s) aiming to defend/restore heat balance.

1.4 Onset, Sensitivity, and Maximum of Thermoeffector Responses

Autonomic [13, 144–149] and behavioural [34–36, 150–152] responses are known to be affected by both skin and core temperatures (though with potentially different weighing for each response). Thus, effector responses are often represented as a function of the change in mean body temperature—calculated by appropriately weighing skin and core temperatures [5]. In this light, each response is activated at an onset threshold of mean body temperature and increases proportionally thereafter until reaching a maximum [153, 154]. This is illustrated in Fig. 1.3, where mean body temperature increases (e.g. during exercise in the heat) and, at a given onset threshold, the heat loss response is activated (Fig. 1.3: point 1) Once activated, the intensity of the heat loss response increases linearly with the increase in mean body temperature, indicating the sensitivity of the response (i.e. the degree of change in the thermoeffector response for every unit of increase in mean body temperature;

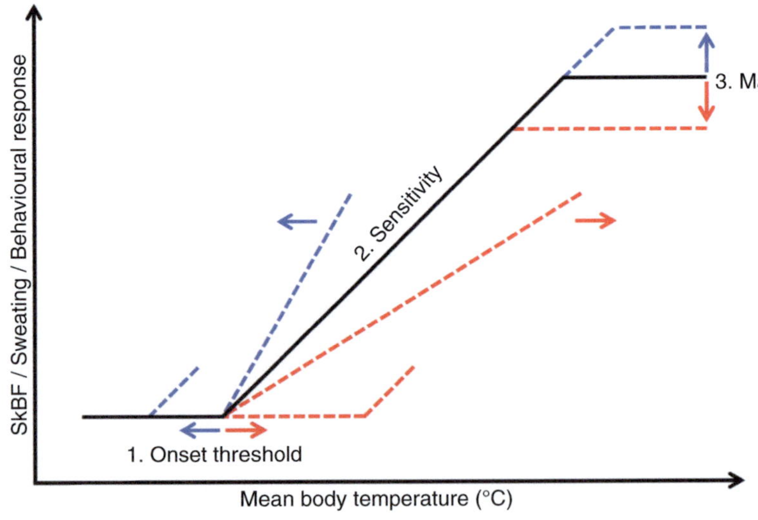

Fig. 1.3 A conceptual model of the relationship between thermoeffector responses and mean body temperature during heat stress. The model has been previously proposed by Gagnon and Kenny [154] and appears here redrawn and slightly modified. The black line illustrates the typical relationship. The dotted lines indicate enhanced (blue; e.g. observed in young and/or acclimatized individuals) or attenuated/delayed (red; e.g. observed in older and/or no-acclimatized individuals) onset, sensitivity, and maximum output of the thermoeffector responses

Fig. 1.3: point 2) If mean body temperature continues to increase, the heat loss response reaches a maximum whereby no further amplification occurs despite further increases in mean body temperature (Fig. 1.3: point 3) Previous research suggested that the range of temperature change from the onset of thermoeffector response to reaching a maximum is approximately 2–4 °C [155, 156]. Regarding sensitivity, a linear relationship between temperature change and effector response is typically reported [155, 156], with animal studies showing sensitivity coefficients of −4 to −6 W kg^{-1} $°C^{-1}$ of metabolic heat generation in response to cooling, as well as evaporative heat loss coefficients in the order of +1 to +3 W kg^{-1} $°C^{-1}$ in response to heating [155]. It has been suggested that the sensitivity values in humans are within the same ranges [156].

During exercise in the heat, heat loss effector responses adapt to changes in heat production, albeit with a brief delay [157]. This is illustrated in Fig. 1.4, showing data for heat production and loss from eight healthy male adults who spent 3 h in a hot (35 °C) dry (30% relative humidity) environment while wearing cotton shorts and t-shirt (Flouris, unpublished data). During the 1st hour, participants rested in a seated position, while during the 2nd and 3rd hour they were requested to perform cycling exercise at fixed rates of metabolic heat production equal to 300 W and 450 W, respectively. Exercise is a wonderful stimulus to study such changes in heat production and loss because it provides a large change, abrupt enough to be considered a 'step change'. When an individual begins to walk, run, or cycle, the increased

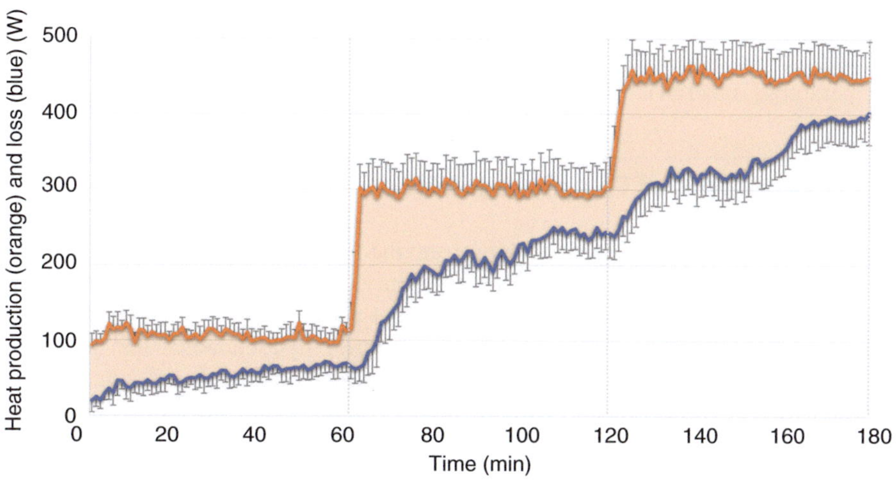

Fig. 1.4 Partitional calorimetry data (mean ± SD) for the rate of net heat production (metabolic heat production minus work; M-W) and loss from eight healthy male adults who spent 3 h in a hot (35 °C) dry (30% relative humidity) environment while wearing cotton shorts and t-shirt (Flouris, unpublished data). During the 1st hour, participants rested in a seated position, while during the 2nd and 3rd hour they were requested to perform cycling exercise at fixed rates of metabolic heat production equal to 300 W and 450 W, respectively. The shaded area represents the rate of residual body heat storage, while vertical dotted lines identify changes in work rate reflecting the protocol timeline

activity of muscle requires more oxidation of fuel and more removal of heat and waste. Within seconds, heart rate and oxygen uptake start to increase, and within 2–3 min new steady states are achieved [158]. As illustrated in Fig. 1.4, following the onset of exercise, heat loss is much slower to increase. This delay, sometimes called inertia [159] or temporal dissociation [158], is followed by a gradual increase in heat loss via eccrine sweating and vasodilation mechanisms until a new plateau is finally reached in 45–60 min. The greater the magnitude and duration of the mismatch between the rate of heat production and the rate of heat loss, the greater the change in body heat content and, thus, the increase in core temperature. As described in the previous paragraph, the duration of the mismatch is influenced by the onset, the sensitivity, and the maximum of the individual's heat loss effector responses. If the conditions are compensable (e.g. heat loss not restricted due to protective garments and/or high humidity), heat balance will be achieved and rate of heat storage will return to zero. Heat balance is not achieved in the example provided in Fig. 1.4, yet heat loss has almost matched heat production towards the end of each 1-h stage. Had the stages been longer (e.g. 2 h), heat balance may had been reached. In that case, core temperature would be maintained at an elevated albeit steady-state value so long as sweat production would not be compromised due to dehydration [5]. However, if heat dissipation from the body is restricted (e.g. high humidity conditions, insulative effect of clothing), this will lead to uncompensable heat stress and a more pronounced increase in body heat storage despite maximum activation of heat loss responses. It is also important to note that factors such as age and/or disease (both acute and chronic) which have been shown to cause impairments in thermoeffector activity can compromise the body's ability to dissipate heat thereby resulting a prolonged thermal imbalance and progressive increase in mean body temperature [160–164].

1.5 Concepts of Thermoregulation

The genesis of intensive experimentation and theory regarding the functional architecture of the human thermoregulatory system is marked [165] by the influential experiments of Barbour [166], who showed that cooling a rabbit brainstem region near to what is termed today the preoptic anterior hypothalamus had profound effects on rectal temperature and that the animal demonstrated analogous behavioural responses. This discovery fitted exquisitely within the theory of 'brain centres'—that is, specific areas in the brain eliciting topical control of somatosensory perception and somatomotor activity—which prevailed during the beginning of the twentieth century and determined the direction of mainstream research on thermoregulation for the next 50 years. Since the 1960s, there have been four major concepts of human thermoregulation and they will be presented in detail in the following pages. To date, none of these models have been unanimously accepted [167]. Each one has inherent advantages and limitations in their attempts to explain the various phenomena of human nature and, after several decades of continuous research, the precise mechanisms of human thermoregulation remain elusive [167].

1.5.1 Hypothalamic Proportional Control Around an Adjustable Set Point

In 1960, Hammel and colleagues conducted hypothalamic cooling and heating in conscious dogs and found significant effects on core temperature, vasoconstriction, shivering, and panting [95]. Interestingly, although alternating or sustained hypothalamic heating and cooling demonstrated qualitative consistency, it revealed quantitative variation with the induced thermoeffector response resulting in changes in core temperature opposite to that of hypothalamic temperature [95]. Based on these and other findings [96, 168, 169], Hammel's group proposed the concept of hypothalamic proportional control of body temperature around an adjustable set point. The set point, in this case, is viewed as a complex variable reflecting the activity of hypothalamic neurons as a function of temperature at specific body regions (or the entire body) [169]. In this light, thermosensors convert temperatures of the different regions of the body into impulses in the form of a neural code, conveying the information of objects to the brain where these impulses are integrated by a separate network to form a mean temperature. Thereafter, this integrating network compares the mean temperature with a reference signal—that is, a set point—and through various processes (which have yet to be elucidated) orders are sent to thermoeffectors to evoke appropriate thermoregulatory responses. Hammel's paper evaluated precisely the skin cold and warm sensors as thermal inputs influencing the temperature set point and showed, for the first time, that it was possible to quantify their systematic input through mean skin temperature [96] and, therefore, present the temperature set point as a function (amongst other variables) of mean skin temperature [170].

The eloquent model of proportional hypothalamic control around an adjustable set point quickly received support, but also some criticism, particularly on mean skin temperature and hypothalamic temperature being the only pertinent regions generating thermal inputs. In the years to follow, Hammel's group studied the parameters thought to affect the set point, namely ambient and core temperatures [171], sleep-wakefulness cycle [96], hibernation [172, 173], exercise [174, 175], and fever [175] were evaluated as modulators of the set point with respect to thermoeffector mechanisms.

The idea of a common set point in endothermic temperature regulation has been challenged by experiments on extrahypothalamic thermoregulatory functions, and particularly that of the spinal cord. Several studies examined different deep-body compartments in order to detect extra-cerebral deep-body thermosensitivity based on the initial observation by Simon and colleagues of shivering in response to cooling of the peridural space of the vertebral canal and of the spinal cord [176]. This notion was in contrast to the popular view which favoured the straightforward and, hence, convenient idea of only one region of deep-body thermosensitivity. Further criticism for the proportional hypothalamic control theory came from more recent experiments demonstrating that the preoptic anterior hypothalamus signals for vasomotion, shivering, and salivation can be dissimilar and functionally independent [177–180]. Arguments against the adjustable set point theory stemmed also from research demonstrating the markedly disparate inter-threshold zones between heat and cold defence activation found in circumstances of thermoregulatory abnormalities [181].

1.5.2 The Comparator Model

Kobayashi argues that, for warm- and cold-sensitive neurons to be sensors, their firing rate (i.e. impulses·sec^{-1}) must be a code by which these neurons convey local temperature information [182] to the central nervous system. For this assumption to be correct the neural code of firing rate must be directly related to temperature and the receivers of these impulses must possess the ability to decipher the code of firing rate and detect the local temperature information. Yet, as demonstrated by Kobayashi [182, 183], no convincing evidence has been presented hitherto confirming either of these assumptions.

Based on the comparator model, heat- and cold-sensitive neurons are comparators of temperature eliciting impulses independently when local temperature surpasses a sensor activation threshold. In turn, these impulses are not a form of neural code (as assumed by the set point model) but only signals to activate thermoeffectors [182–185]. In other words, thermosensory information is relayed to the brain not for integration, computation, or 'decision-making' but solely for sending appropriate orders to the effector organs (since each comparator cannot directly communicate with the relevant thermoeffector).

Regarding the contribution of core and shell temperatures to thermoregulation, the comparator model holds that core temperature is a very stable regulated variable serving as feedback signal, while shell temperatures are highly variable, unregulated parameters serving as feed-forward signals that allow the body to respond to a thermal load in advance to avoid changes in core temperature [186]. Both core and shell temperatures trigger thermoeffector responses in a similar fashion. The combination of core and shell temperature activating a specific thermoeffector depends on the central and peripheral sensory neurons wired to it.

1.5.3 Core Temperature Defence at a Null Zone

Based on the null zone model [153], there must be some degree of variation even in the most simplistic regulatory processes. Inevitably, this will have consequences on the neuronal communication between sensor-to-effector pathways [153]. The basic assumption of null zone model is that the cold- and warm-sensitive neuron inputs are each separately used to produce two thresholds: one for heat production effectors and the other for heat loss effectors [153]. Within the central nervous system these two sensor-to-effector channels are reciprocally crossed with other excitatory and inhibitory inputs from nonthermal channels. As a result, a range of thermoneutral zone is defined—which differs depending on the organism as a consequence of the effect of nonthermal factors on the vasomotor response [186]—where there is no need for any thermoregulatory response to occur. When the capacity of the vasomotor response to maintain a constant core temperature is exceeded, autonomic sweating or shivering is activated. The core temperature at which these effectors are activated is classified as the thermoeffector threshold core temperature [187].

The concept of core temperature defence at a null zone does not support the probability of body temperature regulation at a precise level, that is, a set point. In contrast, it postulates that body temperature regulation is coarse, permitting core temperature under normal physiological conditions to vary within the null zone. Additionally, proponents of this theory believe that "...clear and unequivocal neuronal evidence supportive of the reference signal hypothesis has remained elusive, yet the notion of a stable reference signal with which the variable is being constantly compared remains a popular way of considering the physiology of homeothermy" [153]. Yet, this remains to be confirmed as the null zone model is still more theoretical than proven.

1.5.4 Heat Regulation

The model of heat regulation holds that the human body maintains a heat content equilibrium over a wide range of heat loads by sensing heat flow to/from the body and subsequently defending the body heat content through thermoeffector responses [158, 188]. Therefore, the main difference between the model of heat regulation and the model of temperature set point pertains to the regulated variable based on which thermoeffector responses are triggered. Helped by the easiness to obtain them, temperature data—and the hypothalamic proportional control model—have dominated thermal physiology since the first studies on the topic. As calorimetric techniques are improving, the concept of heat regulation is receiving more attention [189]. However, it is important to note that the heat regulation model is an attempt to explain human (endothermic) thermoregulation employing only physiological canons and thermodynamic principles, and particularly those governing the heat balance equation. Little, or nothing, is mentioned regarding the efferent neural control of thermoeffectors, thus one is left to assume that the hypothalamic proportional control and the heat regulation models share views regarding the efferent pathways to thermoeffectors.

The main driving force behind the heat regulation model has been Paul Webb who wrote several articles in favour of this hypothesis, the most prominent of which was his 1995 review paper [158]. Nevertheless, the concept of heat regulation was proposed several times by others, including Snellen—"...the regulating mechanism operates as if there is heat regulation instead of temperature regulation" [190]—and Adolph—"...in heat regulation, the most general links are between (a) heat content of the body and parts (or net temperature), (b) heat additions, and (c) heat dissipations" [191], as well as others who dealt directly with the concept of heat regulation [159, 192].

The central premise of heat regulation is that the body constantly generates heat and has a characteristic range of temperature. Heat generation varies constantly according to the requirements of daily living, yet a relatively constant temperature implies that heat loss is adjusted to match heat production. Thus, proponents of heat regulation argue that it is more likely that the 'controller' regulates a matching of outflow to inflow of heat than a mean body temperature or some combination of body temperatures [158]. The heat regulation model is exceptionally successful in explaining/predicting thermal responses to some phenomena that are difficult to explain by the hypothalamic proportional control model including thermoeffector responses during: (1) the circadian temperature rhythm [158, 193–198], (2) steady-state exercise [158], (3) incremental exercise [157], (4) cold water immersion [199], (5) heat exposure following precooling [158, 200, 201], (6) hot water immersion following cold water immersion [202–205], (7) core cooling at fixed skin temperature [158], and (8) negative work [158].

Notwithstanding the problems in the hypothalamic proportional control model evaded by the heat regulation model, it is necessary to observe that proponents of the latter are faced with an even more difficult problem, that is, how body heat content as an extensive property might be measured. Based on the model of heat regulation, heat flow in the body is sensed by monitoring transcutaneous thermal gradient via temperature-sensitive neurons located on the skin surface and in different subcutaneous tissue layers [158]. Such an arrangement has been, indeed, detected [188, 200, 206, 207] and changes in the transcutaneous thermal gradient have been related to regulatory processes [207–210], yet it has not been confirmed that they allow the monitoring of heat flow. However, even if one accepts that possibility, the fact that these neurons are not distributed uniformly around the body [211] prohibits a precise assessment of body heat content.

1.6 Conclusion

As shown in this chapter, remarkable progress has been made during the last 100 years regarding the principles of human thermoregulation as well as the relevant afferent and efferent thermoeffector pathways in the central and peripheral nervous systems. Some important concepts do remain elusive, yet it is hoped that the enormous efforts continuously being made by scientists interested in the field of thermoregulation will eventually address the missing knowledge gaps.

References

1. Smith KR, Woodward A, Lemke B, Otto M, Chang CJ, Mance AA, et al. The last summer Olympics? Climate change, health, and work outdoors. Lancet. 2016;388(10045):642–4.
2. Flouris AD, Kenny GP. Heat remains unaccounted for in thermal physiology and climate change research. F1000Research. 2017;6:221.
3. Flouris AD, Piantoni C. Links between thermoregulation and aging in endotherms and ectotherms. Temperature. 2015;2(1):73–85.
4. Burton AC. The clinical importance of the physiology of temperature regulation. Can Med Assoc J. 1956;75(9):715–20.

5. Kenny GP, Flouris AD. The human thermoregulatory system and its response to thermal stress. In: Wang F, Gao C, editors. Protective clothing: managing thermal stress. Cambridge: Woodhead Publishing Limited; 2014.
6. Hardy JD. Physiology of temperature regulation. Physiol Rev. 1961;41:521–606.
7. Hammel HT. Regulation of internal body temperature. Annu Rev Physiol. 1968;30:641–710.
8. Benzinger TH. Heat regulation: homeostasis of central temperature in man. Physiol Rev. 1969;49(4):671–759.
9. Flouris AD, Dinas PC, Valente A, Andrade CMB, Kawashita NH, Sakellariou P. Exercise-induced effects on UCP1 expression in classical brown adipose tissue: a systematic review. Horm Mol Biol Clin Invest. 2017;31:2.
10. Sakellariou P, Valente A, Carrillo AE, Metsios GS, Nadolnik L, Jamurtas AZ, et al. Chronic l-menthol-induced browning of white adipose tissue hypothesis: a putative therapeutic regime for combating obesity and improving metabolic health. Med Hypotheses. 2016;93:21–6.
11. Valente A, Jamurtas AZ, Koutedakis Y, Flouris AD. Molecular pathways linking non-shivering thermogenesis and obesity: focusing on brown adipose tissue development. Biol Rev Camb Philos Soc. 2014;90:77–88.
12. Sato K, Kang WH, Saga K, Sato KT. Biology of sweat glands and their disorders. I. Normal sweat gland function. J Am Acad Dermatol. 1989;20(4):537–63.
13. Shibasaki M, Wilson TE, Crandall CG. Neural control and mechanisms of eccrine sweating during heat stress and exercise. J Appl Physiol. 2006;100(5):1692–701.
14. Sato K. The physiology, pharmacology, and biochemistry of the eccrine sweat gland. Rev Physiol Biochem Pharmacol. 1977;79:51–131.
15. Sato K, Dobson RL. Regional and individual variations in the function of the human eccrine sweat gland. J Invest Dermatol. 1970;54(6):443–9.
16. Weiner JS. The regional distribution of sweating. J Physiol. 1945;104(1):32–40.
17. Hertzman AB, Randall WC, Peiss CN, Seckendorf R. Regional rates of evaporation from the skin at various environmental temperatures. J Appl Physiol. 1952;5(4):153–61.
18. Park SJ, Tamura T. Distribution of evaporation rate on human body surface. Ann Physiol Anthropol. 1992;11(6):593–609.
19. Cotter JD, Taylor NA. The distribution of cutaneous sudomotor and alliesthesial thermosensitivity in mildly heat-stressed humans: an open-loop approach. J Physiol. 2005;565(Pt 1):335–45.
20. Cotter JD, Patterson MJ, Taylor NA. The topography of eccrine sweating in humans during exercise. Eur J Appl Physiol Occup Physiol. 1995;71(6):549–54.
21. Taylor NA, Caldwell JN, Mekjavic IB. The sweating foot: local differences in sweat secretion during exercise-induced hyperthermia. Aviat Space Environ Med. 2006;77(10):1020–7.
22. Machado-Moreira CA, Wilmink F, Meijer A, Mekjavic IB, Taylor NA. Local differences in sweat secretion from the head during rest and exercise in the heat. Eur J Appl Physiol. 2008;104(2):257–64.
23. Havenith G, Fogarty A, Bartlett R, Smith CJ, Ventenat V. Male and female upper body sweat distribution during running measured with technical absorbents. Eur J Appl Physiol. 2008;104(2):245–55.
24. Mack GW, Nadel ER. Body fluid balance during heat stress in humans. In: Fregly MJ, Blatteis CM, editors. Handbook of physiology. Section 4: environmental physiology. New York: Oxford University Press; 1996. p. 187–214.
25. Robinson S, Robinson AH. Chemical composition of sweat. Physiol Rev. 1954;34(2):202–20.
26. Wenger CB. Heat of evaporation of sweat: thermodynamic considerations. J Appl Physiol. 1972;32(4):456–9.
27. Lossius K, Eriksen M, Walløe L. Fluctuations in blood flow to acral skin in humans: connection with heart rate and blood pressure variability. J Physiol. 1993;460:641–55.
28. Brubaker PH, Kaminsky LA, Whaley MH. Coronary artery disease: essentials of prevention and rehabilitation programs. Champaign: Human Kinetics; 2002.
29. Kellogg DL Jr. In vivo mechanisms of cutaneous vasodilation and vasoconstriction in humans during thermoregulatory challenges. J Appl Physiol. 2006;100(5):1709–18.

30. Kellogg DL Jr, Johnson JM, Kosiba WA. Selective abolition of adrenergic vasoconstrictor responses in skin by local iontophoresis of bretylium. Am J Phys. 1989;257(5 Pt 2):H1599–606.
31. Roddie IC, Shepherd JT, Whelan RF. The contribution of constrictor and dilator nerves to the skin vasodilatation during body heating. J Physiol. 1957;136(3):489–97.
32. Edholm OG, Fox RH, Macpherson RK. Vasomotor control of the cutaneous blood vessels in the human forearm. J Physiol. 1957;139(3):455–65.
33. Flouris AD. Functional architecture of behavioural thermoregulation. Eur J Appl Physiol. 2010;111(1):1–8.
34. Schlader ZJ, Stannard SR, Mundel T. Human thermoregulatory behavior during rest and exercise - a prospective review. Physiol Behav. 2010;99(3):269–75.
35. Flouris AD, Schlader ZJ. Human behavioral thermoregulation during exercise in the heat. Scand J Med Sci Sports. 2015;25(Suppl 1):52–64.
36. Flouris AD. Functional architecture of behavioural thermoregulation. Eur J Appl Physiol. 2011;111(1):1–8.
37. MacDougall JD, Reddan WG, Layton CR, Dempsey JA. Effects of metabolic hyperthermia on performance during heavy prolonged exercise. J Appl Physiol. 1974;36(5):538–44.
38. Galloway SD, Maughan RJ. Effects of ambient temperature on the capacity to perform prolonged cycle exercise in man. Med Sci Sports Exerc. 1997;29(9):1240–9.
39. Parkin JM, Carey MF, Zhao S, Febbraio MA. Effect of ambient temperature on human skeletal muscle metabolism during fatiguing submaximal exercise. J Appl Physiol. 1999;86(3):902–8.
40. Gonzalez-Alonso J, Teller C, Andersen SL, Jensen FB, Hyldig T, Nielsen B. Influence of body temperature on the development of fatigue during prolonged exercise in the heat. J Appl Physiol. 1999;86(3):1032–9.
41. Ely MR, Cheuvront SN, Montain SJ. Neither cloud cover nor low solar loads are associated with fast marathon performance. Med Sci Sports Exerc. 2007;39(11):2029–35.
42. Ely MR, Cheuvront SN, Roberts WO, Montain SJ. Impact of weather on marathon-running performance. Med Sci Sports Exerc. 2007;39(3):487–93.
43. Ely MR, Martin DE, Cheuvront SN, Montain SJ. Effect of ambient temperature on marathon pacing is dependent on runner ability. Med Sci Sports Exerc. 2008;40(9):1675–80.
44. Vihma T. Effects of weather on the performance of marathon runners. Int J Biometeorol. 2010;54(3):297–306.
45. Wegelin JA, Hoffman MD. Variables associated with odds of finishing and finish time in a 161-km ultramarathon. Eur J Appl Physiol. 2011;111(1):145–53.
46. Abbiss CR, Burnett A, Nosaka K, Green JP, Foster JK, Laursen PB. Effect of hot versus cold climates on power output, muscle activation, and perceived fatigue during a dynamic 100-km cycling trial. J Sports Sci. 2010;28(2):117–25.
47. Altareki N, Drust B, Atkinson G, Cable T, Gregson W. Effects of environmental heat stress (35 degrees C) with simulated air movement on the thermoregulatory responses during a 4-km cycling time trial. Int J Sports Med. 2009;30(1):9–15.
48. Ely BR, Cheuvront SN, Kenefick RW, Sawka MN. Aerobic performance is degraded, despite modest hyperthermia, in hot environments. Med Sci Sports Exerc. 2010;42(1):135–41.
49. Ely BR, Ely MR, Cheuvront SN, Kenefick RW, Degroot DW, Montain SJ. Evidence against a 40 degrees C core temperature threshold for fatigue in humans. J Appl Physiol. 2009;107(5):1519–25.
50. Periard JD, Cramer MN, Chapman PG, Caillaud C, Thompson MW. Cardiovascular strain impairs prolonged self-paced exercise in the heat. Exp Physiol. 2011;96(2):134–44.
51. Tatterson AJ, Hahn AG, Martin DT, Febbraio MA. Effects of heat stress on physiological responses and exercise performance in elite cyclists. J Sci Med Sport. 2000;3(2):186–93.
52. Tucker R, Rauch L, Harley YX, Noakes TD. Impaired exercise performance in the heat is associated with an anticipatory reduction in skeletal muscle recruitment. Pflugers Arch. 2004;448(4):422–30.
53. Schlader ZJ, Stannard SR, Mundel T. Evidence for thermoregulatory behavior during self-paced exercise in the heat. J Therm Biol. 2011;36:390–6.

54. Schlader ZJ, Stannard SR, Mundel T. Is peak oxygen uptake a determinant of moderate-duration self-paced exercise performance in the heat? Appl Physiol Nutr Metab. 2011;36(6):863–72.
55. Tucker R, Marle T, Lambert EV, Noakes TD. The rate of heat storage mediates an anticipatory reduction in exercise intensity during cycling at a fixed rating of perceived exertion. J Physiol. 2006;574(Pt 3):905–15.
56. Lorenzo S, Halliwill JR, Sawka MN, Minson CT. Heat acclimation improves exercise performance. J Appl Physiol. 2010;109(4):1140–7.
57. Schlader ZJ, Raman A, Morton RH, Stannard SR, Mundel T. Exercise modality modulates body temperature regulation during exercise in uncompensable heat stress. Eur J Appl Physiol. 2011;111(5):757–66.
58. Cheuvront SN, Kenefick RW, Montain SJ, Sawka MN. Mechanisms of aerobic performance impairment with heat stress and dehydration. J Appl Physiol. 2010;109(6):1989–95.
59. Sawka MN, Cheuvront SN, Kenefick RW. High skin temperature and hypohydration impair aerobic performance. Exp Physiol. 2012;97(3):327–32.
60. Sawka MN, Leon LR, Montain SJ, Sonna LA. Integrated physiological mechanisms of exercise performance, adaptation, and maladaptation to heat stress. Compr Physiol. 2011;1(4):1883–928.
61. Periard JD, Racinais S. Self-paced exercise in hot and cool conditions is associated with the maintenance of % VO2 peak within a narrow range. J Appl Physiol. 2015;118(10):1258–65.
62. Pandolf KB. Differentiated ratings of perceived exertion during physical exercise. Med Sci Sports Exerc. 1982;14(5):397–405.
63. Schlader ZJ, Simmons SE, Stannard SR, Mundel T. The independent roles of temperature and thermal perception in the control of human thermoregulatory behavior. Physiol Behav. 2011;103(2):217–24.
64. Cotter JD, Sleivert GG, Roberts WS, Febbraio MA. Effect of pre-cooling, with and without thigh cooling, on strain and endurance exercise performance in the heat. Comp Biochem Physiol A Mol Integr Physiol. 2001;128(4):667–77.
65. Gonzalez-Alonso J, Calbet JAL. Reductions in systemic and skeletal muscle blood flow and oxygen delivery limit maximal aerobic capacity in humans. Circulation. 2003;107(6):824–30.
66. Rowell LB, Marx HJ, Bruce RA, Conn RD, Kusumi F. Reductions in cardiac output, central blood volume, and stroke volume with thermal stress in normal men during exercise. J Clin Invest. 1966;45(11):1801–16.
67. Arngrimsson SA, Petitt DS, Borrani F, Skinner KA, Cureton KJ. Hyperthermia and maximal oxygen uptake in men and women. Eur J Appl Physiol. 2004;92(4-5):524–32.
68. Arngrimsson SA, Stewart DJ, Borrani F, Skinner KA, Cureton KJ. Relation of heart rate to percent VO2 peak during submaximal exercise in the heat. J Appl Physiol. 2003;94(3):1162–8.
69. Wingo JE, Lafrenz AJ, Ganio MS, Edwards GL, Cureton KJ. Cardiovascular drift is related to reduced maximal oxygen uptake during heat stress. Med Sci Sports Exerc. 2005;37(2):248–55.
70. Berry MJ, Weyrich AS, Robergs RA, Krause KM, Ingalls CP. Ratings of perceived exertion in individuals with varying fitness levels during walking and running. Eur J Appl Physiol. 1989;58(5):494–9.
71. Pandolf KB, Billings DS, Drolet LL, Pimental NA, Sawka MN. Differentiated ratings of perceived exertion and various physiological-responses during prolonged upper and lower body exercise. Eur J Appl Physiol. 1984;53(1):5–11.
72. Hartley GL, Flouris AD, Plyley MJ, Cheung SS. The effect of a covert manipulation of ambient temperature on heat storage and voluntary exercise intensity. Physiol Behav. 2012;105(5):1194–201.
73. Barwood MJ, Corbett J, White D, James J. Early change in thermal perception is not a driver of anticipatory exercise pacing in the heat. Br J Sports Med. 2012;46(13):936–42.
74. Fink GD. Hypothesis: the systemic circulation as a regulated free-market economy. A new approach for understanding the long-term control of blood pressure. Clin Exp Pharmacol Physiol. 2005;32(5-6):377–83.

75. Guler AD, Lee H, Iida T, Shimizu I, Tominaga M, Caterina M. Heat-evoked activation of the ion channel, TRPV4. J Neurosci. 2002;22(15):6408–14.
76. Caterina MJ, Schumacher MA, Tominaga M, Rosen TA, Levine JD, Julius D. The capsaicin receptor: a heat-activated ion channel in the pain pathway. Nature. 1997;389(6653):816–24.
77. Feng Q. Temperature sensing by thermal TRP channels: thermodynamic basis and molecular insights. Curr Top Membr. 2014;74:19–50.
78. Caterina MJ. Transient receptor potential ion channels as participants in thermosensation and thermoregulation. Am J Physiol Regul Integr Comp Physiol. 2007;292(1):R64–76.
79. Valente A, Carrillo AE, Tzatzarakis MN, Vakonaki E, Tsatsakis AM, Kenny GP, et al. The absorption and metabolism of a single L-menthol oral versus skin administration: effects on thermogenesis and metabolic rate. Food Chem Toxicol. 2015;86:262–73.
80. Yao J, Liu B, Qin F. Kinetic and energetic analysis of thermally activated TRPV1 channels. Biophys J. 2010;99(6):1743–53.
81. Yao J, Liu B, Qin F. Pore turret of thermal TRP channels is not essential for temperature sensing. Proc Natl Acad Sci U S A. 2010;107(32):E125.
82. Yao J, Liu B, Qin F. Modular thermal sensors in temperature-gated transient receptor potential (TRP) channels. Proc Natl Acad Sci U S A. 2011;108(27):11109–14.
83. Vriens J, Owsianik G, Hofmann T, Philipp SE, Stab J, Chen X, et al. TRPM3 is a nociceptor channel involved in the detection of noxious heat. Neuron. 2011;70(3):482–94.
84. Knowlton WM, Bifolck-Fisher A, Bautista DM, McKemy DD. TRPM8, but not TRPA1, is required for neural and behavioral responses to acute noxious cold temperatures and cold-mimetics in vivo. Pain. 2010;150(2):340–50.
85. Coste B, Mathur J, Schmidt M, Earley TJ, Ranade S, Petrus MJ, et al. Piezo1 and Piezo2 are essential components of distinct mechanically activated cation channels. Science. 2010;330(6000):55–60.
86. Feske S, Gwack Y, Prakriya M, Srikanth S, Puppel SH, Tanasa B, et al. A mutation in Orai1 causes immune deficiency by abrogating CRAC channel function. Nature. 2006;441(7090):179–85.
87. Vig M, Peinelt C, Beck A, Koomoa DL, Rabah D, Koblan-Huberson M, et al. CRACM1 is a plasma membrane protein essential for store-operated Ca2+ entry. Science. 2006;312(5777):1220–3.
88. Yang YD, Cho H, Koo JY, Tak MH, Cho Y, Shim WS, et al. TMEM16A confers receptor-activated calcium-dependent chloride conductance. Nature. 2008;455(7217):1210–5.
89. Caputo A, Caci E, Ferrera L, Pedemonte N, Barsanti C, Sondo E, et al. TMEM16A, a membrane protein associated with calcium-dependent chloride channel activity. Science. 2008;322(5901):590–4.
90. Schroeder BC, Cheng T, Jan YN, Jan LY. Expression cloning of TMEM16A as a calcium-activated chloride channel subunit. Cell. 2008;134(6):1019–29.
91. Craig AD. Significance of the insula for the evolution of human awareness of feelings from the body. Ann N Y Acad Sci. 2011;1225:72–82.
92. Craig AD. How do you feel? Interoception: the sense of the physiological condition of the body. Nat Rev Neurosci. 2002;3(8):655–66.
93. Craig AD. Cooling, pain, and other feelings from the body in relation to the autonomic nervous system. Handb Clin Neurol. 2013;117:103–9.
94. Farrell MJ. Regional brain responses in humans during body heating and cooling. Temperature. 2016;3(2):220–31.
95. Hammel HT, Hardy JD, Fusco MM. Thermoregulatory responses to hypothalamic cooling in unanesthetized dogs. Am J Phys. 1960;198:481–6.
96. Hammel HT, Jackson DC, Stolwijk JA, Hardy JD, Stromme SB. Temperature regulation by hypothalamic proportional control with an adjustable set point. J Appl Physiol. 1963;18:1146–54.
97. Farrell MJ, Trevaks D, McAllen RM. Preoptic activation and connectivity during thermal sweating in humans. Temperature. 2014;1(2):135–41.

98. Farrell MJ, Trevaks D, Taylor NA, McAllen RM. Regional brain responses associated with thermogenic and psychogenic sweating events in humans. J Neurophysiol. 2015;114(5):2578–87.
99. Low PA. Evaluation of sudomotor function. Clin Neurophysiol. 2004;115(7):1506–13.
100. Nakamura K, Matsumura K, Hubschle T, Nakamura Y, Hioki H, Fujiyama F, et al. Identification of sympathetic premotor neurons in medullary raphe regions mediating fever and other thermoregulatory functions. J Neurosci. 2004;24(23):5370–80.
101. Kennedy WR, Wendelschafer-Crabb G, Brelje TC. Innervation and vasculature of human sweat glands: an immunohistochemistry-laser scanning confocal fluorescence microscopy study. J Neurosci. 1994;14(11 Pt 2):6825–33.
102. Torres NE, Zollman PJ, Low PA. Characterization of muscarinic receptor subtype of rat eccrine sweat gland by autoradiography. Brain Res. 1991;550(1):129–32.
103. Kondo N, Shibasaki M, Aoki K, Koga S, Inoue Y, Crandall CG. Function of human eccrine sweat glands during dynamic exercise and passive heat stress. J Appl Physiol. 2001;90(5):1877–81.
104. Shibasaki M, Crandall CG. Effect of local acetylcholinesterase inhibition on sweat rate in humans. J Appl Physiol. 2001;90(3):757–62.
105. Welch G, Foote KM, Hansen C, Mack GW. Nonselective NOS inhibition blunts the sweat response to exercise in a warm environment. J Appl Physiol. 2009;106(3):796–803.
106. Nagashima K, Nakai S, Tanaka M, Kanosue K. Neuronal circuitries involved in thermoregulation. Auton Neurosci. 2000;85(1-3):18–25.
107. Romanovsky AA. Thermoregulation: some concepts have changed. Functional architecture of the thermoregulatory system. Am J Physiol Regul Integr Comp Physiol. 2007;292(1):R37–46.
108. Bamshad M, Song CK, Bartness TJ. CNS origins of the sympathetic nervous system outflow to brown adipose tissue. Am J Phys. 1999;276(6 Pt 2):R1569–78.
109. Cano G, Passerin AM, Schiltz JC, Card JP, Morrison SF, Sved AF. Anatomical substrates for the central control of sympathetic outflow to interscapular adipose tissue during cold exposure. J Comp Neurol. 2003;460(3):303–26.
110. Oldfield BJ, Giles ME, Watson A, Anderson C, Colvill LM, McKinley MJ. The neurochemical characterisation of hypothalamic pathways projecting polysynaptically to brown adipose tissue in the rat. Neuroscience. 2002;110(3):515–26.
111. Kamijo Y, Lee K, Mack GW. Active cutaneous vasodilation in resting humans during mild heat stress. J Appl Physiol. 2005;98(3):829–37.
112. Kenney WL, Johnson JM. Control of skin blood flow during exercise. Med Sci Sports Exerc. 1992;24(3):303–12.
113. Pawelcyzk JA. Neural control of skin and muscle blood flow during exercise and thermal stress. In: Gisolfi CV, Lamb DR, Nadel ER, editors. Perspectives in exercise science and sports medicine. Volume 6: exercise, heat, and thermoregulation. Dubuque: WCB Brown and Benchmark; 1993. p. 119–77.
114. Wilkins BW, Holowatz LA, Wong BJ, Minson CT. Nitric oxide is not permissive for cutaneous active vasodilatation in humans. J Physiol. 2003;548(Pt 3):963–9.
115. Kellogg DL Jr, Zhao JL, Friel C, Roman LJ. Nitric oxide concentration increases in the cutaneous interstitial space during heat stress in humans. J Appl Physiol. 2003;94(5):1971–7.
116. Kellogg DL Jr, Crandall CG, Liu Y, Charkoudian N, Johnson JM. Nitric oxide and cutaneous active vasodilation during heat stress in humans. J Appl Physiol. 1998;85(3):824–9.
117. Kellogg DL Jr, Liu Y, Kosiba IF, O'Donnell D. Role of nitric oxide in the vascular effects of local warming of the skin in humans. J Appl Physiol. 1999;86(4):1185–90.
118. Shibasaki M, Wilson TE, Cui J, Crandall CG. Acetylcholine released from cholinergic nerves contributes to cutaneous vasodilation during heat stress. J Appl Physiol. 2002;93(6):1947–51.
119. McCord GR, Cracowski JL, Minson CT. Prostanoids contribute to cutaneous active vasodilation in humans. Am J Physiol Regul Integr Comp Physiol. 2006;291(3):R596–602.
120. Bennett LA, Johnson JM, Stephens DP, Saad AR, Kellogg DL Jr. Evidence for a role for vasoactive intestinal peptide in active vasodilatation in the cutaneous vasculature of humans. J Physiol. 2003;552(Pt 1):223–32.

121. Wong BJ, Minson CT. Neurokinin-1 receptor desensitization attenuates cutaneous active vasodilatation in humans. J Physiol. 2006;577(Pt 3):1043–51.
122. Wong BJ, Wilkins BW, Minson CT. H1 but not H2 histamine receptor activation contributes to the rise in skin blood flow during whole body heating in humans. J Physiol. 2004;560(Pt 3):941–8.
123. Heimer L, Van Hoesen GW. The limbic lobe and its output channels: implications for emotional functions and adaptive behavior. Neurosci Biobehav Rev. 2006;30(2):126–47.
124. Nunneley SA, Martin CC, Slauson JW, Hearon CM, Nickerson LD, Mason PA. Changes in regional cerebral metabolism during systemic hyperthermia in humans. J Appl Physiol. 2002;92(2):846–51.
125. Kalivas PW, Nemeroff CB, Miller JS, Prange AJ Jr. Microinjection of neurotensin into the ventral tegmental area produces hypothermia: evaluation of dopaminergic mediation. Brain Res. 1985;326(2):219–27.
126. Smith JE, Jansen AS, Gilbey MP, Loewy AD. CNS cell groups projecting to sympathetic outflow of tail artery: neural circuits involved in heat loss in the rat. Brain Res. 1998;786(1-2):153–64.
127. Zhang YH, Hosono T, Yanase-Fujiwara M, Chen XM, Kanosue K. Effect of midbrain stimulations on thermoregulatory vasomotor responses in rats. J Physiol. 1997;503(Pt 1): 177–86.
128. Fechir M, Klega A, Buchholz HG, Pfeifer N, Balon S, Schlereth T, et al. Cortical control of thermoregulatory sympathetic activation. Eur J Neurosci. 2010;31(11):2101–11.
129. McAllen RM, Tanaka M, Ootsuka Y, McKinley MJ. Multiple thermoregulatory effectors with independent central controls. Eur J Appl Physiol. 2010;109(1):27–33.
130. Farrell MJ, Trevaks D, Taylor NA, McAllen RM. Brain stem representation of thermal and psychogenic sweating in humans. Am J Physiol Regul Integr Comp Physiol. 2013;304(10):R810–7.
131. Davison MA, Koss MC. Brainstem loci for activation of electrodermal response in the cat. Am J Phys. 1975;229(4):930–4.
132. McAllen RM. Action and specificity of ventral medullary vasopressor neurones in the cat. Neuroscience. 1986;18(1):51–9.
133. Shafton AD, McAllen RM. Location of cat brain stem neurons that drive sweating. Am J Physiol Regul Integr Comp Physiol. 2013;304(10):R804–9.
134. Egan GF, Johnson J, Farrell M, McAllen R, Zamarripa F, McKinley MJ, et al. Cortical, thalamic, and hypothalamic responses to cooling and warming the skin in awake humans: a positron-emission tomography study. Proc Natl Acad Sci U S A. 2005;102(14):5262–7.
135. Farrell MJ, Johnson J, McAllen RM, Zamarripa F, Denton DA, Fox PT, et al. Brain activity associated with ratings of the hedonic component of thermal sensation during whole-body warming and cooling. J Therm Biol. 2011;36:57–63.
136. Rolls ET, Grabenhorst F, Parris BA. Warm pleasant feelings in the brain. NeuroImage. 2008;41(4):1504–13.
137. Kulkarni B, Bentley DE, Elliott R, Youell P, Watson A, Derbyshire SW, et al. Attention to pain localization and unpleasantness discriminates the functions of the medial and lateral pain systems. Eur J Neurosci. 2005;21(11):3133–42.
138. Fredrikson M, Furmark T, Olsson MT, Fischer H, Andersson J, Langstrom B. Functional neuroanatomical correlates of electrodermal activity: a positron emission tomographic study. Psychophysiology. 1998;35(2):179–85.
139. Nagai Y, Critchley HD, Featherstone E, Trimble MR, Dolan RJ. Activity in ventromedial prefrontal cortex covaries with sympathetic skin conductance level: a physiological account of a "default mode" of brain function. NeuroImage. 2004;22(1):243–51.
140. Patterson JC, Ungerleider LG, Bandettini PA. Task-independent functional brain activity correlation with skin conductance changes: an fMRI study. NeuroImage. 2002;17(4):1797–806.
141. Williams LM, Brammer MJ, Skerrett D, Lagopolous J, Rennie C, Kozek K, et al. The neural correlates of orienting: an integration of fMRI and skin conductance orienting. Neuroreport. 2000;11(13):3011–5.

142. Rothwell JC. Overview of neurophysiology of movement control. Clin Neurol Neurosurg. 2012;114(5):432–5.
143. Baker SN. The primate reticulospinal tract, hand function and functional recovery. J Physiol. 2011;589(Pt 23):5603–12.
144. Wissler EH. A quantitative assessment of skin blood flow in humans. Eur J Appl Physiol. 2008;104(2):145–57.
145. Wenger CB, Bailey RB, Roberts MF, Nadel ER. Interaction of local and reflex thermal effects in control of forearm blood flow. J Appl Physiol. 1985;58(1):251–7.
146. Jessen C. Interaction of body temperatures in control of thermoregulatory effector mechanisms. In: Fregly MJ, Blatteis CM, editors. Handbook of physiology. Section 4: environmental physiology. New York: Oxford University Press; 1996. p. 127–38.
147. Nadel ER, Mitchell JW, Saltin B, Stolwijk JA. Peripheral modifications to the central drive for sweating. J Appl Physiol. 1971;31(6):828–33.
148. Nadel ER, Bullard RW, Stolwijk JA. Importance of skin temperature in the regulation of sweating. J Appl Physiol. 1971;31(1):80–7.
149. Yoshida T, Nakai S, Yorimoto A, Kawabata T, Morimoto T. Effect of aerobic capacity on sweat rate and fluid intake during outdoor exercise in the heat. Eur J Appl Physiol. 1995;71:235–9.
150. Flouris AD, Cheung SS. Human conscious response to thermal input is adjusted to changes in mean body temperature. Br J Sports Med. 2008;43(3):199–203.
151. Schlader ZJ, Prange HD, Mickleborough TD, Stager JM. Characteristics of the control of human thermoregulatory behavior. Physiol Behav. 2009;98(5):557–62.
152. Schlader ZJ, Simmons SE, Stannard SR, Mundel T. Skin temperature as a thermal controller of exercise intensity. Eur J Appl Physiol. 2011;11:1631–9.
153. Bligh J. A theoretical consideration of the means whereby the mammalian core temperature is defended at a null zone. J Appl Physiol. 2006;100(4):1332–7.
154. Gagnon D, Kenny GP. Does sex have an independent effect on thermoeffector responses during exercise in the heat? J Physiol. 2012;590(Pt 23):5963–73.
155. Simon E, Pierau FK, Taylor DC. Central and peripheral thermal control of effectors in homeothermic temperature regulation. Physiol Rev. 1986;66(2):235–300.
156. Simon E. The enigma of deep-body thermosensory specificity. Int J Biometeorol. 2000;44(3):105–20.
157. Flouris AD, Cheung SS. Thermometry and calorimetry assessment of sweat response during exercise in the heat. Eur J Appl Physiol. 2010;108(5):905–11.
158. Webb P. The physiology of heat regulation. Am J Phys. 1995;268(4 Pt 2):R838–50.
159. Grucza R. Body heat balance in man subjected to endogenous and exogenous heat load. Eur J Appl Physiol Occup Physiol. 1983;51(3):419–33.
160. Kenny GP, Groeller H, McGinn R, Flouris AD. Age, human performance, and physical employment standards. Appl Physiol Nutr Metab. 2016;41(6 Suppl 2):S92–S107.
161. Kenny GP, Sigal RJ, McGinn R. Body temperature regulation in diabetes. Temperature. 2016;3(1):119–45.
162. Carrillo AE, Flouris AD, Herry CL, Poirier MP, Boulay P, Dervis S, et al. Heart rate variability during high heat stress: a comparison between young and older adults with and without Type 2 diabetes. Am J Physiol Regul Integr Comp Physiol. 2016;311(4):R669–R75.
163. Stapleton JM, Poirier MP, Flouris AD, Boulay P, Sigal RJ, Malcolm J, et al. Aging impairs heat loss, but when does it matter? J Appl Physiol. 2015;118(3):299–309.
164. McGinn R, Carter MR, Barrera-Ramirez J, Sigal RJ, Flouris AD, Kenny GP. Does type 1 diabetes alter post-exercise thermoregulatory and cardiovascular function in young adults? Scand J Med Sci Sports. 2015;25(5):e504–14.
165. Nagashima K. Central mechanisms for thermoregulation in a hot environment. Ind Health. 2006;44(3):359–67.
166. Barbour HG. The effect of direct warming and cooling of the thermoregulatory centers on body temperature (translated from the German title: Die Wirkung unmittelbarer Erwärmung und Abkühlung der Wärmezentra auf die Körpertemperatur). Naunyn-Schmiedebergs Arch. 1912;70:1–26.

167. Flouris AD. Shaping our understanding of endothermic thermoregulation. Temperature. 2015;2(3):328–9.
168. Hammel HT, Stromme S, Cornew RW. Proportionality constant for hypothalamic proportional control of metabolism in unanesthetized dog. Life Sci. 1963;12:933–47.
169. Nakayama T, Hammel HT, Hardy JD, Eisenman JS. Thermal stimulation of electrical activity of single units of the preoptic region. Am J Phys. 1963;204:1122–6.
170. Hammel HT. Concept of the adjustable set temperature. In: Hardy JD, Gagge AP, Stolwijk JAJ, editors. Physiological and behavioral temperature regulation. Springfield: Charles C. Thomas; 1970. p. 676–83.
171. Hellstrøm B, Hammel HT. Some characteristics of temperature regulation in unanesthetized dog. Am J Phys. 1967;213:547–56.
172. Heller HC, Hammel HT. CNS control of body temperature during hibernation. Comp Biochem Physiol A. 1972;41(2):349–59.
173. Hammel HT, Heller HC, Sharp FR. Probing the rostral brainstem of anesthetized, unanesthetized, and exercising dogs and of hibernating and euthermic ground squirrels. Fed Proc. 1973;32(5):1588–97.
174. Jackson DC, Hammel HT. Hypothalamic "set" temperature decreased in exercising dog. Life Sci. 1963;8:554–63.
175. Sharp FR, Hammel HT. Effects of fever on salivation response in the resting and exercising dog. Am J Phys. 1972;223(1):77–82.
176. Simon E, Rautenberg W, Thauer R, Iriki M. Ausloesung thermoregulatorischer Reaktionen durch lokale Kuehlung im Vertebralkanal (Stimulation of thermoregulatory reactions by local cooling within the vertebral canal). Naturwissenschaften. 1963;50:337.
177. Kanosue K, Nakayama T, Tanaka H, Yanase M, Yasuda H. Modes of action of local hypothalamic and skin thermal stimulation on salivary secretion in rats. J Physiol. 1990;424:459–71.
178. Kanosue K, Yanase-Fujiwara M, Hosono T. Hypothalamic network for thermoregulatory vasomotor control. Am J Phys. 1994;267(1 Pt 2):R283–8.
179. Kanosue K, Zhang YH, Yanase-Fujiwara M, Hosono T. Hypothalamic network for thermoregulatory shivering. Am J Phys. 1994;267(1 Pt 2):R275–82.
180. Kanosue K, Yoshida K, Maruyama M, Nagashima K. The central organization of the thermoregulatory system. In: Kosaka M, Sugahara T, Schmidt KL, Simon E, editors. Thermotherapy for neoplasia, inflammation, and Pain. Tokyo: Springer; 2001. p. 2–11.
181. Romanovsky AA. Do fever and anapyrexia exist? Analysis of set point-based definitions. Am J Physiol Regul Integr Comp Physiol. 2004;287(4):R992–5.
182. Kobayashi S. Temperature-sensitive neurons in the hypothalamus: a new hypothesis that they act as thermostats, not as transducers. Prog Neurobiol. 1989;32(2):103–35.
183. Kobayashi S, Okazawa M, Hori A, Matsumura K, Hosokawa H. Paradigm shift in sensory system - animals do not have sensors. J Therm Biol. 2006;31:19–23.
184. Kobayashi S. Warm- and cold-sensitive neurons inactive at normal core temperature in rat hypothalamic slices. Brain Res. 1986;362(1):132–9.
185. Okazawa M, Takao K, Hori A, Shiraki T, Matsumura K, Kobayashi S. Ionic basis of cold receptors acting as thermostats. J Neurosci. 2002;22(10):3994–4001.
186. Romanovsky AA, Ivanov AI, Shimansky YP. Selected contribution: ambient temperature for experiments in rats: a new method for determining the zone of thermal neutrality. J Appl Physiol. 2002;92(6):2667–79.
187. Mekjavic IB, Eiken O. Contribution of thermal and nonthermal factors to the regulation of body temperature in humans. J Appl Physiol. 2006;100(6):2065–72.
188. Bazett HC. Theory of reflex controls to explain regulation of body temperature at rest and during exercise. J Appl Physiol. 1951;4(4):245–62.
189. Meade RD, Poirier MP, Flouris AD, Hardcastle SG, Kenny GP. Do the threshold limit values for work in hot conditions adequately protect workers? Med Sci Sports Exerc. 2016;48(6):1187–96.
190. Snellen JW. Set point and exercise. In: Bligh J, Moore RE, editors. Essays on temperature regulation. Amsterdam: North Holland; 1972. p. 139–48.

191. Adolph EF. Look at physiological integration. Am J Phys. 1979;237(5):R255–9.
192. Houdas Y, Sauvage A, Bonaventure M, Ledru C, BGuieu JD. Thermal control in man: regulation of central temperature or adjustments of heat exchanges by servomechanisms? J Dyn Syst Meas Control. 1973;95:331–5.
193. Aschoff J, Heise A. Thermal conductance in man: its dependence on time of day and on ambient temperature. In: Itoh S, Ogata K, Yoshimura H, editors. Advances in climatic physiology. Tokyo: Igaku Shoin; 1972. p. 334–48.
194. Krauchi K, Wirz-Justice A. Circadian rhythm of heat production, heart rate, and skin and core temperature under unmasking conditions in men. Am J Phys. 1994;267(3 Pt 2):R819–29.
195. Robinson EL, Demaria-Pesce VH, Fuller CA. Circadian rhythms of thermoregulation in the squirrel monkey (Saimiri sciureus). Am J Phys. 1993;265(4 Pt 2):R781–5.
196. Aschoff J, Pohl H. Thermal conductance in man: its dependence on time of day and on ambient temperature. In: Itoh S, Ogata K, Yoshimura H, editors. Advances in climatic physiology. Tokyo: Igaku Shoin; 1972. p. 334–48.
197. Webb P. Continuous thermal comfort in a suit calorimeter. In: Durand J, Raynaud J, editors. Thermal comfort: physiological and psychological bases. Paris: INSERM; 1979. p. 177–85.
198. Webb P, Annis JF, Troutman SJ Jr. Energy balance in man measured by direct and indirect calorimetry. Am J Clin Nutr. 1980;33(6):1287–98.
199. Tikuisis P. Heat balance precedes stabilization of body temperatures during cold water immersion. J Appl Physiol. 2003;95(1):89–96.
200. Kerslake DM. The stress of hot environments. Cambridge: Cambridge University Press; 1972.
201. Veghte JH, Webb P. Body cooling and response to heat. J Appl Physiol. 1961;16:235–8.
202. Webb P. Rewarming after diving in cold water. Aerosp Med. 1973;44(10):1152–7.
203. Ereth MH, Lennon RL, Sessler DI. Limited heat transfer between thermal compartments during rewarming in vasoconstricted patients. Aviat Space Environ Med. 1992;63(12):1065–9.
204. Kruk B, Pekkarinen H, Harri M, Manninen K, Hanninen O. Thermoregulatory responses to exercise at low ambient temperature performed after precooling or preheating procedures. Eur J Appl Physiol Occup Physiol. 1990;59(6):416–20.
205. Webb P. Heat storage and body temperature during cooling and rewarming. Eur J Appl Physiol Occup Physiol. 1993;66(1):18–24.
206. Ebbecke U. Uber die Temperaturempfindungen in ihrer Abhängigkeit von der Hautdurchblutung und von den Reflexzentren. Pfluegers Arch. 1917;169:385–462.
207. Nielsen B. Thermoregulation in rest and exercise. Acta Physiol Scand Suppl. 1969;323:1–74.
208. Webb P. Temperatures of skin, subcutaneous tissue, muscle and core in resting men in cold, comfortable and hot conditions. Eur J Appl Physiol Occup Physiol. 1992;64(5):471–6.
209. Ivanov KP. The location and function of different skin thermoreceptors. In: Bligh J, Voigt K, editors. Thermoreception and temperature regulation. Berlin: Springer; 1990. p. 37–43.
210. Ivanov KP, Dymnikova L, Danilova N. The hypothalamus neuron's response to the signals from surface and deep skin thermoreceptors. J Therm Biol. 1987;12:289–93.
211. Simon E. Deep-body thermosensitivity: Another look. A tribute to Harold Theodore (Ted) Hammel (1921–2005). J Therm Biol. 2006;31:4–18.

The Biophysics of Human Heat Exchange

2

Nicholas Ravanelli, Coen C. W. G. Bongers, and Ollie Jay

2.1 Introduction

The ability to maintain body temperature within a narrow range during acute or chronic exposure to environmental extremes is paramount for optimal human performance, and ultimately, survival. Muscle contractions during different sporting activities can result in a greatly elevated internal heat production. The subsequent changes in body temperature are managed to an extent by physiologically modulating heat exchange between the skin surface and the surrounding environment via sensible (convection (C), radiation (R) and conduction (K)) and insensible (evaporation (E)) heat transfer. However, the net heat dissipation via these heat transfer avenues is also strongly determined by the physical characteristics of the thermal environment that the sport is performed in. To optimally assess the risk of thermal stress for an athlete performing a given sport in a particular environment, the biophysical processes that govern the dynamic balance between internal heat production and skin surface heat dissipation must therefore be fully considered.

N. Ravanelli
School of Human Kinetics, University of Ottawa, Ottawa, ON, Canada

Thermal Ergonomics Laboratory, Faculty of Health Sciences, The University of Sydney, Sydney, NSW, Australia

C. C. W. G. Bongers
Thermal Ergonomics Laboratory, Faculty of Health Sciences, The University of Sydney, Sydney, NSW, Australia

O. Jay (✉)
School of Human Kinetics, University of Ottawa, Ottawa, ON, Canada

Thermal Ergonomics Laboratory, Faculty of Health Sciences, The University of Sydney, Sydney, NSW, Australia

The University of Sydney, Charles Perkins Centre, Sydney, NSW, Australia
e-mail: ollie.jay@sydney.edu.au

J. D. Périard, S. Racinais (eds.), *Heat Stress in Sport and Exercise*,
https://doi.org/10.1007/978-3-319-93515-7_2

2.2 Human Heat Balance

The fundamental law of human heat balance illustrates that internal metabolic heat production (M–W) must be balanced by an equal rate of net heat dissipation, i.e. combined sensible and insensible heat losses from the skin (sk) and respiratory tract (res) to the surrounding environment to ensure a rate of body storage (*S*) of zero (i.e. heat balance):

$$(M-W)=(\pm K_{sk}\pm C_{sk}\pm R_{sk})+(C_{res}+E_{res})+E_{sk}\pm S\quad \left[\text{in W m}^{-2}\right] \tag{2.1}$$

2.2.1 Metabolic Heat Production (M–W)

Metabolic heat production (M–W) is the difference between metabolic rate (*M*) and the external work performed (*W*). In its most basic form, *M* is the amount of energy released by hydrolysing adenosine triphosphate (ATP) into adenosine diphosphate (ADP) and an inorganic phosphate molecule. It follows that a proportion of the energy released from this process is then utilised to create *W*; however, the human body is quite inefficient and about 75–95% of *M* does not ultimately contribute to *W* but instead is liberated internally as heat [1–3]. Road cycling is one of the most efficient sporting activities (~30% of *M* is used for *W* [4]), so at an external work load of 240 W a metabolic rate of ~840 W is required, with ~600 W of this energy released as heat (Fig. 2.1). Running and walking are among the least efficient activities, especially on a flat surface, where effectively no external work is performed and all metabolic energy is converted to heat [5, 6].

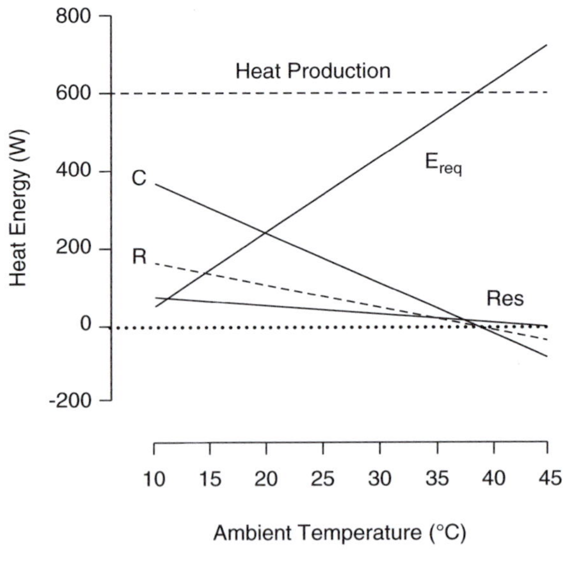

Fig. 2.1 An example of partitional heat exchange for an exercising individual on an upright ergometer. E_{req} evaporative requirements for heat balance, *C* convection, *R* radiation, *Res* respiratory heat loss

Carbohydrates and lipids are the two main substrates utilised by the body to produce ATP, and although ATP can be produced both aerobically and anaerobically within a cell, oxygen consumption is required to restore ATP pools. Thus, M can be estimated [7] by measuring the rate of oxygen consumption and carbon dioxide production using:

$$M = VO_2 \frac{\left[\left(\left(\frac{RER-0.7}{0.3}\right)\cdot e_c\right)+\left(\left(\frac{1.0-RER}{0.3}\right)\cdot e_f\right)\right]}{60} 1000 \quad [\text{in W}] \tag{2.2}$$

where VO_2 is the rate of oxygen consumption in L min^{-1}, RER is the ratio of carbon dioxide production to oxygen consumption, e_c is the caloric equivalent per litre of oxygen for the oxidation of carbohydrates (21.13 kJ) and e_f is the caloric equivalent per litre of oxygen for the oxidation of lipids (19.62 kJ). To normalise $M–W$ in W m^{-2}, it must be divided by the body surface area (BSA) of the individual using the Dubois and Dubois equation [8]:

$$BSA = 0.202 \times mass^{0.425} \times height^{0.725} \quad \left[\text{in m}^2\right] \tag{2.3}$$

where mass of the person is in kg and the height of the person is in m.

2.2.2 Sensible Heat Transfer from the Skin ($\pm K_{sk} \pm C_{sk} \pm R_{sk}$)

Sensible Heat Transfer from the Skin ($\pm K_{sk} \pm C_{sk} \pm R_{sk}$) is the sum of conduction (K_{sk}), convection (C_{sk}) and radiation (R_{sk}). These three avenues of heat transfer abide by the second law of thermodynamics, whereby heat energy moves from an area of high concentration to low concentration (e.g. from high to low temperature). During active or passive heat stress, the prevailing temperature gradients for sensible heat transfer may be minimal or even negative, which leads to sensible heat gain through one or more avenue at ambient temperatures above skin temperature (i.e. 35–36 °C) (Fig. 2.1).

2.2.2.1 Conduction (K_{sk})

Conduction (K_{sk}) is the transfer of heat energy through direct contact between the skin and a solid object. From a whole-body heat balance perspective, particularly human heat stress conditions, K is generally assumed to be negligible, with the primary means for sensible heat transfer via convection and radiation. However, when a solid object is in direct contact with the skin (e.g. a cold metallic wall), conductive heat transfer can be calculated as:

$$K = kA(T_2 - T_1)/L \quad \left[\text{in W m}^{-2}\right] \tag{2.4}$$

where k is the estimated thermal conductivity of the object in contact with the skin, A is the total surface area of contact between the skin and solid (in m^2), $(T_2 - T_1)$ is the absolute temperature difference between the skin and the solid's external surface and L is the thickness of the solid object in contact with the skin surface.

2.2.2.2 Radiation (R_{sk})

Heat exchange by radiation is the electromagnetic energy transfer between a relatively cool and warm body. Radiative heat loss from the skin for a nude person can be derived using:

$$R_{sk} = h_r \left(T_{sk} - T_r\right) \quad \left[\text{in W m}^{-2}\right] \tag{2.5}$$

where T_{sk} is mean skin temperature (in °C), T_r is mean radiant temperature (in °C) and h_r is the radiative heat transfer coefficient (in W m^{-2} K^{-1}), which is estimated using:

$$h_r = 4\varepsilon\sigma \frac{A_r}{A_D}\left[273.2 + \frac{T_{sk} + T_r}{2}\right]^3 \quad \left[\text{in W m}^{-2}\ \text{K}^{-1}\right] \tag{2.6}$$

where ε is the emissivity of the body surface (usually assumed to be 0.95), σ is the Stefan–Boltzmann constant (5.67×10^{-8} W m^{-2} K^{-4}), A_r/A_D is the effective radiative area of the body (m^2) which can be estimated as 0.70 or 0.73 for a seated or standing person [9], respectively, and $T_{sk} + T_r$ is the sum of mean skin temperature and mean radiant temperature. Mean radiant temperature is assumed to be equal to ambient air temperature when indoors without any substantial sources of radiation. However, in other environments, e.g. outdoor sun exposure, mean radiant temperature of the environment must be estimated using black globe temperature (T_g) measured with a standard 150-mm diameter black globe thermometer placed in a similar location as the exposed individual (e.g. in direct sunlight). T_g will vary depending on the time of day and year due to differences in the angle between the sun and the horizon. However, when interested in calculating radiative heat transfer for an individual wearing clothing, a black globe thermometer may overestimate the effect of a radiative source (particularly the sun) and should therefore be similar in colour to the clothing worn by the individual. Lastly, air velocity (v) in m/s near the black globe thermometer must be measured as greater air flow will alter T_g for a given amount of radiant heat energy. According to ISO 7726:1998 [10], mean radiant temperature (T_r) can be derived as follows:

If $v < 0.15$ m/s:

$$T_r = \left[\left(T_g + 273\right)^4 + \frac{0.25 \times 10^8}{\varepsilon} \cdot \left[\frac{T_g + T_a}{d}\right]^{0.25} \left(T_g - T_a\right)\right]^{0.25} - 273 \quad \left[\text{in}^{\circ}\text{C}\right] \tag{2.7}$$

where d is black globe diameter (in cm).

If $v \geq 0.15$ m/s:

$$T_r = \left[\left(T_g + 273\right)^4 + \frac{1.1 \times 10^8 v^{0.6}}{0.44} \cdot \left(T_g - T_a\right)\right]^{0.25} - 273 \quad \left[\text{in}^{\circ}\text{C}\right] \tag{2.8}$$

2.2.2.3 Convection (C_{sk})

Convection (C_{sk}) is the transfer of heat to a moving gas (air) or liquid (water), which is increased by the movement of the body in air or water or movement of air or water across the skin. It is directly proportional to the temperature difference between the skin surface and the ambient environment, and air velocity passing across the skin. A warm surface such as the skin can also produce natural convection when a person is still, where the boundary layer movement is a result of differing air densities arising from a temperature gradient (e.g. warm air rises). Alternatively, and more commonly, forced convection pushes air across the skin surface (e.g. a fan) or convection is self-generated as a person moves through an air mass. Convective heat transfer for a nude person can be estimated using [11]:

$$C_{sk} = h_c \left(T_{sk} - T_a\right) \quad \left[\text{in W m}^{-2}\right] \tag{2.9}$$

where T_{sk} is mean skin temperature (°C), T_a is ambient air temperature (°C) and h_c is the convective heat transfer coefficient (in W m^{-2} K^{-1}). For natural convection in still conditions, this value can be assumed to be 3.1 W m^{-2} K^{-1} [12]. If air velocity is >0.2 m/s, but <4.0 m/s, the convective heat transfer coefficient can be estimated using:

$$h_c = 8.3v^{0.6} \quad \left[\text{in W m}^{-2}\,\text{K}^{-1}\right] \tag{2.10}$$

where v is the mean air velocity around the body in m/s. During physical activity, it may be more practical to consider the mean net air flow across the body surface rather than just the mean ambient air velocity as this accounts for the path of movement relative to wind direction. Indeed, it is evident that the magnitude of self-generated air flow can influence the convective heat transfer coefficient. For example, independent of clothing and equipment, the higher heat strain for American football lineman compared to non-lineman has been attributed to the more static nature of their position-specific activities (e.g. blocking vs. running routes) [13, 14]. Alternatively, self-generated convection during outdoor cycling (>20 km/h) will in most cases be far greater than in laboratory settings [15]. As such, specific equations have been derived for estimating the convective heat transfer coefficient during different modalities of human movement (Table 2.1).

Table 2.1 Estimations of the convective heat transfer coefficient (h_c) for common modalities of exercise

Exercise modality	Equation/constant h_c (W m^{-2} K^{-1})	Comments
Stationary cycle ergometer (50 RPM)	5.4	Ambient air flow <0.2 m/s [16]
Stationary cycle ergometer (60 RPM)	6.0	Ambient air flow <0.2 m/s [16]
Outdoor cycling	$h_c = 8.4\text{v}_{\text{speed}}^{0.84}$	v_{speed}: cycling velocity (m/s) [15]
Walking/Running	$h_c = 8.3\text{v}_{\text{loc}}^{0.531}$	v_{loc}: speed of locomotion (m/s) [16]
Treadmill exercise	$h_c = 8.3\text{v}_{\text{loc}}^{0.391}$	v_{loc}: speed of locomotion (m/s) [16]

All convective heat transfer coefficients presented have been developed for thermal stress at approximately sea level. The relationship between barometric pressure (P_b) and convective heat transfer can be integrated into Eq. (2.9) as follows [17]:

$$C_{\mathrm{sk}} = h_c\left(T_{\mathrm{sk}} - T_a\right)\left(P_b / 760\right)^{0.55} \quad \left[\mathrm{in\ W\,m^{-2}}\right] \tag{2.11}$$

If clothing is worn, combined sensible heat transfer via convection and radiation ($C_{\mathrm{sk}} + R_{\mathrm{sk}}$) can be estimated using:

$$C_{\mathrm{sk}} + R_{\mathrm{sk}} = \frac{\left(T_{\mathrm{sk}} - T_o\right)}{\left(R_{\mathrm{cl}} + \dfrac{1}{h \cdot f_{\mathrm{cl}}}\right)} \quad \left[\mathrm{in\ W\,m^{-2}}\right] \tag{2.12}$$

where T_o is operative temperature (in °C):

$$T_o = \frac{\left(h_r T_r + h_c T_a\right)}{\left(h_r + h_c\right)} \quad \left[\mathrm{in\ ^\circ C}\right] \tag{2.13}$$

and h is the combined heat transfer coefficient (in W m^{-2} K^{-1}), i.e. $h_c + h_r$; and f_{cl} is the clothing area factor defined as the surface area of the clothed body divided by the surface area of the nude body and estimated using [18]:

$$f_{\mathrm{cl}} = 1 + \left[\frac{0.31 \cdot R_{\mathrm{cl}}}{0.155}\right] \quad [\mathrm{ND}] \tag{2.14}$$

where R_{cl} is the dry heat transfer resistance of clothing (in m^2 °C^{-1} W^{-1}), which can be obtained from normative tables [18, 19] such as the International Standardisation Organisation (ISO) 9920 standard.

Additionally, convective heat loss by water movement is important for example in swimmers. The convective heat loss is, in contradiction to convective heat loss by air movement, not a function of the water velocity [20]. Due to water turbulence created during swimming in a swimming pool, the effective water velocity around a swimmer does not differ irrespective of swimming speed. As a result, convective heat exchange for swimmers is predominantly determined by water temperature, whereby heat is lost if water temperature is lower than skin temperature, and vice versa. Due to a higher density, specific heat capacity and thermal conductivity [21], convective heat loss is much greater in water than in air [20]. Given exercise is typically performed on land, and detailed equations for convective heat exchange in water are beyond the scope of this chapter. However, for a detailed description we refer readers to the article of Brandt and Pichowsky [22].

2.2.3 Respiratory Heat Exchange (C_{res} + E_{res})

Respiratory heat exchange occurs through the convective heat transfer (C_{res}) between inhaled air and the lungs, and evaporative heat loss from the respiratory tract (E_{res}) due to the saturation of air with water vapour when entering the lungs. Net respiratory heat exchange can be estimated using (2.2):

$$C_{\text{res}} + E_{\text{res}} = \left[0.0014M \cdot (34 - T_a)\right] + \left[0.0173M \cdot (5.87 - P_a)\right] \quad \left[\text{in W m}^{-2}\right] \tag{2.15}$$

where M is metabolic rate in W m^{-2}, T_a is air temperature in °C and P_a is the ambient water vapour pressure in kPa.

The rate of respiratory heat loss is dependent on the temperature and humidity of inspired air [23, 24] and minute ventilation [25, 26]. As such, the amount of convective heat transfer through respiration during exercise in the heat compared to the cold is minimal due to the small temperature gradient between ambient and core temperature. Additionally, the amount of evaporative heat loss via respiration is dependent on the humidity gradient between the lungs and the air, and the rate of ventilation which is assumed to have a linear relationship with the rate of metabolic rate (up to 80% of maximum oxygen consumption [26]).

2.2.4 Evaporation from Skin Surface (E_{sk})

The evaporation of sweat (or water) from the skin surface is the largest modifiable avenue of heat loss from the body. During heat stress, sweat evaporation becomes the predominant factor for determining whether heat balance is achieved, and when air temperature equals skin temperature and dry heat loss is eliminated, evaporation becomes the only avenue for dissipating metabolic heat at the skin surface [27]. The latent heat lost for every gram of sweat that completely evaporates from the skin is 2.426 kJ [28]. As such, evaporative heat loss can be estimated using body mass changes corrected for metabolic and respiratory mass losses, as well as any ingested fluids, but only under conditions that permit complete evaporation [29]. Arguably, the most accurate method for estimating evaporative heat loss is direct calorimetry, which measures the difference in absolute water vapour pressure between influent and effluent of an enclosed air space [30]. However, once again the complete evaporation of all sweat from the skin is a necessity and is typically achieved in a calorimeter by ensuring a high and turbulent air mass flow [31].

Under combinations of climate and activity that yield incomplete sweat evaporation from the skin surface, evaporative efficiency (i.e. the proportion of secreted sweat that actually evaporates [32]) can be roughly estimated. It is known that as the sweat saturation level of the skin reaches a maximum, evaporative efficiency rapidly declines [32–34]. First described by Gagge [35], sweat saturation levels can be expressed as a “skin wettedness” value (ω), which is physiologically defined as the fraction of the skin surface that is covered in sweat. It follows that reductions in evaporative efficiency have been reported when $\omega > 0.50$ during passive heat stress [36], and when $\omega > 0.30$ during upright cycling [32]; meaning that while greater levels of skin wettedness permit greater rates of evaporation, this comes at the expense of a disproportionately greater rate of sweating. Mathematically, the ω value required (for heat balance; ω_{req}) is defined as the ratio of the evaporative requirement to maintain heat balance (E_{req}) relative to the maximum evaporative capacity in the ambient environment (E_{max}):

$$\omega_{req} = E_{req} / E_{max} \quad [\text{ND}] \tag{2.16}$$

By rearranging the conceptual heat balance equation (Eq. (2.1)), and assuming a rate of body heat storage (S) of zero, E_{req} can be estimated as follows:

$$E_{req} = (M - W) - (\pm K_{sk} \pm C_{sk} \pm R_{sk}) - (C_{res} + E_{res}) \quad [\text{in W m}^{-2}] \tag{2.17}$$

E_{max} is determined by the water vapour pressure gradient between the skin and the air, as well as air speed, clothing properties and the maximum proportion of the skin that can be physiologically saturated with sweat (ω_{max}):

$$E_{max} = \omega_{max} \frac{(P_{sk,sat} - P_a)}{\left(R_{e,cl} + \frac{1}{h_e \cdot f_{cl}}\right)} \quad [\text{in W m}^{-2}] \tag{2.18}$$

where ω_{max} is maximum skin wettedness, which can reach 1.00 for a fully heat acclimated person but only 0.72 in an untrained, non-heat acclimated individual [37]; $P_{sk,sat}$ is the saturated water vapour pressure at skin temperature (in kPa); P_a is the water vapour pressure measured in ambient air (in kPa); $R_{e,cl}$ is the evaporative heat transfer resistance of clothing (in m^2 kPa W^{-1}); f_{cl} is the clothing area factor (Eq. (2.14)) and h_e is the evaporative heat transfer coefficient (in W m^{-2} kPa^{-1}).

Values for $P_{sk,sat}$ can be derived using Antoine’s equation [38] as follows:

$$P_{sk} = \text{EXP}\left[18.956 - \frac{4030.18}{T_{sk} + 235}\right] \quad [\text{in kPa}] \tag{2.19}$$

Values for the h_e can be estimated using h_c (from Eq. (2.10)/Table 2.1) as follows:

$$h_e = 16.5 h_c \quad [\text{in W} \cdot \text{m}^{-2}\,\text{kPa}^{-1}] \tag{2.20}$$

Values for evaporative efficiency (E_{eff}) (i.e. as a fraction of secreted sweat that evaporates from the skin) can be subsequently estimated for a given level of ω_{req} using [39]:

$$E_{eff} = 1 - \frac{\omega_{req}^{\,2}}{2} \quad [\text{ND}] \tag{2.21}$$

Evaporative efficiency can also be estimated by directly measuring the mass of dripped sweat trapped in an oil pan placed on scale directly underneath the participant. However, this technique had been primarily reported during passive heating [36] and is difficult to implement during exercise.

Evaporative heat loss from the skin surface (E_{sk}) can then be estimated using:

$$E_{sk} = (\text{WBSL} \times 2.426) \times E_{eff} \quad [\text{in kJ}] \tag{2.22}$$

where WBSL is whole-body sweat loss over a fixed exercise time (in g).

It is important to acknowledge that the approach described above is especially limited for individuals wearing layered clothing outfits. While trapped sweat can indeed still evaporate, the effective latent heat of vaporisation of this sweat (which is usually assumed to be 2.426 kJ g^{-1}) has been shown to decline dramatically (by up to ~80%) depending on the material properties and most importantly the number of clothing layers [40]. As such, E_{sk} from measured sweat losses, even if sweat trapped in clothing is accounted for, could be overestimated by more than fourfold.

2.2.5 Heat Storage (*S*)

Heat Storage (S) occurs when an imbalance arises between metabolic heat production and the parallel rate of net heat dissipation via sensible and evaporative heat transfer. Typically, at rest in a temperate environment, humans are in heat balance (i.e. $S = 0$) as heat loss from sensible heat exchange via convection and radiation matches resting metabolic rate without any requirement for evaporation other than passively through respiration. However, elevated rates of heat production following the onset of exercise under nearly all environmental conditions lead to a positive rate of heat storage. On the other hand, cold exposure without sufficient clothing insulation can cause high rates of convective and radiative heat loss that exceed metabolic heat production leading to a heat imbalance and thus a negative rate of heat storage. Cumulatively over time, sustained rates of positive or negative heat storage result in changes in internal (i.e. core) body temperature, which if left unchecked can become detrimental to human performance and ultimately health.

The change in heat storage required to alter core temperature is dependent on biophysical factors. Firstly, the body mass of an individual represents their heat sink, meaning that changes in core temperature for an absolute amount of heat stored in the body are negatively correlated, i.e. a smaller rise in core temperature is observed with a larger body mass for a fixed heat storage [41, 42]. Secondly, large differences in the specific heat of the tissues of the body (C_p) caused by marked differences in body composition can also alter core temperature despite a similar heat storage. A C_p of 3.47 kJ kg^{-1} $°C^{-1}$ is assumed for the average person [43]. However, owing to the different C_p of fat tissue (2.97 kJ kg^{-1} $°C^{-1}$) and lean mass (3.64 kJ kg^{-1} $°C^{-1}$) overall C_p can vary depending on adiposity. While small differences in C_p do not seem to meaningfully influence core temperature, it has been recently demonstrated that a ~20% difference in body fat percentage is sufficient to independently yield ~0.2 to 0.3 °C greater rises in core temperature during moderate exercise at a fixed metabolic heat production of 6 W/kg of total body mass in healthy males (mean body fat % of 10.8 versus 32.0%) in a 28 °C environment [44].

2.2.6 Temporal Changes in Human Heat Balance

Reflex physiological mechanisms as described in Chap. 1 aid the maintenance of body temperature within the prescribed limits for human health by modifying heat balance. Autonomic increases in vascular conductance of the skin mediated by a cutaneous vasodilatation, and eccrine sweating, are observed in proportion to elevations in skin and/or core temperature during exercise and/or with heat exposure. Similarly, in the cold, a vasoconstriction response and shivering thermogenesis occur in proportion to reductions in skin and/or core temperature [45–47].

While at rest, mean T_{sk} in a temperate and thermally comfortable environment is typically 33–34 °C [11, 48]. During heat stress, an initial vasodilatation causes an increase in T_{sk}, which alters the temperature difference between the skin surface and the ambient environment and thus increases sensible heat loss (or decreases sensible heat gain if $T_a > T_{sk}$) via convection (Eq. (2.9)) and radiation (Eq. (2.5)) [27]. If net heat loss via convection and radiation (and the small heat losses via respiration) are not sufficient to balance the rate of internal heat production via metabolism (Eq. (2.2)), eccrine sweating must be initiated to enhance evaporative heat loss from the skin. Once sweat is secreted from eccrine sweat glands and reaches the skin surface, the area of skin directly under the sweat is considered to be 100% saturated with water vapour [49]. As such, the gradient between the partial pressure of water vapour at the skin surface (P_{sk}) and in ambient air (P_a), and therefore the rate of evaporative heat loss, is increased by sweating. As sweat gland output increases, skin wettedness (ω) increases until reaching a maximum theoretical value of 1.00 when the entire surface area of the body that is available for evaporation (typically equal to total body surface area in healthy humans) is completely covered in sweat. This level of ω is only possible in fully heat acclimated individuals [39] and has been recently shown to be as low as 0.72 in an untrained, non-heat acclimated individual, and 0.84 in trained but non-heat acclimated people [37]. The ability to saturate ~15 to 25% more of the skin surface following heat acclimation permits an extension of the range of compensable conditions for a given ambient temperature and humidity (Fig. 2.2). Whole-body sweat rate is regulated to ensure that, and if possible, a steady-state core temperature is attained [50]. It follows that for thermal equilibrium to be possible a rate of heat storage of zero must be achieved. As such, whole-body sweat rate is effectively controlled to ensure heat balance, or more specifically the evaporative heat requirement for heat balance (E_{req}) (Eq. (2.17)) [49]. The relationship between whole-body sweat rate and E_{req} becomes non-linear however, once decrements in evaporative efficiency (Eq. (2.21)) are observed; that is, when E_{req} is approximately greater than 50% of E_{max} (Eq. (2.18)).

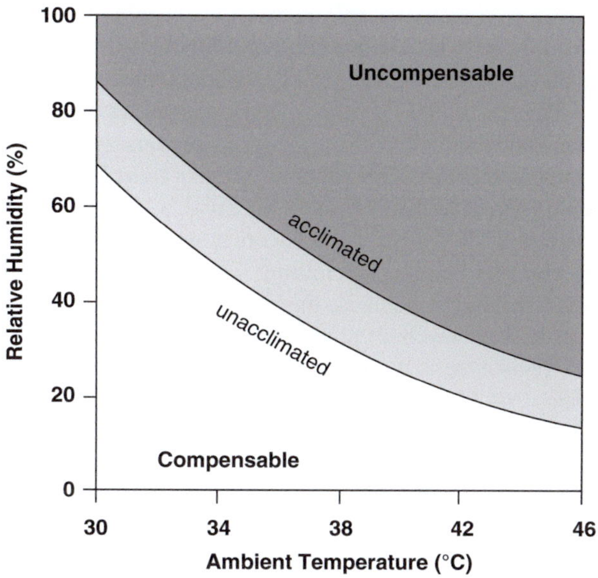

Fig. 2.2 The upper limits of compensability for an exercising individual at 600 W of heat production who is unacclimated (white area) or fully heat acclimated (light grey area). As depicted, the ability to attain a greater maximum skin wettedness following complete heat acclimation permits the maintenance of heat balance in more humid environments (for a given ambient temperature) in comparison to unacclimated

As skin surface sweating is autonomically controlled via a feedback loop using afferent signals from thermoreceptors throughout the body [51], sweating cannot commence without a "load error", i.e. a rise in internal temperature [45, 52], which in almost all circumstances requires heat storage. As a result, the time course of the activation of physiologically mediated changes in skin surface heat dissipation is longer relative to the almost immediate increase in heat production following the onset of exercise leading to a transient heat imbalance. The duration of this imbalance is determined by: (1) the rate at which sweating and skin blood flow increase relative to the rise in core and skin temperature and (2) the maximum physiological capacity to increase sweat production and skin blood flow. The longer this imbalance between heat production and net heat dissipation lasts, the greater S will be for a particular individual, and the greater the rise in internal tissue temperatures.

2.2.7 Compensable and Uncompensable Heat Stress

The risk associated with heat stress or hyperthermia, in terms of the magnitude of rise in core temperature, is greatly dependent on the physiological compensability of the individual in a given environment. While body temperature will eventually plateau if heat balance is attainable (compensable), it will continue to rise without a plateau occurring if heat balance is not possible (uncompensable). Uncompensable heat stress can occur because: (1) metabolic heat production is too high, and/or (2) the physiological capacity to sweat has been reached, and/or (3) the environment/clothing prevents a sufficiently high rate of heat dissipation from the skin. In the context of the previously described heat balance components, whether a given heat exposure is compensable or not is determined by: (a) the amount of evaporation required for heat balance (E_{req}) and (b) the maximal evaporative capacity of environment (E_{max}): that is, if $E_{req} \leq E_{max}$ = Compensable and if $E_{req} > E_{max}$ = Uncompensable.

The E_{req} and E_{max} for a given exposure are determined by both environmental conditions and physiological characteristics. A lower E_{req} is observed as: (1) T_a and T_r become lower, (2) v becomes greater apart from when $T_a > T_{sk}$, (3) M–W is lower and (4) T_{sk} is higher. On the other hand, a higher E_{max} is observed as: (1) P_a becomes lower and thus drier, (2) v becomes greater, (3) body surface area of the person is greater and (4) $R_{e,cl}$ of clothing worn is lower. Therefore, the cooler, windier and drier an environment is, the more likely it will lead to compensable heat stress, especially if levels of physical activity are low and/or clothing with a low evaporative resistance is worn. Nevertheless, numerous combinations of activities and climates can yield uncompensable heat stress. Even activities with a low metabolic heat production can result in uncompensable heat stress if the climate is sufficiently hot, humid and still. Similarly, activities with high rates of metabolic heat production can result in uncompensable heat stress even in relatively temperate climates. In sum, for a fixed set of environmental characteristics, the more skin temperature can be increased through elevations in skin blood flow, and the greater the skin wettedness that can be achieved, the more likely an individual will (a) avoid uncompensable heat stress and a continued increase in core temperature and (b) limit the magnitude of heat storage and therefore the increase in core temperature during a compensable heat stress exposure.

2.2.8 Cold Stress

Although the current chapter focuses primarily on heat stress, it will conclude with a brief comment on the biophysical processes associated with cold stress as they follow identical principles. In the cold, large temperature gradients between the cold ambient air and the warmer skin cause extensive sensible heat loss, primarily via convection and radiation. A reduction in skin blood flow via sympathetic vasoconstriction causes a concomitant decrease in T_{sk} [53], and subsequently reduces the temperature gradient between the skin surface and the ambient environment and therefore blunts sensible heat loss for a given air temperature and air velocity. Restricting blood flow to the skin and maintaining blood flow to the body core ensures that the heat produced via metabolism remains close to the deep visceral organs and the brain. As a result, a substantial temperature gradient develops between the body core and peripheral tissues. If skin blood flow does not sufficiently limit dry heat loss and a negative rate of body heat storage persists, shivering thermogenesis will be instigated to increase the rate of metabolic heat production via the asynchronous firing of muscle fibres to produce heat without external work. Heat production during maximal shivering can reach up to 5–6 times resting metabolic rate [54]. Primary input for the magnitude of shivering thermogenesis appears to come from deep body thermoreceptors (i.e. spinal cord, intestines and brain), whereas the onset threshold for shivering is modified by skin temperature [55, 56]. While shivering thermogenesis is an effective means of compensating for a negative rate of body heat storage, shivering has been shown to interfere with the performance of fine motor tasks [57, 58].

2.2.9 Summary

This chapter describes the fundamental factors that influence heat exchange between the human body and its surrounding environment. The bulk of heat exchange takes place at the skin surface via sensible heat transfer (i.e. convection and radiation) and evaporation. With increasing ambient temperature, the gradient for sensible heat transfer declines, meaning that the human body becomes increasingly dependent on the evaporation of sweat for heat dissipation. If the combination of climate (air temperature, radiant temperature, humidity and air velocity) and clothing permit a sufficient level of heat dissipation to counterbalance the rate of internal heat production, elevations in core temperature are moderated (i.e. compensable heat stress). However, if heat production exceeds the upper capacity to lose heat from the skin surface due to high ambient temperatures, humidity, low wind speeds or high evaporative resistance of clothing, a continuous increase in core temperature occurs (i.e. uncompensable heat stress).

References

1. Edwards RH, Ekelund LG, Harris RC, Hesser CM, Hultman E, Melcher A, Wigertz O. Cardiorespiratory and metabolic costs of continuous and intermittent exercise in man. J Physiol. 1973;234(2):481–97.
2. Gladden LB, Welch HG. Efficiency of anaerobic work. J Appl Physiol Respir Environ Exerc Physiol. 1978;44(4):564–70.
3. Wasserman K, Van Kessel AL, Burton GG. Interaction of physiological mechanisms during exercise. J Appl Physiol. 1967;22(1):71–85.
4. Whipp BJ, Wasserman K. Efficiency of muscular work. J Appl Physiol. 1969;26(5):644–8.
5. Margaria R. Positive and negative work performances and their efficiencies in human locomotion. Int Z Angew Physiol. 1968;25:339–51.
6. Snellen JW. External work in level and grade walking on a motor-driven treadmill. J Appl Physiol. 1960;15(5):759–63.
7. Nishi Y. Measurement of thermal balance in man. In: Bioengineering, thermal physiology and comfort. New York: Elsevier; 1981.
8. Du Bois D, Du Bois EF. A formula to estimate surface area if height and weight are known. Arch Intern Med. 1916;5:303–11.
9. Fanger PO. Calculation of thermal comfort: introduction of a basic comfort equation. ASHRAE Trans. 1967;73:III4.1.
10. International Organisation for Standardisation. Ergonomics of the thermal environment. Instruments for measuring physical quantities. ISO 7726:1998, Geneva, 1998.
11. Parsons K. Human thermal environments: the effects of hot, moderate, and cold environments on human health, comfort and performance. 2nd ed. New York: CRC Press; 2002.
12. Mitchell D. Convective heat loss from man and other animals. In: Convective heat loss from man and other animals. London: Butterworths; 1974.

13. Deren TM, Coris EE, Casa DJ, DeMartini JK, Bain AR, Walz SM, Jay O. Maximum heat loss potential is lower in football linemen during an NCAA summer training camp because of lower self-generated air flow. J Strength Cond Res. 2014;28(6):1656–63.
14. Godek SF, Bartolozzi AR, Godek JJ. Sweat rate and fluid turnover in American football players compared with runners in a hot and humid environment. Br J Sports Med. 2005;39(4):205–11.
15. Defraeye T, Blocken B, Koninckx E, Hespel P, Carmeliet J. Computational fluid dynamics analysis of drag and convective heat transfer of individual body segments for different cyclist positions. J Biomech. 2011;44(9):1695–701.
16. Nishi Y, Gagge AP. Direct evaluation of convective heat transfer coefficient by naphthalene sublimation. J Appl Physiol. 1970;29(6):830–8.
17. Gagge APN, Nishi Y. Heat exchange between human skin surface and thermal environment. In: A. C. Society, editor. Comprehensive physiology. Hoboken: Wiley; 1977. p. 69–92.
18. McCullough EA, Jones EW, Huck J. A comprehensive database for estimating clothing insulation. ASHRAE Trans U S. 1985;91:29–47.
19. International Organisation for Standardisation. Ergonomics of the thermal environment - estimation of thermal insulation and water vapour resistance of a clothing ensemble. ISO 9920:2007 Geneva, 2004.
20. Nadel ER, Holmer I, Bergh U, Astrand PO, Stolwijk JA. Energy exchanges of swimming man. J Appl Physiol. 1974;36(4):465–71.
21. Lemmon EWH. Thermophysical properties of water and steam. In: Handbook of chemistry and physics. New York: CRC Press; 2015. p. 1–3.
22. Brandt RA, Pichowsky MA. Conservation of energy in competitive swimming. J Biomech. 1995;28(8):925–33.
23. Mc CJ, Taylor CL. Respiratory heart exchange with varying temperature and humidity of inspired air. J Appl Physiol. 1951;4(2):121–35.
24. Walker JE, Wells RE Jr, Merrill EW. Heat and water exchange in the respiratory tract. Am J Med. 1961;30:259–67.
25. Brebbia DR, Goldman RF, Buskirk ER. Water vapor loss from the respiratory tract during outdoor exercise in the cold. J Appl Physiol. 1957;11(2):219–22.
26. Mitchell JW, Nadel ER, Stolwijk JA. Respiratory weight losses during exercise. J Appl Physiol. 1972;32(4):474–6.
27. Nielsen M. Die regulation der körpertemperatur bei muskelarbeit. Skand Arch Physiol. 1938;79:193–230.
28. Wenger CB. Heat of evaporation of sweat: thermodynamic considerations. J Appl Physiol. 1972;32(4):456–9.
29. Bain AR, Lesperance NC, Jay O. Body heat storage during physical activity is lower with hot fluid ingestion under conditions that permit full evaporation. Acta Physiol. 2012;206(2):98–108.
30. Snellen JW, Chang KS, Smith W. Technical description and performance characteristics of a human whole-body calorimeter. Med Biol Eng Comput. 1983;21(1):9–20.
31. Reardon FD, Leppik KE, Wegmann R, Webb P, Ducharme MB, Kenny GP. The Snellen human calorimeter revisited, re-engineered and upgraded: design and performance characteristics. Med Biol Eng Comput. 2006;44(8):721–8.
32. Alber-Wallerstrom B, Holmer I. Efficiency of sweat evaporation in unacclimatized man working in a hot humid environment. Eur J Appl Physiol Occup Physiol. 1985;54(5):480–7.
33. Candas V, Libert JP, Vogt JJ. Human skin wettedness and evaporative efficiency of sweating. J Appl Physiol Respir Environ Exerc Physiol. 1979;46(3):522–8.
34. Givoni B. Man, climate and architecture. London: Applied Science Publ.; 1976.
35. Gagge AP. A new physiological variable associated with sensible and insensible perspiration. Am J Phys. 1937:277–87.
36. Candas V, Libert JP, Vogt JJ. Influence of air velocity and heat acclimation on human skin wettedness and sweating efficiency. J Appl Physiol Respir Environ Exerc Physiol. 1979;47(6):1194–200.
37. Ravanelli N, Coombs GB, Imbeault P, Jay O. Maximum skin wettedness after aerobic training with and without heat acclimation. Med Sci Sports Exerc. 2018;50(2):299–307.

38. Thomson GW. The Antoine equation for vapor-pressure data. Chem Rev. 1946;38:1–39.
39. International Organisation for Standardisation. Ergonomics of the thermal environment; Analytical determination and interpretation of heat stress using calculation of the predicted heat strain. ISO 7933:2004, Geneva, 2004.
40. Havenith G, Brode P, den Hartog E, Kuklane K, Holmer I, Rossi RM, Richards M, Farnworth B, Wang XX. Evaporative cooling: effective latent heat of evaporation in relation to evaporation distance from the skin. J Appl Physiol. 2013;114(6):778–85.
41. Cramer MN, Jay O. Selecting the correct exercise intensity for unbiased comparisons of thermoregulatory responses between groups of different mass and surface area. J Appl Physiol. 2014;116(9):1123–32.
42. Ravanelli N, Cramer M, Imbeault P, Jay O. The optimal exercise intensity for the unbiased comparison of thermoregulatory responses between groups unmatched for body size during uncompensable heat stress. Physiol Rep. 2017;5(5):e13099.
43. Geddes LA. The specific resistance of biological material - a compendium of data for the biomedical engineer and physiologist. Med Biol Eng. 1967;5:271–93.
44. Dervis S, Coombs GB, Chaseling GK, Filingeri D, Smoljanic J, Jay O. A comparison of thermoregulatory responses to exercise between mass-matched groups with large differences in body fat. J Appl Physiol. 2016;120(6):615–23.
45. Hammel HT. Regulation of internal body temperature. Annu Rev Physiol. 1968;30:641–710.
46. Hammel HT, Jackson DC, Stolwijk JA, Hardy JD, Stromme SB. Temperature regulation by hypothalamic proportional control with an adjustable set point. J Appl Physiol. 1963;18:1146–54.
47. Werner J. System properties, feedback control and effector coordination of human temperature regulation. Eur J Appl Physiol. 2010;109(1):13–25.
48. Bierman W. The temperature of the skin surface. J Am Med Assoc. 1936;106:1158–62.
49. Gagnon D, Jay O, Kenny GP. The evaporative requirement for heat balance determines whole-body sweat rate during exercise under conditions permitting full evaporation. J Physiol. 2013;591(11):2925–35.
50. Werner J. Control aspects of human temperature regulation. Automatica. 1981;17(2):351–62.
51. Cabanac M. Regulation and modulation in biology - a reexamination of temperature regulation. Ann N Y Acad Sci. 1997;813:21–31.
52. Cabanac M. Adjustable set point: to honor Harold T. Hammel. J Appl Physiol. 2006;100(4):1338–46.
53. Jessen C. Interaction of body temperatures in control of thermoregulatory effector mechanisms. In: Comprehensive physiology. Hoboken: Wiley; 1996. p. 127–38.
54. Eyolfson DA, Tikuisis P, Xu X, Weseen G, Giesbrecht GG. Measurement and prediction of peak shivering intensity in humans. Eur J Appl Physiol. 2001;84(1-2):100–6.
55. Benzinger TH. Heat regulation: homeostasis of central temperature in man. Physiol Rev. 1969;49(4):671–759.
56. Euler v. Physiology and pharmacology of temperature regulation. Pharmacol Rev. 1961;13:361–98.
57. Meigal AY, Oksa J, Gerasimova LI, Hohtola E, Lupandin YV, Rintamaki H. Force control of isometric elbow flexion with visual feedback in cold with and without shivering. Aviat Space Environ Med. 2003;74(8):816–21.
58. Meigal AY, Oksa J, Hohtola E, Lupandin YV, Rintamaki H. Influence of cold shivering on fine motor control in the upper limb. Acta Physiol Scand. 1998;163(1):41–7.

Integrative Human Cardiovascular Responses to Hyperthermia

3

Scott T. Chiesa, Steven J. Trangmar, Kazuhito Watanabe, and José González-Alonso

3.1 Introduction

When ambient temperature exceeds that of the skin, heat is transferred from the surrounding environment to the cooler body tissues, and internal body temperature begins to rise. With an increase in core temperature of as little as 3–4 °C potentially proving fatal [1], the presence of raised tissue temperatures results in profound changes in vascular tone and cardiac output, alongside a redistribution of blood flow from core to peripheral sites in order to promote heat loss to the environment [2]. Even at rest, the magnitude of increase in cardiac output is striking, with elevations of 5–8 l/min reported during prolonged severe passive heating, an increase second only to that witnessed during aerobic exercise [3, 4]. This increase in blood flow is directed predominantly to the cutaneous (skin) circulation, aided by hyperthermia-induced vasoconstriction in splanchnic, renal, and cerebral vascular beds. Recent evidence suggests that skeletal muscle, fat, and bone blood flow may also contribute

S. T. Chiesa (✉)
Institute of Cardiovascular Science, University College London, London, UK
e-mail: s.chiesa@ucl.ac.uk

S. J. Trangmar
Department of Life Sciences, University of Roehampton, London, UK

K. Watanabe
Centre for Human Performance, Exercise and Rehabilitation, Brunel University London, Uxbridge, UK

J. González-Alonso
Centre for Human Performance, Exercise and Rehabilitation, Brunel University London, Uxbridge, UK

Division of Sport, Health and Exercise Sciences, Department of Life Sciences, Brunel University London, Uxbridge, UK

J. D. Périard, S. Racinais (eds.), *Heat Stress in Sport and Exercise*,
https://doi.org/10.1007/978-3-319-93515-7_3

to the observed hyperemic response, although the magnitude of this effect appears relatively modest.

The superimposition of exercise onto a hyperthermic individual results in one of the greatest challenges faced by the human cardiovascular system. Even when normothermic, the profound ability of skeletal muscle to vasodilate during intense whole-body exercise [5] exceeds the capacity of the heart to maintain the level of blood flow required for metabolic demand [6, 7]. When exercise is performed in a hyperthermic state, the combined demand of both skeletal muscle and skin vascular beds for O_2 delivery and heat dissipation—alongside a host of other vascular, hormonal, and neural alterations—can rapidly push the cardiovascular system to its limits [8, 9]. Despite this, however, mean arterial pressure is only slightly reduced, and cardiac output and active skeletal muscle blood flow are maintained or may even be increased at all but the very highest intensities of whole-body exercise. This chapter aims to document the integrative cardiovascular response to heat stress, firstly during progressive increases in body temperature at rest (i.e. passive heat stress, induced through thermal interventions such as wearing a suit that circulates hot water, exposure to sauna, or hot water baths), followed by cardiovascular adjustments across a spectrum of exercise intensities (i.e. exercise heat stress, primarily via the use of water-perfused suit while cycling or exposure to very hot environmental temperatures). As the majority of human research studies in this field have been carried out in young healthy adults, it is important to note that the absolute cardiovascular responses described in this chapter likely apply to this population alone, and should not be considered representative of other groups such as clinical patients or the elderly. Likewise, conditions such as orthostatic stress, dehydration, or haemorrhage—which may act to further compromise cardiovascular function in the heat—are outside the scope of this chapter, and so will not be addressed.

3.2 Respiratory and Cardiovascular Responses to Passive Heat Stress

The overall respiratory and cardiovascular adjustments observed during passive heat stress are at first glance similar to that observed during exercise, comprising an increase in respiratory rate, an increase in cardiac output, a decrease in peripheral resistance, and a selective elevation and redistribution of blood flow from the core to the extremities (Fig. 3.1). In contrast to exercise, however, these changes are largely driven by thermoregulatory as opposed to metabolic requirements, and are therefore underpinned by distinct respiratory, cardiac, and vascular adjustments. These act to maximise convective heat transfer within the body and evaporative heat dissipation to the surrounding environment while preserving mean arterial and perfusion pressure and increasing O_2 delivery in many tissues above the metabolic demand.

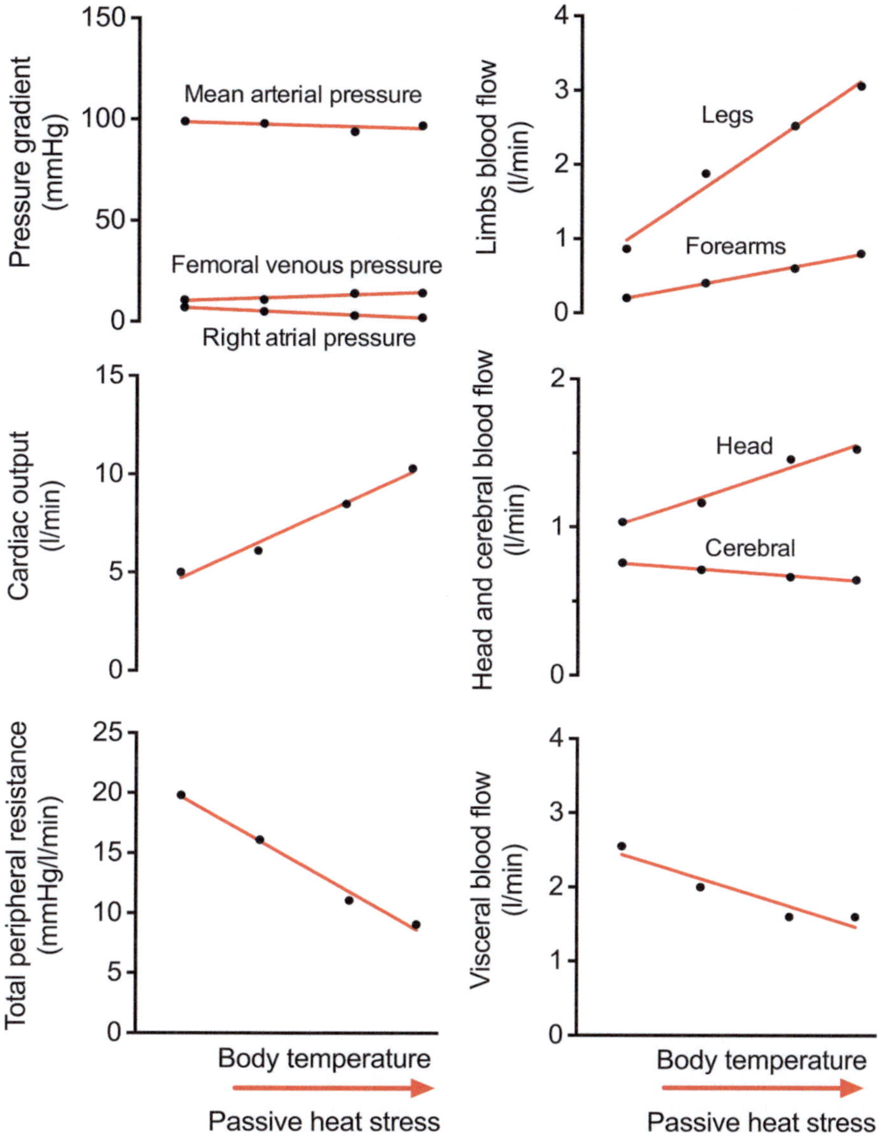

Fig. 3.1 Systemic and regional haemodynamic responses to progressive passive heat stress-induced hyperthermia. Data for mean arterial pressure, femoral venous pressure, cardiac output, and total peripheral resistance graphs are redrawn from [10] and [22]. The limbs blood flow graph is based on data reported by Pearson et al. [10] (legs) and Kalsi et al. [45] (forearms), whereas the head and cerebral blood flow graph is from [12]. Lastly, the visceral blood flow and right atrial pressure graphs are based on data from [4]. The increase in blood flow in the legs, arms, and head accounts for the majority (>80%) of the increase in cardiac output

3.2.1 Respiratory Responses to Passive Heat Stress

Elevations in core temperature >1 °C induce a hyperventilatory response in humans, despite a small absolute increase in metabolic requirements and O_2 consumption ($\Delta\dot{V}O_2$ ~0.15 l/min or 70% increase in resting metabolic rate) [10]. The potential mechanisms controlling the hyperventilatory response to hyperthermia are outside the scope of this chapter, but are discussed in detail in a recent review [11]. In hyperthermic but euhydrated individuals, arterial O_2 content is unchanged or increased, and mixed venous O_2 is significantly increased due to the significant increase in blood flow to the cutaneous and deep tissue circulations for thermoregulatory purposes [10]. As a result, even at the very highest levels of heat stress, O_2 delivery to both thermoregulatory and metabolically active tissues is elevated well in excess of that required to maintain the modest increases in $\dot{V}O_2$, and systemic and limb O_2 extraction is therefore markedly blunted (see Fick principle in Fig. 3.2). In contrast to its negligible effect on arterial O_2 content, however, the effect of hyperventilation on intravascular CO_2 levels is pronounced [12]. This has potential implications for brain blood flow and cerebral O_2 delivery, as will be discussed later in the chapter.

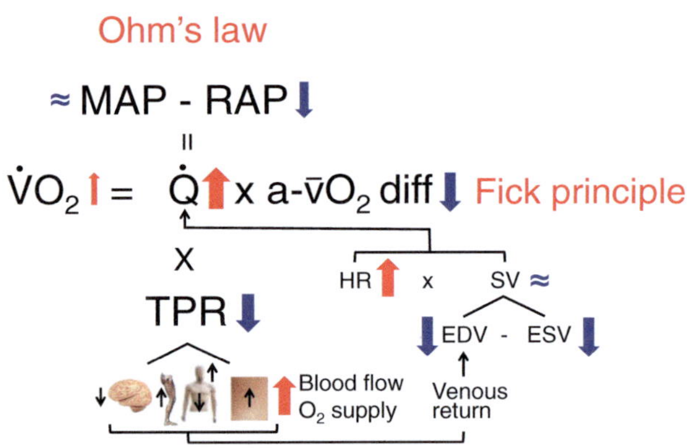

Fig. 3.2 Schematic illustration of the impact of severe whole-body heat stress on physiological function according to Ohm's law and the Fick principle. Ohm's law states that perfusion pressure equals flow x resistance, with pressure being the force that drives flow and resistance the force that opposes flow. The Fick principle, in turn, determines the rate of oxygen consumption by the human body, an organ, limb, or tissue. This is equal to the product of blood flow and the arterial-venous oxygen content differences. Note that blood flow is the common denominator in both equations. $\dot{Q}$ cardiac output, *TPR* total peripheral resistance, *MAP* mean arterial pressure, *RAP* right atrial pressure, $\dot{V}O_2$ systemic oxygen consumption, *a-vO_2 diff* arterio-mixed venous oxygen content differences, *HR* heart rate, *SV* stroke volume, *EDV* end-diastolic volume, *ESV* end-systolic volume

3.2.2 Cardiac Responses to Passive Heat Stress

At the most fundamental physical level, the cardiovascular system can be thought of as a variation of Ohm's law, where pressure (in this case the pressure gradient between the arterial circulation and the right side of the heart) is the product of flow x resistance. Therefore, if blood pressure is to be maintained in the face of substantial reductions in peripheral resistance (due to profound vasodilation within the peripheral tissues), blood flow must also increase in a proportionate manner (Fig. 3.1). During passive heat stress, this is achieved through elevations in cardiac output, which increases by ~3 l/min for every 1 °C increase in core body temperature [4, 10, 12–19], and can result in elevations in cardiac output of >5 l/min during severe whole-body hyperthermia. In basic terms, these profound increases in cardiac output can be mediated by one of the three mechanisms: an increase in heart rate, an increase in stroke volume, or a combination of the two.

Heart rate: Heart rate increases at a rate of ~35 beats/min/°C core temperature [3, 4, 12, 14–16, 20] and is indisputably the primary determinant of cardiac output during prolonged passive heat stress. The mechanisms underlying this increase are multifactorial, with roughly equal contributions from both neural pathways (60%) and a direct effect of temperature on the pacemaker of the heart itself (40%) [21]. Around 75% of neural-mediated increases result from the combined effects of increased sympathetic nerve activity and elevated levels of circulating catecholamines, altering the sino-atrial node pacemaker current and causing an increase in heart rate of ~7 beats/min for every 1 °C increase in mixed venous blood temperature.

Stroke volume: In contrast to heart rate, stroke volume has been found to remain unchanged or to even slightly increase during passive heat stress [4, 13, 22]. This finding is remarkable given that whole-body hyperthermia results in a profound reduction in central blood volume [23], cardiac filling time, cardiac filling pressure, and end-diastolic volume [3, 4, 22, 24–26], thereby decreasing the amount of blood available for ejection during each ventricular contraction. In spite of this, the preservation of stroke volume observed in the majority of studies to date suggests that compensatory adjustments to cardiac function are able to effectively counteract this challenge, either through improvements in diastolic function, systolic function, or both. Multiple studies have indicated that, despite decreases in end-diastolic volume, diastolic function may in fact be enhanced during hyperthermia [27, 28], thereby protecting stroke volume in the face of the decreased filling pressures previously described [22, 29]. Systolic function is also enhanced, as decreases in end-systolic volume mirror the drop in end-diastolic volume [22]. Although the cellular and molecular mechanisms underlying these regulatory adjustments are still not fully characterised and understood, potential explanations may involve the interaction of numerous functionally intertwined factors, including an enhanced intrinsic myocardial contractility [22], an increased rate of myocyte relaxation [29], and improved cardiac 'suction' during ventricular filling as a result of an increased left ventricular untwisting velocity [22, 28].

3.2.3 Vascular Responses to Passive Heat Stress

Skin blood flow: The skin—with its large surface area (1.5–2.0 m^2) and abundant sweat glands (~2 million)—provides a 30 μm thick interface between the human body and external environment, and is therefore the most important tissue for thermoregulatory control during heat exposure. As a result, the majority of the increase in cardiac output and blood flow 'redistribution' from the splanchnic, renal, and cerebral organs observed during prolonged passive heating is likely directed towards this vascular bed (Fig. 3.1). Although the exact magnitude is unknown because direct measures at whole-body level has never been obtained, whole-body skin blood flow (estimated from the increase in cardiac output and decline in visceral blood flow) is routinely reported to be in the region of 7–8 l/min at the most extreme levels of resting heat stress. Early findings which led to the adoption of this figure [3, 13, 30], however, have rarely—if ever—been replicated, with levels closer to 4–5 l/min more commonly reported [4, 10, 12, 14–19] (Fig. 3.1). The control of skin blood flow during heat stress is a complex process involving both central and local thermoregulatory networks, and can be additionally modified by numerous non-thermoregulatory feedback mechanisms such as exercise [31], circadian rhythms [32], metaboreceptors [33], and baroreceptors [34]. Core temperature is often considered the major driving stimulus behind changes in skin blood flow at rest, as a 1 °C increase in core temperature has been demonstrated to have a ninefold greater effect on skin blood flow than the same increase in skin temperature. It should be noted, however, that the capacity to alter skin temperature during prolonged or severe heating is substantially greater than that possible for core temperature (~15 °C compared to ~3 °C), resulting in the potential for a much greater contribution from local temperature-sensitive mechanisms than is often appreciated. In addition, with these peripheral mechanisms able to induce a maximal cutaneous vasodilation even when core temperature remains unchanged [35–39], passive heat stress via exposure to high skin temperatures therefore has the capacity to elevate skin blood flow through predominantly local mechanisms prior to the development of increased core temperatures [17, 40]. Although increases in cardiac output are sufficient to offset this profound cutaneous vasodilatory response at rest, the superimposition of additional stresses such as dehydration, orthostatic stress, or—as will be discussed subsequently—exercise, may have implications for the maintenance of perfusion pressure.

Muscle blood flow: The question of whether increases in skin blood flow account for the full hyperaemic response during passive heating has remained highly controversial for many years, mainly due to the difficulty in separating skin (average depth

~30 μm) and underlying skeletal muscle tissue perfusion in vivo and the variety of experimental techniques employed among studies. With early research concluding that peripheral vasodilation was confined to the skin alone [41–43], the widely held view has been that skeletal muscle tissue is unresponsive to increases in deep tissue and blood temperature, which themselves can increase up to 5 °C during local and whole-body heating [10, 17, 42, 44, 45]. However, a number of later studies have since cast doubt on these findings. Firstly, substantial elevations in thigh skeletal muscle perfusion have been observed in studies using high frequency microwave diathermy to achieve intense local deep muscle heating [46–50]. While it could be argued that the extreme stimulus (muscle temperature 40–45 °C) employed during this technique is outside normal physiological limits, later studies involving techniques such as contrast-enhanced ultrasound [51], near infra-red spectroscopy [52], Xe^{133} clearance [53], laser Doppler flowmetry [54], and duplex Doppler ultrasound [10, 17, 55] have since supported these findings using more traditional and physiologically relevant methods of passive heating. Collectively, these data—together with the observation that resting metabolic rate in heat-stressed humans is modestly elevated—indicate that skeletal muscle blood flow may make a modest yet meaningful contribution to heat-stress hyperemia, particularly when considering its large contribution to body mass. The mechanisms underlying this response remain unclear, but appear to be solely mediated at the local level and—much like in the skin—are the net result of the augmented vasodilator activity overriding the parallel increase in vasoconstrictor activity [45, 53, 55–59]. Multiple vascular, interstitial, and intracellular signalling mechanisms sensitive to increases in temperature are likely involved in this response [56, 58–62]. One potential intravascular mechanism for control of tissue blood flow with alterations in local temperature could be the recently proposed temperature-dependent release of ATP from the circulating erythrocytes [45, 58]. This is an attractive possibility because of ATP's strong vasodilator and sympatholytic properties [45, 63–66] and the direct relationship between blood temperature and plasma ATP during passive and active hyperaemic conditions [10, 45, 58, 67].

Visceral blood flow: In direct contrast to the peripheral circulation, splanchnic (i.e. stomach, liver, spleen, pancreas, intestine) and renal (kidney) blood flow is substantially reduced in the hyperthermic state (Fig. 3.1). This phenomenon most likely occurs due to temperature-driven reflex vasoconstriction within the abdominal vascular beds, thereby contributing to the redistribution of blood flow and volume to the peripheral vasculature and aiding in the maintenance of mean arterial pressure [68]. The presence of this response was first suggested in classic studies in the 1960s, where central blood volume and hepatic blood flow were estimated to decrease in some [30, 68], but not all studies [3]. Revisiting this question in more recent times, estimates of visceral blood flow have demonstrated decreases in the region of 30% during severe heat stress [4], while estimates of blood volume in different regions of the human torso have suggested reductions of between 14% and 23% in the heart, central vasculature, thorax, inferior vena cava, and liver [23].

These findings, combined with the well-documented decrease in central venous (right atrial) pressure during heat stress, highlight the ability of hyperthermia to elicit selective vasoconstriction within the central vasculature, thereby aiding the redistribution of up to ~1 l/min blood flow to the periphery independently of the much larger increase in tissue blood flow and cardiac output (~5 l/min).

Cerebral blood flow: Passive elevations in core body temperature are associated with progressive reductions in cerebral (CBF) blood flow (~0.1 l/min or 15% fall in CBF; Fig. 3.1) [12, 69], with these effects occurring largely as a function of a hyperventilation-induced reduction in the potent cerebral vasodilator CO_2 [70]. Increases in body core temperature by as little as 1 °C can result in changes in minute ventilation and CO_2 of +40% and −10%, respectively [12], while further increases >1.5 °C result in marked hyperventilation and a decrease in CBF to levels ~20% below that seen in thermoneutral conditions. In contrast to blood perfusion in the brain-supplying internal carotid and vertebral arteries, blood flow in the external carotid artery, which perfuses the face and neck [12, 69], increases in a linear fashion with body temperature—leading to an overall elevation in blood flow to the head of ~0.5 l/min or 48% [12] (Fig. 3.1). This response appears to act independently of CBF [71], and is likely driven by thermoregulatory (thermosensitive) rather than blood gas or pH-related mechanisms [12, 69].

3.2.4 Summary

Passive heat stress results in a decrease in peripheral resistance, an increase in cardiac output, and a selective elevation and redistribution of blood flow from the core to the extremities, head, and superficial tissues of the torso. Despite modest increases in aerobic metabolism, this response is predominantly driven by thermosensitive mechanisms, resulting in significant cardiovascular strain even at rest, alongside tissue blood flows well in excess of that required to meet additional metabolic requirements.

3.3 Respiratory and Cardiovascular Responses to Exercise Heat Stress

The combination of exercise and heat stress places additional functional and regulatory demands on the cardiovascular system as both the metabolic and thermoregulatory demands of the body must be met. A long-standing question is whether the human circulation can appropriately meet the demands for blood flow of both the active skeletal muscles and skin when exercise heat stress engaging a large muscle mass is performed (e.g. cycling, running, rowing, swimming). The answer to this question is a complex one, with evidence suggesting different circulatory responses depending on both the magnitude of hyperthermia experienced and the type of exercise performed (i.e. small vs. large muscle-mass exercise, prolonged submaximal exercise vs. incremental maximal exercise, etc.). While circulatory adjustments may be sufficient to support blood perfusion during small muscle-mass exercise and during early whole-body exercise at low-to-moderate intensities, the superimposition of heat stress during exhaustive whole-body exercise can rapidly push the cardiovascular system to its regulatory limits, where both peripheral blood flow and cardiac output become impaired [72, 73]. Under these circumstances, aerobic power and endurance capacity may be hindered due to the development of severe hyperthermia alongside an attenuated O_2 delivery to the active skeletal muscles, heart, and brain. In this section, we will explore how global cardiovascular function, O_2 delivery to regional tissues and organs, and removal of heat from exercising limbs are altered by exercise heat stress at different exercise intensities. Building upon our previous findings in resting conditions, we will begin with small muscle-mass exercise in the heat (when overall metabolic demands are low), before progressing to whole-body exercise at both submaximal and maximal levels. As the focus of this chapter is on whether the cardiovascular system is capable of sustaining dual metabolic and thermoregulatory requirements of exercise in the heat, we will address these questions using data from uncompensable heat-stress conditions, where skin temperature is held at very high levels (39–41 °C) while core temperature increases to different degrees (38.5–40 °C). As such, we will not address data pertaining to compensable heat-stress conditions, where core temperature can increase to 39–40 °C while becoming significantly dehydrated, but skin temperature is stable around 32–35 °C. Decrements in endurance capacity in these conditions are associated with the physiological and perceptual strain caused by dehydration and core hyperthermia, which are discussed elsewhere [74].

3.3.1 Respiratory Responses to Exercise Heat Stress

Oxygen transport from the atmospheric air to the skeletal muscle mitochondria (where the majority of ATP and heat are produced during exercise [75]) is a major determinant of aerobic power and endurance capacity, regardless of environmental conditions. As the lungs are the first potential site limiting oxygen transport through their impact on alveolar O_2 diffusion and—according to the Fick principle—arterial O_2 content, we will briefly discuss below whether or not respiratory responses may limit aerobic exercise capacity in the heat.

Submaximal exercise:Similar to observations during passive heat stress, prolonged submaximal exercise in the heat is accompanied by a ventilatory response higher than that required to meet metabolic demand [76–78]. This hyperventilation is characterised by an increase in breath frequency rather than tidal volume, and appears dependent solely on the rate of rise in core temperature, as skin temperature plays a negligible role [78].

Maximal exercise: Minute ventilation is also elevated during the submaximal stages of incremental exercise in the heat [76], before peaking at similar values to exercise in control conditions at maximum exercise intensities [76, 79]. In contrast to prolonged exercise, however, the ventilatory response during graded exercise is not solely dependent on thermal stimuli, as neural, metabolic, and reflex mechanisms may all play a role [80]. Regardless of the intensity of the exercise performed or the mechanisms involved, however, O_2 content is either unaffected or elevated even at the very highest levels of hyperthermia and/or exercise intensity, suggesting that decrements in O_2 transport and aerobic capacity during strenuous exercise in the heat do not arise from direct respiratory limitations negatively impacting blood O_2 content. In contrast, reductions in arterial CO_2 due to a hyperthermia-induced hyperventilatory response may have implications for the control of CBF, as will be discussed later in the chapter.

3.3.2 Cardiac and Peripheral Skeletal Muscle, Skin, and Visceral Responses to Exercise Heat Stress

Oxygen transport via the systemic circulation (i.e. convective O_2 delivery = blood flow x arterial O_2 content) is the next potential limiting step in oxygen utilisation by tissues and organs. In the following sections, we will review the systemic, muscle, skin, visceral, and cerebral vascular responses and their consequences on systemic and regional O_2 utilisation with emphasis on the influence of exercise type and intensity.

Submaximal exercise: Studies utilising small muscle-mass exercise (e.g. single-leg knee-extensor exercise) demonstrate an elevation in both leg blood flow and cardiac output during exercise heat stress—suggesting an additive effect of metabolic and thermoregulatory blood flow requirements when whole-body physiological demand is very low (power output 10–50 W; Figs. 3.3 and 3.4). Even at these low absolute exercise intensities, however, this effect is lower than would be expected given the magnitude of hyperemia observed when heat stress and exercise are studied in isolation. As an example, while severe passive hyperthermia results in elevations in leg blood flow and cardiac output of ~1 l/min and ~5 l/min, respectively, the superimposition of the same thermal stimulus onto single-leg knee-extensor exercise in the same individuals results in increases of only ~0.5 l/min and ~3 l/min—an attenuation of 40–50% of the response seen at rest [10, 55, 81]. When a larger muscle mass is engaged (e.g. cycling), the hyperthermia-induced augmentation in exercising limb blood flow disappears altogether (Fig. 3.4) [82–85], suggesting that the much greater metabolic and thermoregulatory effects of whole-body exercise on exercising leg blood flow (~16 to 19 l/min at peak exercise) cancel out the independent influence of hyperthermia observed during passive heating or small muscle-mass exercise (Figs. 3.1 and 3.4). This response is likely due in part to an attenuation in the skin circulation, as reduced cutaneous vasodilator activity during exercise has been shown to sharply curtail skin blood flow (compared to resting

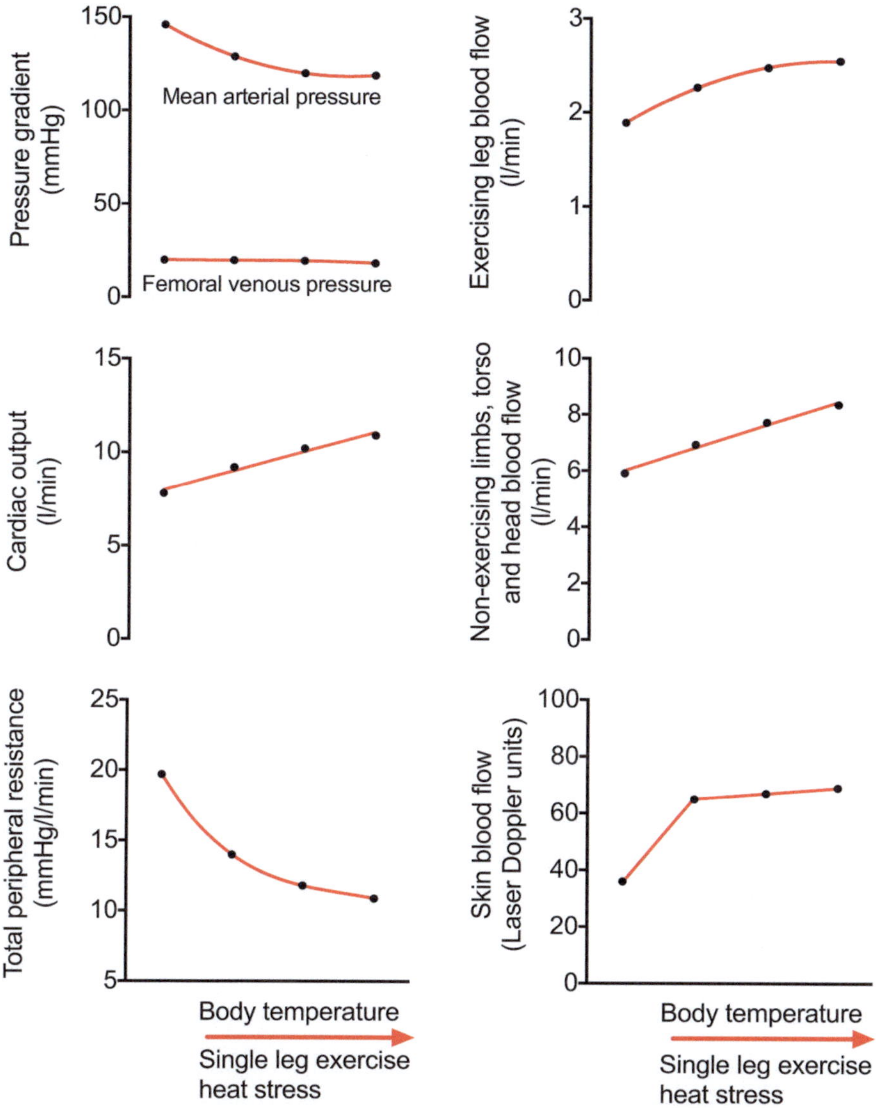

Fig. 3.3 Systemic and regional haemodynamic responses to combined constant-load single-leg knee-extensor exercise and graded heat stress. Redrawn from [10]

values) when core temperature increases >38 °C [86]. Although restrictions in skin blood flow may be expected to compromise thermoregulation, a number of factors likely help to offset this apparent insufficient vasodilatory response. Firstly, exercise reduces the core temperature at which sweating is initiated, thereby facilitating heat loss at an earlier stage [87]. Secondly, active muscle blood flow plays a key role in

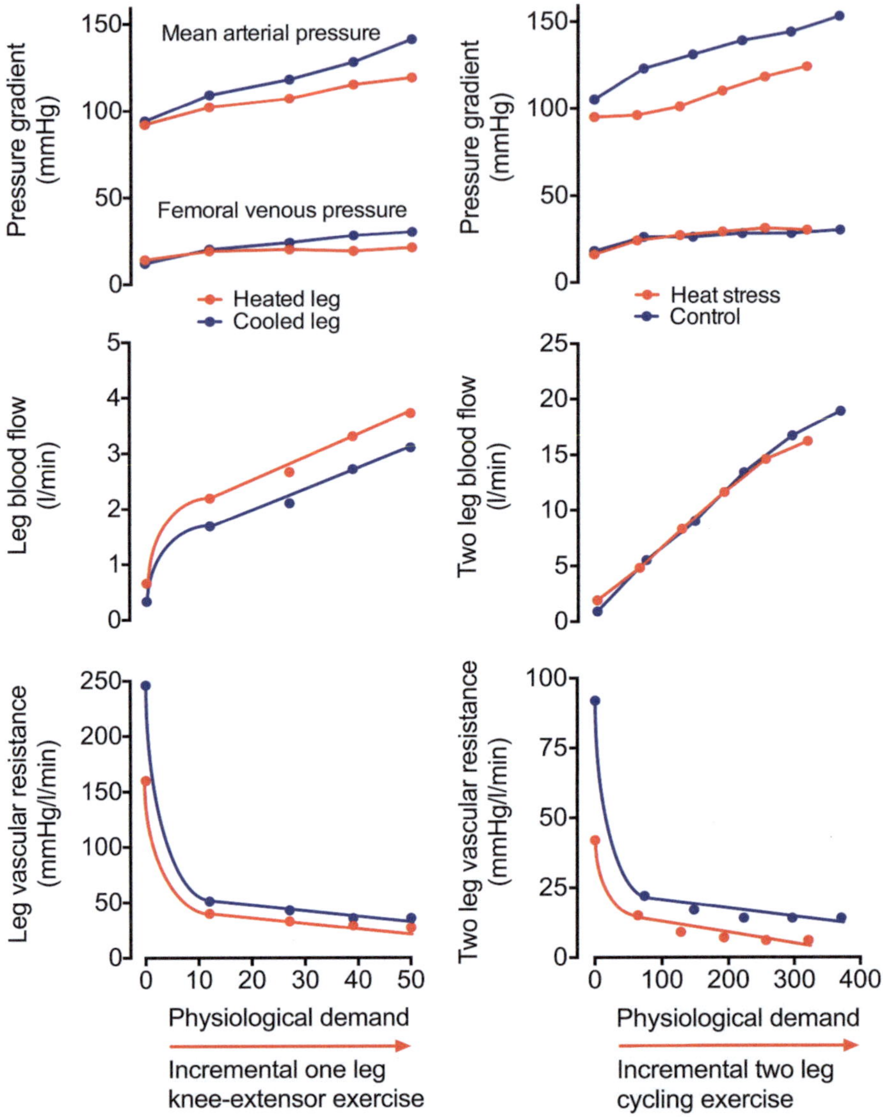

Fig. 3.4 Leg haemodynamic responses to incremental knee-extensor exercise to volitional exhaustion and incremental cycling to voluntary exhaustion with exposure to heat stress (induced by 1 h of whole-body exposure to heat stress with a water-perfused suit prior to exercise and maintained during exercise) and control conditions. Redrawn from [55, 85], respectively

thermoregulation during exercise, aiding in the convective transfer of heat from the interior of the exercising muscles to the overlying skin. This heat transfer pathway can account for the dissipation of >60% of the leg energy turnover during both prolonged [88] and incremental exercise [67], and thus may explain why exercise heat stress has no additive effects on locomotor limb blood flow during whole-body exercise. It may also help to explain why cardiac output only shows small increases

(1–3 l/min) compared to passive heat stress or exercise in control conditions [10, 82–85], as increases in exercising limb blood flow account for the majority of the increase in cardiac output during constant-load and incremental exercise [6, 7, 64, 72, 89]. Despite these haemodynamic adjustments, progressive increases in core temperature will not be prevented during uncompensable heat-stress conditions, and whole-body hyperthermia therefore becomes an important factor in the development of fatigue during prolonged exercise in hot environments [90].

Maximal exercise: The combination of heat stress and whole-body exercise at near-maximal intensities (e.g. 80–100% $\dot{V}O_{2max}$) poses one of the greatest challenges to cardiovascular control. Severe hyperthermia results in a quicker onset of fatigue due to a more rapid reduction in mean arterial blood pressure, cardiac output, and blood flow to the working skeletal muscles of the locomotor limbs compared to exercise at temperate ambient temperatures [72, 73] (Fig. 3.5). This fall in cardiac output occurs due to a decline in stroke volume, as heart rate is higher over

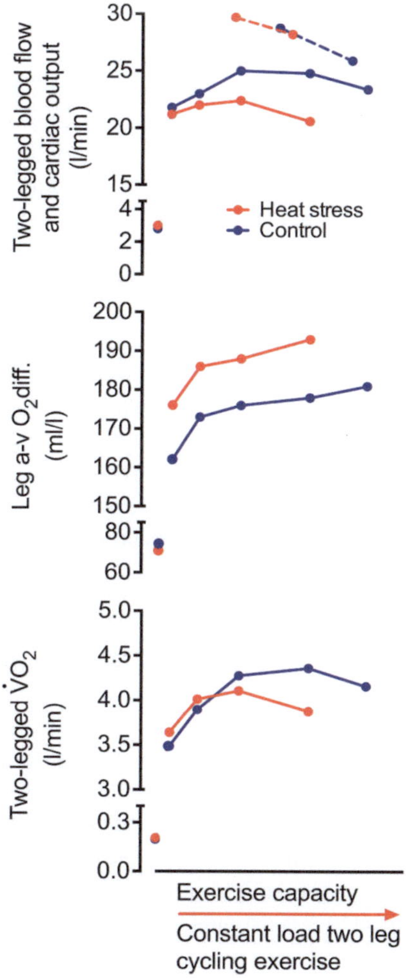

Fig. 3.5 Cardiac output (dashed lines) and leg haemodynamic and metabolic responses (solid lines) to constant-load cycling to voluntary exhaustion with exposure to heat stress (induced by 1 h of whole-body exposure to heat stress with a water-perfused suit prior to exercise and maintained during exercise) and control conditions. Redrawn from [72]

time and attains similar maximal values compared to temperate conditions [72, 73, 85, 91]. Similar to passive heat stress, incremental exercise results in selective reductions in visceral blood flow (up to 20–40% of resting values). The impact of exercise on visceral vasoconstriction appears stronger than that observed during graded passive heat stress [92, 93], and as a result, the addition of heat stress onto exercise does not seem to change the rate of decline in visceral blood flow as exercise intensity increases. It is therefore unlikely that redistribution of blood flow from splanchnic and renal organs contributes significantly to the peripheral blood flow responses during exercise heat stress. Ultimately, the fall in exercising leg blood flow appears to be associated with reductions in perfusion pressure secondary to the fall in cardiac output, rather than an active vasoconstriction process in the exercising limb itself, as limb vascular conductance is largely stable throughout exercise [85]. Although this accelerated decline in flow is often portrayed as an inability of the heart to supply both skin and muscle circulations for metabolic and thermoregulatory purposes [8], two major observations argue against a role of skin hyperperfusion per se as the limiting factor affecting maximal endurance capacity in hot environments. Firstly, stroke volume is reduced to a greater extent during maximal exercise in temperate conditions, compared to the same exercise performed in the heat [72]. Secondly, despite decreases in both stroke volume and cardiac output immediately prior to exhaustion in a hot environment, their absolute values remain elevated compared to those observed in temperature conditions [72]. Instead, a shortened cardiac cycle—perpetuated by hyperthermia-induced increases in heart rate—likely compounds the fall in cardiac filling and end-diastolic volume [22, 72, 94]. In these non-steady state physiological conditions, maximal heart rate is attained at a faster rate than in comparable exercise in a temperate environment and, immediately prior to exhaustion, is accompanied by falls in stroke volume, cardiac output, perfusion pressure, leg blood flow, and eventually, O_2 delivery to the working skeletal muscle [72, 90] (Fig. 3.5). This reduction in convective O_2 delivery reduces limb aerobic metabolism, because maximal functional O_2 extraction of the exercising limb tissues (~90 to 95%) is achieved in the early stages of severe-intensity exercise [45, 72]. Ultimately, these hemodynamic alterations compromise active limb and systemic oxygen uptake and, at least in part, contribute to the reduced aerobic power and endurance capacity when exercising in a hyperthermic state (Fig. 3.6).

3.3.3 Cerebrovascular Responses to Exercise Heat Stress

The cerebral vascular bed is small in comparison to other peripheral tissues, and is strongly regulated by autoregulation and arterial CO_2 levels. As such, its response to both heat stress and exercise is distinct to that observed in other peripheral vascular beds, and may independently contribute to impaired exercise capacity. Dynamic exercise requires the activation of motor and cardiorespiratory neurons, and is generally thought to necessitate increases in CBF and O_2 supply in order to meet the presumed greater 'brain metabolic activity' [95, 96]. As a result, any

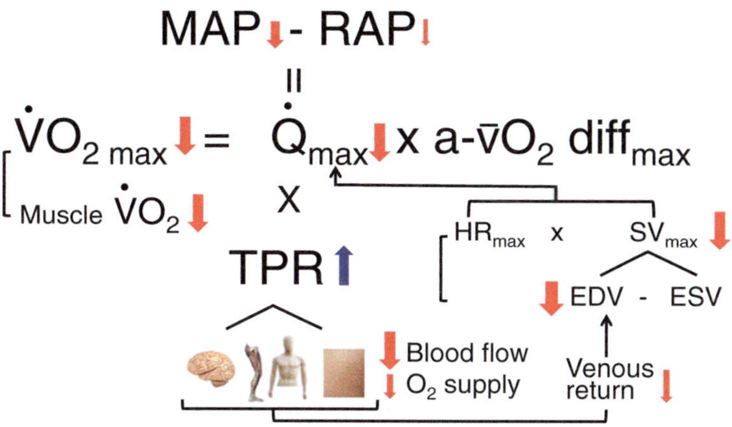

Fig. 3.6 Schematic illustration of the impact of severe whole-body exercise heat stress on physiological function according to Ohm's law and the Fick principle. Note that impaired aerobic capacity is associated with blunted peripheral blood flow and cardiac output and enhanced peripheral vascular resistance, preferentially affecting active skeletal muscle. Diminished venous return and filling of the heart is proposed as a primary factor reducing stroke volume, and thus cardiac output

factors compromising CBF may potentially compromise motor neurone activation and exercise capacity.

Submaximal exercise: At the onset of prolonged submaximal exercise (<60% $\dot{V}O_{2max}$), CBF increases by ~20% above baseline values, irrespective of the ambient conditions [76, 97–100]. If submaximal exercise is continued in a cool environment, CBF remains stable as long as the rise in internal body temperature and associated ventilatory and blood CO_2 responses are prevented [76, 91, 101]. However, the development of hyperthermia during exercise heat stress results in a relative hyperventilation (discussed previously), an accompanying reduction in the potent cerebral vasodilator CO_2—even at submaximal exercise intensities—and ultimately a progressive reduction of CBF back towards resting baseline values [76, 100, 101].

Maximal exercise: When incremental exercise is performed to near-maximal levels in temperate conditions, the augmented minute ventilation observed during the later stages (i.e. above the ventilatory threshold) also results in a fall in CBF towards baseline values [97, 98, 102]. When exercising in a hyperthermic state, this fall in CBF occurs both at a lower absolute work rate during incremental protocols [85, 99] and sooner during severe constant-load exercise [73], suggesting that augmented core temperature plays a role in the differing CBF dynamics. Similar to passive heat stress, blood flow to other tissues across the head remains high, with external carotid artery blood flow shown to more than double from resting values (up to ~700 ml/min) [73, 99, 100, 103]. This substantial increase in blood flow is related to the rise in body temperature, which is perhaps unsurprising as the external carotid arteries in part perfuse the cutaneous circulation of the face and neck. While this association has led some to hypothesise that the rise in external carotid artery blood flow may 'steal' blood destined for the cerebral vasculature [98], this is unlikely as changes in blood flow in these different tissues appear unrelated [69]. Despite the marked fall in CBF, however, global aerobic metabolism of the brain does not appear to be compromised at any level of exercise intensity due to compensatory elevations in brain O_2 extraction [73, 99]. Thus, reduced brain oxygen consumption is unlikely to contribute to the fall in aerobic power and endurance capacity during exercise heat stress (Fig. 3.6).

3.3.4 Summary

Exercise heat stress can place substantially greater functional and regulatory demands on the cardiovascular system compared to heat stress alone. Although the magnitude of cardiovascular response differs depending on the overall metabolic demand, the independent circulatory effects of heat stress and exercise are not additive. On the contrary, exercising limb blood flow during moderate to heavy intensity whole-body exercise is actually not different from that seen in normal conditions. The superimposition of heat stress onto maximal whole-body exercise, however, more rapidly pushes the cardiovascular system to its regulatory limits, where systemic and regional tissue and organ blood perfusion become restricted. While some organs can cope better with the decline in convective O_2 supply (e.g. the brain), a compromised active skeletal muscle O_2 metabolism is a major factor explaining why aerobic capacity is reduced in the heat-stressed human.

References

1. Crandall CG, González-Alonso J. Cardiovascular function in the heat-stressed human. Acta Physiol. 2010;199(4):407–23.
2. Tipton CM. Medicine AC of S. ACSM's advanced exercise physiology. Philadelphia: Lippincott Williams & Wilkins; 2006.
3. Rowell LB, Murray JA, Brengelmann GL, Kraning KK. Human cardiovascular adjustments to rapid changes in skin temperature during exercise. Circ Res. 1969;24(5): 711–24.
4. Minson CT, Wladkowski SL, Cardell AF, Pawelczyk JA, Kenney WL. Age alters the cardiovascular response to direct passive heating. J Appl Physiol. 1998;84(4):1323–32.
5. Andersen P, Saltin B. Maximal perfusion of skeletal muscle in man. J Physiol. 1985;366(1):233–49.
6. Mortensen SP, Dawson EA, Yoshiga CC, Dalsgaard MK, Damsgaard R, Secher NH, et al. Limitations to systemic and locomotor limb muscle oxygen delivery and uptake during maximal exercise in humans. J Physiol. 2005;566(1):273–85.
7. Mortensen SP, Damsgaard R, Dawson EA, Secher NH, González-Alonso J. Restrictions in systemic and locomotor skeletal muscle perfusion, oxygen supply and VO2 during high-intensity whole-body exercise in humans. J Physiol. 2008;586(10):2621–35.
8. Rowell LB. Human cardiovascular adjustments to exercise and thermal stress. Physiol Rev. 1974;54(1):75–159.
9. Kenney WL, Stanhewicz AE, Bruning RS, Alexander LM. Blood pressure regulation III: what happens when one system must serve two masters: temperature and pressure regulation? Eur J Appl Physiol. 2014;114(3):467–79.

10. Pearson J, Low DA, Stöhr E, Kalsi K, Ali L, Barker H, et al. Hemodynamic responses to heat stress in the resting and exercising human leg: insight into the effect of temperature on skeletal muscle blood flow. AJP Regul Integr Comp Physiol. 2011;300(3):R663–73.
11. White MD. Components and mechanisms of thermal hyperpnea. J Appl Physiol. 2006;101(2):655–63.
12. Ogoh S, Sato K, Okazaki K, Miyamoto T, Hirasawa A, Morimoto K, et al. Blood flow distribution during heat stress: cerebral and systemic blood flow. J Cereb Blood Flow Metab. 2013;33(12):1915–20.
13. Rowell LB, Brengelmann GL, Murray JA. Cardiovascular responses to sustained high skin temperature in resting man. J Appl Physiol. 1969;27(5):673–80.
14. Bonde-Petersen F, Schultz-Pedersen L, Dragsted N. Peripheral and central blood flow in man during cold, thermoneutral, and hot water immersion. Aviat Space Environ Med. 1992;63(5):346–50.
15. Niimi Y, Matsukawa T, Sugiyama Y, Shamsuzzaman ASM, Ito H, Sobue G, et al. Effect of heat stress on muscle sympathetic nerve activity in humans. J Auton Nerv Syst. 1997;63(1):61–7.
16. Ganio MS, Overgaard M, Seifert T, Secher NH, Johansson PI, Meyer MAS, et al. Effect of heat stress on cardiac output and systemic vascular conductance during simulated hemorrhage to presyncope in young men. AJP Hear Circ Physiol. 2012;302(8):H1756–61.
17. Chiesa ST, Trangmar SJ, González-Alonso J. Temperature and blood flow distribution in the human leg during passive heat stress. J Appl Physiol. 2016;120(9):1047–58.
18. Peters JK, Nishiyasu T, Mack GW. Reflex control of the cutaneous circulation during passive body core heating in humans. J Appl Physiol. 2000;88(5):1756–64.
19. Nelson MD, Altamirano-Diaz LA, Petersen SR, DeLorey DS, Stickland MK, Thompson RB, et al. Left ventricular systolic and diastolic function during tilt-table positioning and passive heat stress in humans. Am J Physiol Circ Physiol. 2011;301(2):H599–608.
20. Wyss CR, Brengelmann GL, Johnson JM, Rowell LB, Silverstein D. Altered control of skin blood flow at high skin and core temperatures. J Appl Physiol. 1975;38(5):839–45.
21. Gorman AJ, Proppe DW. Mechanisms producing tachycardia in conscious baboons during environmental heat stress. J Appl Physiol. 1984;56:441–6.
22. Stöhr EJ, González-Alonso J, Pearson J, Low DA, Ali L, Barker H, et al. Effects of graded heat stress on global left ventricular function and twist mechanics at rest and during exercise in healthy humans. Exp Physiol. 2011;96(2):114–24.
23. Crandall CG, Wilson TE, Marving J, Vogelsang TW, Kjaer A, Hesse B, et al. Effects of passive heating on central blood volume and ventricular dimensions in humans. J Physiol. 2008;586(1):293–301.
24. Keller DM, Low DA, Wingo JE, Brothers RM, Hastings J, Davis SL, et al. Acute volume expansion preserves orthostatic tolerance during whole-body heat stress in humans. J Physiol. 2009;587(5):1131–9.
25. Wilson TE, Brothers RM, Tollund C, Dawson EA, Nissen P, Yoshiga CC, et al. Effect of thermal stress on Frank-Starling relations in humans. J Physiol. 2009;587(13):3383–92.
26. Wilson TE, Tollund C, Yoshiga CC, Dawson EA, Nissen P, Secher NH, et al. Effects of heat and cold stress on central vascular pressure relationships during orthostasis in humans. J Physiol. 2007;585(1):279–85.
27. Brothers RM, Bhella PS, Shibata S, Wingo JE, Levine BD, Crandall CG. Cardiac systolic and diastolic function during whole body heat stress. Am J Physiol Circ Physiol. 2009;296(4):H1150–6.
28. Nelson MD, Haykowsky MJ, Petersen SR, DeLorey DS, Cheng-Baron J, Thompson RB. Increased left ventricular twist, untwisting rates, and suction maintain global diastolic function during passive heat stress in humans. Am J Physiol Circ Physiol. 2010;298(3): H930–7.
29. Brothers RM, Bhella PS, Shibata S, Wingo JE, Levine BD, Crandall CG. Cardiac systolic and diastolic function during whole body heat stress. Am J Physiol Heart Circ Physiol. 2009;296(4):H1150–6.

30. Rowell LB, Brengelmann GL, Blackmon JR, Murray JA. Redistribution of blood flow during sustained high skin temperature in resting man. J Appl Physiol. 1970;28(4):415–20.
31. Taylor WF, Johnson JM, Kosiba WA, Kwan CM. Graded cutaneous vascular responses to dynamic leg exercise. J Appl Physiol. 1988;64(5):1803–9.
32. Stephenson LA, Wenger CB, O'Donovan BH, Nadel ER. Circadian rhythm in sweating and cutaneous blood flow. Am J Phys. 1984;246(3):R321–4.
33. Crandall CG, Stephens DP, Johnson JM. Muscle metaboreceptor modulation of cutaneous active vasodilation. Med Sci Sports Exerc. 1998;30(4):490–6.
34. Jackson DN, Kenny GP. Upright LBPP application attenuates elevated postexercise resting thresholds for cutaneous vasodilation and sweating. J Appl Physiol. 2003;95(1):121–8.
35. Kellogg DL Jr, Liu Y, Kosiba IF, O'Donnell D. Role of nitric oxide in the vascular effects of local warming of the skin in humans. J Appl Physiol. 1999;86(4):1185–90.
36. Kellogg DL, Zhao JL, Wu Y. Endothelial nitric oxide synthase control mechanisms in the cutaneous vasculature of humans in vivo. Am J Physiol Heart Circ Physiol. 2008;295(1):H123–9.
37. Johnson JM, Kellogg DL Jr. Local thermal control of the human cutaneous circulation. J Appl Physiol. 2010;109(4):1229–38.
38. Minson CT, Berry LT, Joyner MJ. Nitric oxide and neurally mediated regulation of skin blood flow during local heating. J Appl Physiol. 2001;91(4):1619–26.
39. Black MA, Green DJ, Cable NT. Exercise prevents age-related decline in nitric-oxide-mediated vasodilator function in cutaneous microvessels. J Physiol. 2008;586(14):3511–24.
40. Taylor WF, Johnson JM, O'Leary D, Park MK. Effect of high local temperature on reflex cutaneous vasodilation. J Appl Physiol. 1984;57(1):191–6.
41. Johnson JM, Brengelmann GL, Rowell LB. Interactions between local and reflex influences on human forearm skin blood flow. J Appl Physiol. 1976;41(6):826–31.
42. Detry JM, Brengelmann GL, Rowell LB, Wyss C. Skin and muscle components of forearm blood flow in directly heated resting man. J Appl Physiol. 1972;32(4):506–11.
43. Hales JRS, Rowell LB, King RB. Regional distribution of blood flow in awake heat-stressed baboons. Am J Phys. 1979;237(6):H705–12.
44. Abramson DI, Kahn A, Tuck S Jr, Turman GA, Rejal H, Fleischer CJ. Relationship between a range of tissue temperature and local oxygen uptake in the human forearm. I. Changes observed under resting conditions. J Clin Invest. 1958;37(7):1031–8.
45. Kalsi KK, Chiesa ST, Trangmar SJ, Ali L, Lotlikar MD, González-Alonso J. Mechanisms for the control of local tissue blood flow during thermal interventions: influence of temperature-dependent ATP release from human blood and endothelial cells. Exp Physiol. 2017;102(2):228–44.
46. Wyper DJ, McNiven DR. The effect of microwave therapy upon muscle blood flow in man. Br J Sports Med. 1976;10(1):19–21.
47. Lehmann JF, Guy AW, Stonebridge JB, DeLateur BJ. Evaluation of a therapeutic direct-contact 915-MHz microwave applicator for effective deep-tissue heating in humans. IEEE Trans Microw Theory Tech. 1978;26(8):556–63.
48. Sekins KM, Lehmann JF, Esselman P, Dundore D, Emery AF, DeLateur BJ, et al. Local muscle blood flow and temperature responses to 915MHz diathermy as simultaneously measured and numerically predicted. Arch Phys Med Rehabil. 1984;65(1):1–7.
49. Song CW. Effect of local hyperthermia on blood flow and microenvironment: a review. Cancer Res. 1984;44(S10):4721.
50. Giombini A, Giovannini V, Di Cesare A, Pacetti P, Ichinoseki-Sekine N, Shiraishi M, et al. Hyperthermia induced by microwave diathermy in the management of muscle and tendon injuries. Br Med Bull. 2007;83(1):379–96.
51. Akyürekli D, Gerig LH, Raaphorst GP. Changes in muscle blood flow distribution during hyperthermia. Int J Hyperth. 1997;13(5):481–96.
52. Okada K, Yamaguchi T, Minowa K, Inoue N. The influence of hot pack therapy on the blood flow in masseter muscles. J Oral Rehabil. 2005;32(7):480–6.

53. Keller DM, Sander M, Stallknecht B, Crandall CG. α-Adrenergic vasoconstrictor responsiveness is preserved in the heated human leg. J Physiol. 2010;588(19):3799–808.
54. Binzoni T, Tchernin D, Richiardi J, Van De Ville D, Hyacinthe J-N. Haemodynamic responses to temperature changes of human skeletal muscle studied by laser-Doppler flowmetry. Physiol Meas. 2012;33(7):1181–97.
55. Chiesa ST, Trangmar SJ, Kalsi KK, Rakobowchuk M, Banker DS, Lotlikar MD, et al. Local temperature-sensitive mechanisms are important mediators of limb tissue hyperemia in the heat-stressed human during rest and small muscle mass exercise. Am J Physiol Circ Physiol. 2015;309:H369–80.
56. Heinonen I, Brothers RM, Kemppainen J, Knuuti J, Kalliokoski KK, Crandall CG. Local heating, but not indirect whole body heating, increases human skeletal muscle blood flow. J Appl Physiol. 2011;111(3):818–24.
57. Ives SJ, Andtbacka RH, Kwon SH, Shiu YT, Ruan T, Noyes RD, et al. Heat and alpha1-adrenergic responsiveness in human skeletal muscle feed arteries: the role of nitric oxide. J Appl Physiol. 2012;113(11):1690–8.
58. Kalsi KK, González-Alonso J. Temperature-dependent release of ATP from human erythrocytes: mechanism for the control of local tissue perfusion. Exp Physiol. 2012;97(3):419–32.
59. Ives SJ, Andtbacka RHI, Noyes RD, McDaniel J, Amann M, Witman MAH, et al. Human skeletal muscle feed arteries studied in vitro: the effect of temperature on α1-adrenergic responsiveness. Exp Physiol. 2011;96(9):907–18.
60. Heinonen I, Wendelin-Saarenhovi M, Kaskinoro K, Knuuti J, Scheinin M, Kalliokoski KK. Inhibition of alpha-adrenergic tone disturbs the distribution of blood flow in the exercising human limb. Am J Physiol Circ Physiol. 2013;305(2):H163–72.
61. Kluess HA, Buckwalter JB, Hamann JJ, Clifford PS. Elevated temperature decreases sensitivity of P2X purinergic receptors in skeletal muscle arteries. J Appl Physiol. 2005;99(3): 995–8.
62. Fujii N, McGinn R, Halili L, Singh MS, Kondo N, Kenny GP. Cutaneous vascular and sweating responses to intradermal administration of ATP: a role for nitric oxide synthase and cyclooxygenase? J Physiol. 2015;593(11):2515–25.
63. González-Alonso J, Olsen DB, Saltin B. Erythrocyte and the regulation of human skeletal muscle blood flow and oxygen delivery role of circulating ATP. Circ Res. 2002;91(11):1046–55.
64. González-Alonso J, Mortensen SP, Jeppesen TD, Ali L, Barker H, Damsgaard R, et al. Haemodynamic responses to exercise, ATP infusion and thigh compression in humans: insight into the role of muscle mechanisms on cardiovascular function. J Physiol. 2008;586(9):2405–17.
65. Rosenmeier JB, Hansen J, González-Alonso J. Circulating ATP-induced vasodilatation overrides sympathetic vasoconstrictor activity in human skeletal muscle. J Physiol. 2004;558(1):351–65.
66. Rosenmeier JB, Yegutkin GG, González-Alonso J. Activation of ATP/UTP-selective receptors increases blood flow and blunts sympathetic vasoconstriction in human skeletal muscle. J Physiol. 2008;586(20):4993–5002.
67. González-Alonso J, Calbet JAL, Boushel R, Helge JW, Søndergaard H, Munch-Andersen T, et al. Blood temperature and perfusion to exercising and non-exercising human limbs. Exp Physiol. 2015;100(10):1118–31.
68. Rowell LB, Detry JR, Profant GR, Wyss C. Splanchnic vasoconstriction in hyperthermic man--role of falling blood pressure. J Appl Physiol. 1971;31(6):864–9.
69. Bain AR, Smith KJ, Lewis NC, Foster GE, Wildfong KW, Willie CK, et al. Regional changes in brain blood flow during severe passive hyperthermia: effects of PaCO2 and extracranial blood flow. J Appl Physiol. 2013;115(5):653–9.
70. Willie CK, Tzeng Y-C, Fisher JA, Ainslie PN. Integrative regulation of human brain blood flow. J Physiol. 2014;592(5):841–59.
71. Miyazawa T, Horiuchi M, Ichikawa D, Subudhi AW, Sugawara J, Ogoh S. Face cooling with mist water increases cerebral blood flow during exercise: effect of changes in facial skin blood flow. Front Physiol. 2012;3:308.

72. González-Alonso J, Calbet JAL. Reductions in systemic and skeletal muscle blood flow and oxygen delivery limit maximal aerobic capacity in humans. Circulation. 2003;107(6): 824–30.
73. González-Alonso J, Dalsgaard MK, Osada T, Volianitis S, Dawson EA, Yoshiga CC, et al. Brain and central haemodynamics and oxygenation during maximal exercise in humans. J Physiol. 2004;557(1):331–42.
74. Trangmar SJ, González-Alonso J. New insights into the impact of dehydration on blood flow and metabolism during exercise. Exerc Sport Sci Rev. 2017;45(3):146–53.
75. González-Alonso J, Quistorff B, Krustrup P, Bangsbo J, Saltin B. Heat production in human skeletal muscle at the onset of intense dynamic exercise. J Physiol. 2000;524(2):603–15.
76. Nybo L, Nielsen B. Middle cerebral artery blood velocity is reduced with hyperthermia during prolonged exercise in humans. J Physiol. 2001;534(1):279–86.
77. Fujii N, Honda Y, Hayashi K, Soya H, Kondo N, Nishiyasu T. Comparison of hyperthermic hyperpnea elicited during rest and submaximal, moderate-intensity exercise. J Appl Physiol. 2008;104(4):998–1005.
78. Hayashi K, Honda Y, Ogawa T, Kondo N, Nishiyasu T. Relationship between ventilatory response and body temperature during prolonged submaximal exercise. J Appl Physiol. 2006;100(2):414–20.
79. Keiser S, Flück D, Stravs A, Hüppin F, Lundby C. Restoring heat stress-associated reduction in middle cerebral artery velocity does not reduce fatigue in the heat. Scand J Med Sci Sports. 2015;25(1):145–53.
80. Forster HV, Haouzi P, Dempsey JA. Control of breathing during exercise. Compr Physiol. 2012;2(1):743–77.
81. Ferguson RA, Krustrup P, Kjaer M, Mohr M, Ball D, Bangsbo J. Effect of temperature on skeletal muscle energy turnover during dynamic knee-extensor exercise in humans. J Appl Physiol. 2006;101(1):47–52.
82. Nielsen B, Savard G, Richter EA, Hargreaves M, Saltin B. Muscle blood flow and muscle metabolism during exercise and heat stress. J Appl Physiol. 1990;69(3):1040–6.
83. Nielsen B, Hales JR, Strange S, Christensen NJ, Warberg J, Saltin B. Human circulatory and thermoregulatory adaptations with heat acclimation and exercise in a hot, dry environment. J Physiol. 1993;460(1):467–85.
84. Nielsen B, Strange S, Christensen NJ, Warberg J, Saltin B. Acute and adaptive responses in humans to exercise in a warm, humid environment. Pflugers Arch - Eur J Physiol. 1997;434(1):49–56.
85. Trangmar SJ, Chiesa ST, Kalsi KK, Secher NH, González-Alonso J. Whole body hyperthermia, but not skin hyperthermia, accelerates brain and locomotor limb circulatory strain and impairs exercise capacity in humans. Physiol Rep. 2017;5(1):e13108.
86. Kellogg DL, Johnson JM, Kenney WL, Pergola PE, Kosiba WA. Mechanisms of control of skin blood flow during prolonged exercise in humans. Am J Physiol Circ Physiol. 1993;265(2):562–8.
87. Shibasaki M, Wilson TE, Crandall CG. Neural control and mechanisms of eccrine sweating during heat stress and exercise. J Appl Physiol. 2006;100:1692–701.
88. González-Alonso J, Calbet JA, Nielsen B. Metabolic and thermodynamic responses to dehydration-induced reductions in muscle blood flow in exercising humans. J Physiol. 1999;520:577–89.
89. Calbet JA, González-Alonso J, Helge JW, Sondergaard H, Munch-Andersen T, Boushel R, et al. Cardiac output and leg and arm blood flow during incremental exercise to exhaustion on the cycle ergometer. J Appl Physiol. 2007;103(3):969–78.
90. González-Alonso J, Teller C, Andersen SL, Jensen FB, Hyldig T, Nielsen B. Influence of body temperature on the development of fatigue during prolonged exercise in the heat. J Appl Physiol. 1999;86(3):1032–9.
91. Nybo L, Møller K, Volianitis S, Nielsen B, Secher NH, Moller K, et al. Effects of hyperthermia on cerebral blood flow and metabolism during prolonged exercise in humans. J Appl Physiol. 2002;93(1):58–64.

92. Rowell LB. Human circulation: regulation during physical stress. Oxford: Oxford University Press; 1986. p. 416.
93. Rowell LB. Cardiovascular adjustments to thermal stress. Comprehensive physiology. Hoboken: Wiley; 2011.
94. Trinity JD, Pahnke MD, Lee JF, Coyle EF. Interaction of hyperthermia and heart rate on stroke. J Appl Physiol. 2011;111(3):891–7.
95. Buxton RB, Frank LR. A model for the coupling between cerebral blood flow and oxygen metabolism during neural stimulation. J Cereb Blood Flow Metab. 1997;17(1):64–72.
96. Secher NH, Seifert T, Van Lieshout JJ. Cerebral blood flow and metabolism during exercise: implications for fatigue. J Appl Physiol. 2008;104(1):306–14.
97. Hellstrom G, Fischer-Colbrie W, Wahlgren NG, Jogestrand T. Carotid artery blood flow and middle cerebral artery blood flow velocity during physical exercise. J Appl Physiol. 1996;81(1):413–8.
98. Sato K, Ogoh S, Hirasawa A, Oue A, Sadamoto T. The distribution of blood flow in the carotid and vertebral arteries during dynamic exercise in humans. J Physiol. 2011;589(11):2847–56.
99. Trangmar SJ, Chiesa ST, Stock CG, Kalsi KK, Secher NH, González-Alonso J. Dehydration affects cerebral blood flow but not its metabolic rate for oxygen during maximal exercise in trained humans. J Physiol. 2014;592(14):3143–60.
100. Trangmar SJ, Chiesa ST, Llodio I, Garcia B, Kalsi KK, Secher NH, et al. Dehydration accelerates reductions in cerebral blood flow during prolonged exercise in the heat without compromising brain metabolism. Am J Physiol Heart Circ Physiol. 2015;309(9):H1598–607.
101. Périard JD, Racinais S. Heat stress exacerbates the reduction in middle cerebral artery blood velocity during prolonged self-paced exercise. Scand J Med Sci Sports. 2015;25(S1):135–44.
102. Moraine JJ, Lamotte M, Berré J, Niset G, Leduc A, Naeijel R. Relationship of middle cerebral artery blood flow velocity to intensity during dynamic exercise in normal subjects. Eur J Appl Physiol Occup Physiol. 1993;67(1):35–8.
103. Sato K, Oue A, Yoneya M, Sadamoto T, Ogoh S. Heat stress redistributes blood flow in the arteries of the brain during dynamic exercise. J Appl Physiol. 2016;120(7):766–73.

Neural and Muscular Function in the Heat

4

Sébastien Racinais, Mohammed Ihsan, and Julien D. Périard

4.1 Introduction

Body temperature, muscle temperature in particular, modulates performance during short duration explosive exercise (e.g., sprinting and jumping) through adjustments in neuromuscular function [1]. For instance, muscle power production has been shown to be reduced following the administration of lower body cold water immersion, and in contrast shown to be enhanced following hot water immersion of the legs [2, 3]. Hence there is a positive relationship between muscle temperature and contractile performance, and depending on contraction type and velocity, a 2–5% change in performance can be expected per degree of variation in temperature [4]. However, this benefit tends to dissipate when sprints are repeated [5] and hyperthermia develops [6]. Indeed, the development of whole-body hyperthermia may lead to various alterations impairing neural drive [4].

The integrity of the neural system in the heat depends on both central (i.e., supraspinal) and peripheral (i.e., spinal and peripheral neural system) responses. The first animal studies (in anesthetised cats) examining the influence of hyperthermia on central nervous system (CNS) activity reported altered cortical excitability following the attainment of central temperatures of 44–45 °C [7]. The effects of hyperthermia have since been investigated in exercising humans on many facets of the CNS, including cognitive function, perceptual responses, circulatory adjustments, neural activity, and muscle activation. At the peripheral level (i.e., spinal modulation and proprioception) it has been shown that an increase in muscle temperature can modify contractile properties, oxidative capacity, and substrate utilisation. There is also

S. Racinais (✉) · M. Ihsan
Athlete Health and Performance Research Centre, Aspetar, Doha, Qatar
e-mail: sebastien.racinais@aspetar.com

J. D. Périard
University of Canberra Research Institute for Sport and Exercise (UCRISE), Canberra, ACT, Australia

J. D. Périard, S. Racinais (eds.), *Heat Stress in Sport and Exercise*, https://doi.org/10.1007/978-3-319-93515-7_4

emerging research demonstrating enhanced signalling within the myofibrillar and mitochondrial regulating pathways following heat exposure. This chapter will present the effect of heat on the entire neuromuscular system, including the central and peripheral nervous systems, and skeletal muscle.

4.2 Cerebral Responses

4.2.1 Cognitive Functions

The effect of hyperthermia on the CNS can be evidenced through an impairment in cognitive function [8, 9], with complex tasks shown to be more affected by hyperthermia than simple tasks [10–13] (Fig. 4.1). It is purported that thermal stress induces pleasure or displeasure depending on the change favouring or perturbing homeostasis [14]. As such, the alliesthesial change accompanying compensatory physiological responses to hot environmental conditions may impair complex task performance, especially during the dynamic phase of temperature change [12].

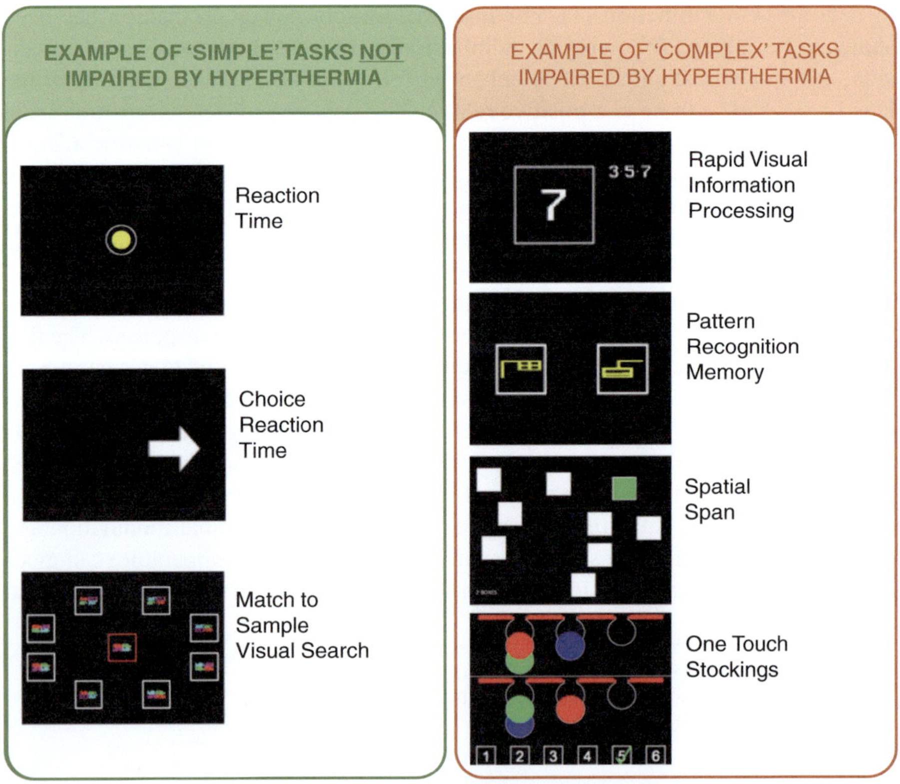

Fig. 4.1 Example of cognitive tasks being impaired or not by hyperthermia. Example of test from the CANTABeclipse battery (Cambridge Cognition, Cambridge, UK)

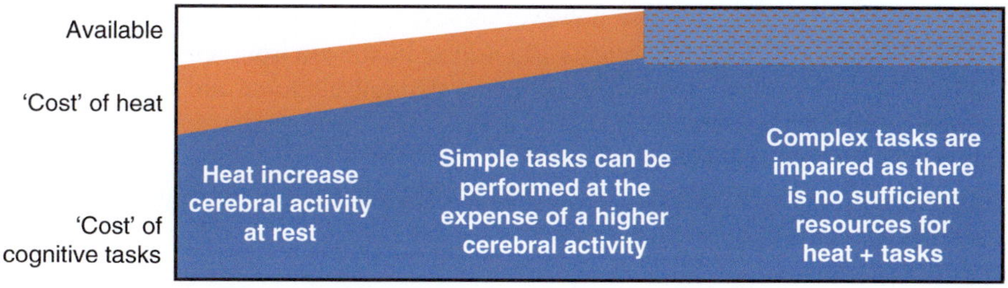

Fig. 4.2 Increasing task complexity increases cerebral activity such as theta wave (blue). However, hyperthermia (orange) also increases cerebral activity, limiting the amount of available resource (white)

Thus, thermal stress and strain represent a cognitive load, potentially limiting the resources available to process complex tasks [9]. For example, whilst theta activity increases with cognitive task complexity, it also increases when exposed to heat stress [15]. Indeed, it was shown that despite brain activity being higher at rest in the heat, a simple planning task could still be performed, albeit at the cost of a higher activity, but a complex planning task requiring more cognitive resources was impaired [15] (Fig. 4.2). This model may also partly explain why cooling the head and reducing thermal discomfort protect the ability to perform some complex cognitive tasks (e.g., short-term memory) in hot environments [16, 17]. Of note, hyperthermia may also increase impulsivity [13], as suggested by an increase in the rate of false alarms during a sustained attention task [17], or by the faster but false responses recorded during a complex planning task [12].

Heat acclimation appears to protect cognitive function in the heat [18–20]. Repeated passive heat exposure has been observed to negate the hyperthermia-induced decrement in complex cognitive function (i.e., planning task) and allow for a similar performance in hyperthermic compared with normothermic individuals. The similarity in performance occurred despite thermal sensation, thermal discomfort and negative to positive affects ratio, remaining increased in the hyperthermic state [20].

4.2.2 Perceptual Responses

Whilst the effects of heat perception on behavioural thermoregulation are detailed in Chap. 1, it should also be acknowledged that the increased 'cognitive load' induced by heat exposure may be partly related to the sense of displeasure when homeostasis is perturbed [14]. Passive hyperthermia has been shown to increase negative affect (with or without change in positive affects) as estimated by the positive and negative affect schedule (PANAS) [13, 20]. Along the same lines, it has been suggested that hyperthermia may reduce motivation and thus exercise capacity [21, 22]. Others have further suggested that thermal perception in response to a high skin temperature may affect work rate at the onset of exercise [23].

This remains a tenuous premise however as several studies have shown that exercise is initiated at the same intensity regardless of the environmental temperature [24–28], or previous heat exposures [29]; and that muscle force and voluntary activation is decreased by an elevated core temperature rather than skin temperatures [30, 31]. As such, whilst heat perception likely affects cognitive performances, physical performance appears to be more affected by the physiological strain consequent to the heat stress.

4.2.3 Cerebral Circulatory Responses

The development of hyperthermia during constant rate cycling in the heat is associated with a progressive reduction in cerebral blood flow, typically measured as middle cerebral artery mean blood velocity (MCA V_{mean}) [32–35]. This reduction is in contrast to the stable elevation in MCA V_{mean} that is maintained under cool conditions when a physiological steady state occurs [32]. The reduction in cerebral blood flow under heat stress is purported to occur in response to a decrease in cardiac output and arterial blood pressure, and to hyperventilation-induced hypocapnia (i.e., a decrement in arterial carbon dioxide pressure; $PaCO_2$) [32, 33], which accounts for 30–50% of the reduction [32, 34]. The difference in cerebral blood flow observed between constant rate exercise in hot and cool conditions may also stem from difference in relative exercise intensity, induced by the development of thermal and cardiovascular strain in the heat [26]. Indeed, as maximal aerobic capacity decreases with increasing levels of thermal strain [36], adjustments in cardiovascular function occur that result in a progressive increase in relative exercise intensity, leading to a rise in ventilation and a decrease in $PaCO_2$.

In contrast to constant rate exercise, cerebral blood flow increases at the onset of self-paced exercise in both hot and cool environmental conditions, but then progressively decreases [26]. This occurs in response to a relative exercise intensity of around 80–85% VO_{2max} during 40 km time trial efforts [27, 37]. Consequently, a hyperventilation-induced decrease in $PaCO_2$ occurs, contributing to the reduced cerebral blood flow. This reduction is exacerbated when exercising in the heat in part due to a decrease in arterial blood pressure and increase in peripheral blood flow redistribution [26]. Interestingly, decreases in cerebral blood flow are associated with a brain temperature of ~0.2 °C higher than that of the body core during exercise, independent of thermal state (i.e., hyperthermia) [38]. Reductions in cerebral blood flow during strenuous exercise have also been suggested to compromise the ability of the central nervous system to voluntarily activate skeletal muscles, consequent to inadequate oxygen delivery to the brain [32, 39, 40]. The development of fatigue during such high-intensity exercise however, with or without hyperthermia, is associated with an enhanced cerebral metabolism manifested by a compensatory increase in oxygen extraction in the brain [33, 41–43]. Accordingly, the exacerbated reduction in cerebral blood flow observed under heat stress does not appear to mediate an additional decrement in performance via central inhibition.

4.2.4 Cerebral Activity

Electroencephalographic activity (EEG) is a direct real-time measure of neural processing. It allows for assessing the relationship between cerebrocortical activity and behaviour using electrodes placed at specific locations across the scalp. EEG activity is evaluated along various frequency bands such as alpha (α: 8–12 Hz) and beta (β: 12–30 Hz), which are associated with different states of brain functioning. For example, α activity is linked to the capacity to maintain attention, mental readiness, and relaxed focus, whereas β activity is related to wakefulness, mental activity, and cortical arousal [44, 45].

When exhaustive exercise is undertaken in the heat, a decrease in β activity has been reported in the frontal lobe (i.e., F3–F4), in conjunction with the rise in core temperature and ratings of perceived exertion [46–49]. Similar findings were reported following the completion of a self-paced time trial in the heat [50] and suggested to represent a suppression of arousal, with potential links to the development of fatigue and the impairment of performance under heat stress [46–48]. A more recent study has extended these findings by reporting a decrease in both α and β activity in the frontal and central areas during high-intensity self-paced exercise in hot compared with cool conditions [51]. This decrement was associated with an increased metabolic activity and thermal perception (i.e., discomfort), as the strongest predictors of change in β activity were power output, oxygen uptake, and thermal discomfort. The authors also noted that the decrement in α activity in the heat was seemingly time and intensity dependent [51]. These findings are in line with reductions in prefrontal and motor cortex activity in the α, β, and γ frequency bands during incremental exercise in temperate conditions, which occurred upon reaching the respiratory compensation point (~85% VO_{2max}) [52], an intensity typically maintained during prolonged self-paced exercise [27, 28, 37].

To localise the specific areas where changes in cerebrocortical activity occurred, Périard et al. [51] used standardised low-resolution brain electromagnetic tomography (sLORETA). They showed that α and β activity increased at the onset of self-paced exercise in cool conditions in various brain areas (e.g., the primary somatosensory cortex, motor cortices, dorsolateral prefrontal cortex, insula, supramarginal gyrus, cingulate cortex, and frontal gyrus), whilst only β activity increased in the heat (Fig. 4.3). Thereafter, a decrease in activity ensued, with α activity being lower in the somatosensory and somatosensory association cortices under heat stress. This suggests that prolonged high-intensity exercise in the heat is associated with changes in brain activity that are linked to a reduced capacity to inhibit conflicting attentional processing, decreased mental readiness, and suppressed arousal [51].

Of note, following a series of studies, Nybo and Nielsen [47] speculated that a reduction in cerebral blood flow during exercise in the heat may contribute to a decrease in EEG activity in the prefrontal area, consequently increasing perceived exertion and leading to the volitional termination of exercise (i.e., exhaustion). However, Rasmussen et al. [48] examined the changes in cerebral perfusion during constant load exercise to exhaustion in the heat, and concluded that adjustments in

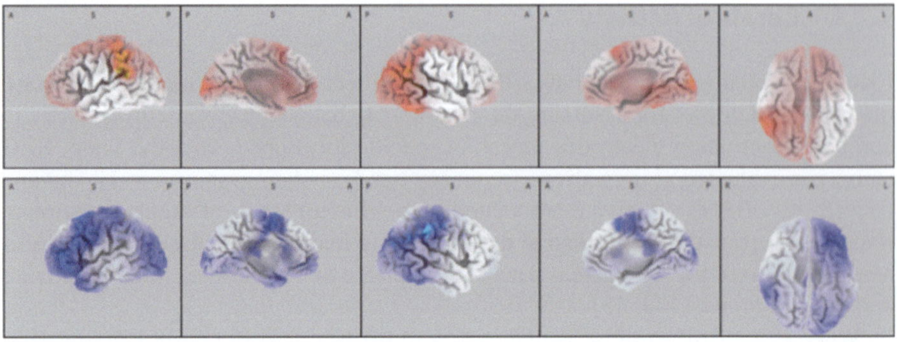

Fig. 4.3 Beta activity increased (red) during the first 10% of a time trial in the heat and then decreased (blue) upon reaching 30%. *A* anterior, *L* left, *P* posterior, *R* right, *S* superior. Reproduced with permission from Periard et al. [51]

cerebrocortical activity were not causally related to changes in cerebral circulation. As such, it would appear that although cerebral blood flow is reduced during intense exercise in the heat, it does not represent a modulating influence on performance, which appears more strongly linked to an inhibition in mental focus and decreased arousal.

4.2.5 Neuromuscular Drive

Electromyographic activity (EMG) is a measure of muscle electrical activity. Some [53–55] but not all [56, 57] studies have reported a decrease in EMG during endurance exercise (mainly cycling) in the heat. It is difficult to draw conclusions from these studies as EMG activity depends on exercise intensity (i.e., maximal vs. sub-maximal) and type (i.e., self-paced vs. fixed intensity). For example, EMG activity has been shown to decrease in conjunction with power output due to fatigue even in temperate environments [58]. Thus, EMG activity is similar in hot and cool conditions at the same power output, but lower at exhaustion in the heat due to the attainment of a lower power output [59].

Several studies have used maximal voluntary isometric contractions (MVC) to measure 'maximal' EMG activity. Interestingly, EMG and force production capacity are not affected by exercise-induced heat stress during brief (~3 to 5 s) MVC [22, 60, 61]. Although EMG activity and force production may [62] or may not [60] decrease when brief MVCs are repeated, EMG and force have been consistently shown to decrease more during a prolonged MVC (120 s) in hyperthermic compared with normothermic states [60]. Of note, an increase in skin temperature can also decrease surface EMG in response to peripheral vasodilation or changes in the skin–electrode properties [63]. As such, data based on voluntary activation measurements obtained by electrical stimulation of the motor nerve during an MVC may represent a better indicator of the effect of hyperthermia on central drive than simple EMG measures.

When an electrical impulse is evoked on a motor nerve, it induces a muscle twitch. The amplitude of this twitch is maximal when the muscle is relaxed and the electrical impulse is (supra)maximal, but decreases when background voluntary force increases until becoming nil when the muscle produces its maximal force [64]. A percentage voluntary activation can be estimated by comparing a superimposed twitch (usually a doublet) to its resting value, albeit the meaning of this percentage has drawn some criticisms [65, 66]. Using this method, heat stress has been shown to reduce voluntary activation and force production capacity of locomotor muscles (e.g., knee extensors) without force production capacity decreasing in non-exercising muscles (e.g., forearm), suggesting a selective alteration in exercising muscles [61, 67]. However, a non-selective effect of hyperthermia has also been suggested, as a decrease in force of both the exercising (knee extensors) and non-exercising (handgrip) muscles has also been reported following exhaustive exercise in the heat [60]. In any case, exercise-induced fatigue has been shown to decrease force, EMG activity, and voluntary activation, even in temperate environments [58, 68], and the alterations following exercise may not differ in the heat and temperate environments following self-paced exercise [69]. To isolate the effect of hyperthermia per se, several studies have used a passive hyperthermia model.

4.2.6 Passive Hyperthermia and Voluntary Activation

Several studies have shown that passive hyperthermia impairs voluntary activation, confirming an effect of temperature per se on the ability to produce force [16, 30, 31, 70]. This decrement has been attributed to an elevated core temperature rather than any local effects of elevated muscle and skin temperatures [30, 31]. This decrement is negligible during brief MVCs [16, 30, 31, 70], but is exacerbated during sustained contractions [16, 20] (Fig. 4.4). The additional decrement when contractions are prolonged was not related to changes in peripherally evoked potentials (i.e., M-wave and H-reflex, see below) and was thus attributed to a failure at the supraspinal level [16]. Such a supraspinal failure was confirmed using transcranial magnetic stimulation (TMS) [71]. In this study, peak muscle relaxation rate following TMS was ~20% faster during hyperthermia, leading the authors to suggest that the motor cortex was unable to compensate for the faster contractile speeds (i.e., faster firing rate) required to reach and maintain the fusion of force [71]. However, another study reported a further increase in relaxation rate from moderate (38.5 °C) to severe (39.5 °C) passive hyperthermia without influencing the loss of voluntary muscle and cortical activation, suggesting that centrally mediated firing rates are sufficient to overcome hyperthermia-induced changes in muscle contractile properties [72]. Thus, the hyperthermia-induced decrease in voluntary activation may not originate from a failure to account for temperature-related adjustments in muscle contractile speed. Although the origin of the supraspinal failure during prolonged contractions is currently unknown, it is important to note that such contractions are difficult to perform and require practise, feedback, and exceptionally high motivation [73]. Importantly, the negative effects of whole-body hyperthermia on the

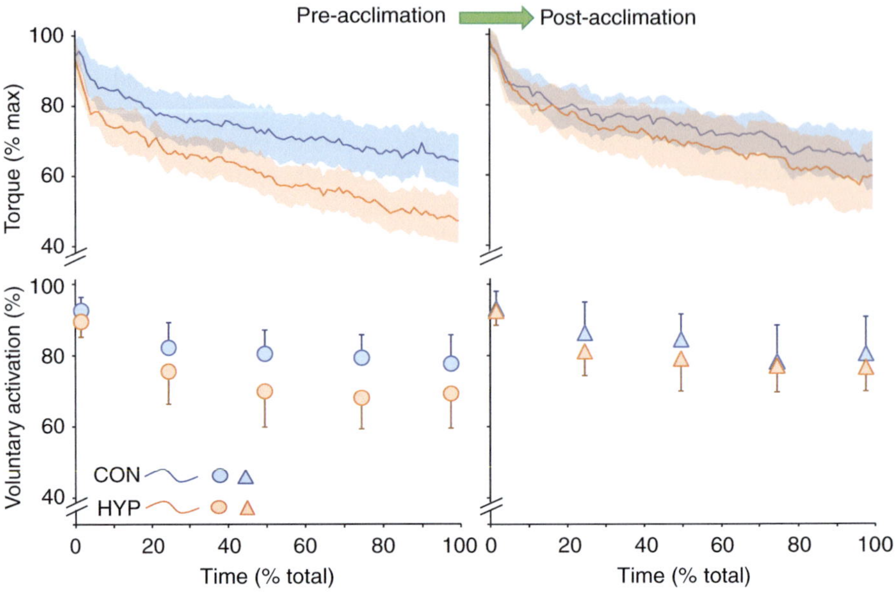

Fig. 4.4 The ability to sustain torque during a maximal contraction is impaired by hyperthermia (orange) as compared to control conditions (blue). The effect of hyperthermia on torque mirrors the effect on voluntary activation. Acclimation partly counteracts those impairments. Reproduced with permission from Racinais et al. [20]

supraspinal ability to maintain voluntary activation during a prolonged contraction can be partly circumvented by heat acclimation [20]. Indeed, the decrease in voluntary activation and torque production noted during a sustained MVC in non-acclimated hyperthermic individuals relative to the normothermic state disappears after acclimation [20] (Fig. 4.4).

4.3 Spinal and Peripheral Responses

4.3.1 Spinal Modulation

Beyond the various CNS alterations discussed above, hyperthermia may also affect spinal and peripheral responses [16]. In temperate conditions, fatigue has been shown to affect the spinal modulation of neural drive, partly through a presynaptic inhibition mediated by group III and IV afferents [68, 74–78]. As these afferents are temperature sensitive, spinal modulation of the neural drive is also likely to occur in response to hyperthermia. Several studies have reported decreased amplitude of the H-reflex, a monosynaptic spinal reflex, following passive hyperthermia [16, 79] (Fig. 4.5). This finding was persistent regardless of whether the H-reflex was expressed in absolute terms or normalised to the M-wave to account for peripheral changes (e.g., at the level of the sarcolemma or neuromuscular junction) [16, 79], and was further confirmed by a decrease in V-wave, an electrophysiological variant

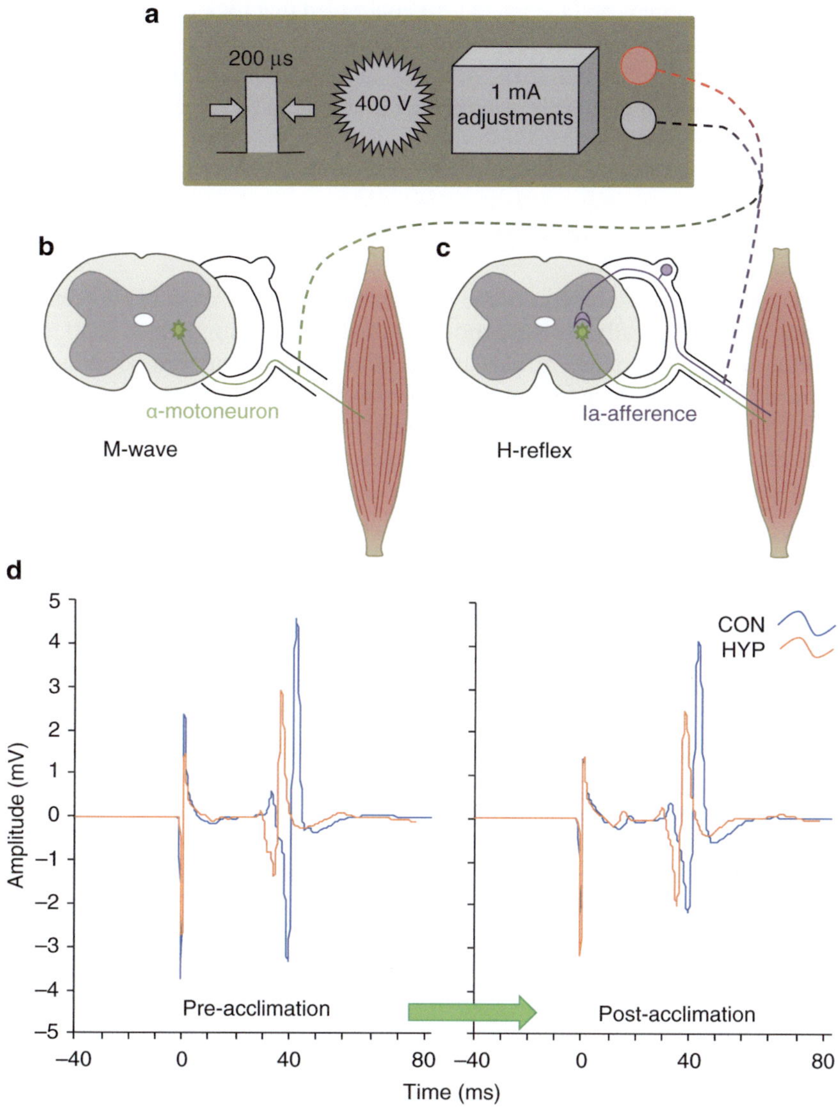

Fig. 4.5 Electrical stimulations (**a**) are applied to the alpha-motoneuron (**b**) and Ia-afferents (**c**) to evoke M-wave and H-reflex, respectively. As displayed in panel (**d**) H-reflex occurs earlier and with a lower amplitude in hyperthermic participants (HYP vs. CON), and this pattern of response is not affected by acclimation. Of note, M-wave shows a similar pattern of response. Modified from Racinais et al. [20]

of the H-reflex [16]. However, the mechanism underpinning the altered reflex response following hyperthermia is uncertain, as it is not clear whether group III and IV muscle afferents directly inhibit motoneurons or act upstream of the motor cortex to impair descending voluntary drive [80]. In addition, using paired pulse H-reflexes to measure the homosynaptic postactivation depression (HPAD) recovery curve, Racinais and Creswell [79] concluded that hyperthermia altered the amplitude of the reflex pathway response, without affecting the synaptic depression from repeated discharges. As such, it is not clear if hyperthermia affects spinal reflexes through spinal inhibition, a synaptic failure, or an increase in nerve conduction velocity.

4.3.2 Motoneuron and Neuromuscular Junction

Under heat stress, the increase in tissue temperature has been shown to decrease the amplitude of an electrically evoked action potential, both in vitro [81–83] and in vivo [16]. Importantly, a decrease in M-wave amplitude in humans can be induced by either locally heating the leg [84] or via whole-body passive hyperthermia [16]. This suggests that the decrement in M-wave amplitude is due to a failure at the peripheral level, rather than a central response. The decrease in amplitude when recorded via surface EMG may partly be related to a higher cutaneous blood flow or alterations in the skin/electrode pairing [63, 85], although intra-muscular EMG recordings have confirmed the existence of neural alterations [20]. Such peripheral neural alterations could be due to changes in the transmission of the neural drive from the α-motoneuron to the sarcolemma (i.e., neuromuscular junction), or from transmission of the action potential through the α-motoneuron and/or the sarcolemma themselves.

A synaptic failure induced by hyperthermia may potentially explain the decrement in both the H-reflex (spinal synapse, see above) and M-wave (neuromuscular junction). High temperatures have been shown to inhibit the transmission of action potentials from the pre- to the post-synaptic element in vitro [86]. Specifically, stimulation of the presynaptic element of a drosophila nerve consistently produced post-synaptic quantal events at 22 °C, but the amplitude of the response declined and failures appeared as temperature was progressively increased, with a complete failure in transmission observed at 35 °C [87]. However, in animal models, such failures have been shown to be attenuated by prior heat exposure, probably through an increase in basal heat shock protein (HSP) expression [86, 88]. In contrast, heat acclimation in humans has been shown to have no influence on M-waves and H-reflexes following heat stress [20], suggesting that these decreases are not linked to synaptic failure. Human neuromuscular junctions have a very high safety factor, with far more acetylcholine released per stimulus than necessary to induce muscle fibres depolarisation [89]. In addition, total synaptic failure has never been demonstrated in hyperthermic humans, despite attaining higher temperatures than previous in vitro studies [72, 79]. Therefore, peripheral alterations are likely related to a gradual effect of temperature on axonal conduction velocity rather than synaptic failure. This is confirmed by the observation that temperature is negatively correlated to the latency, amplitude, duration, and area of the action potential [90]. This can be explained by the rule of Vant-Hoff, stating that there is a two- to fourfold increase in the rate of chemical reactions for every 10 °C increase in temperature. As such, at higher temperatures, the voltage-gated sodium channel remains open for a shorter period of time, decreasing the time for ion movements and consequently decreasing the duration and amplitude of the axon potential [89, 91]. In hyperthermic humans, the peripheral decrease in the amplitude of an action potential is likely a side effect of a faster nerve conduction velocity, which remains independent of acclimation [20] (Fig. 4.5).

4.3.3 Proprioception

Numerous studies have investigated the effect of hyperthermia on the neuromuscular system using isometric contractions coupled with electrically evoked potentials [16, 30, 31, 60]. There is an emerging body of literature looking at the consequences of neural alterations in terms of proprioception, balance, and motor control. It has recently been reported that whole-body hyperthermia impairs proprioceptive acuity during specific movement discrimination tasks, and reduces both dynamic and static postural stability during a Star Excursion Balance test and a single leg stance [92]. These impairments are suggested to be due to hyperthermia rather than an increase in localised skin/muscle temperature, as localised heating has been shown to have no influence on proprioception or balance [93, 94]. Regardless, the influence of hyperthermia on proprioception, balance, and motor control during complex athletic movements has yet to be determined and warrants further investigations.

4.4 Muscle Responses

4.4.1 Muscle Contractile Properties

Tetanic force (i.e., the fusion of several twitches) may be improved by an increase in muscle temperature [95–97], probably due to the improved binding capabilities of contractile proteins [97]. However, the effects of increased temperature in vivo could be lower than previously believed from in vitro experiments, which are routinely performed at sub-physiological temperatures [98]. Indeed, increasing muscle temperature within the physiological range (i.e., 37–43 °C) does not seem to affect muscle fibre force [99], and as such may not modify peak twitch amplitude (Pt) in humans [59, 70, 100] (Fig. 4.6).

A decrease in muscle temperature has been shown to delay chemical reactions [101] and cross-bridge binding [102], as well as decrease in actomyosin sensitivity for calcium [103]. Correspondingly, an increase in muscle temperature will increase the rate of force development of a muscle twitch [95, 104], probably due to the combined effects of increased ATPase activity of the myosin heads [105], and calcium retention by sarcoplasmic reticulum [106]. Thus, as opposed to muscle force, an increase in muscle temperature will increase muscle contraction velocity [4]. The faster contraction time (CT) and half relaxation time (HRT) of a muscle twitch reported in vitro [96] have also been consistently reported in humans, even when Pt was shown to be unchanged [59, 70, 100]. However, it remains unknown if these responses are partly influenced by non-contractile tissues, as it has been suggested

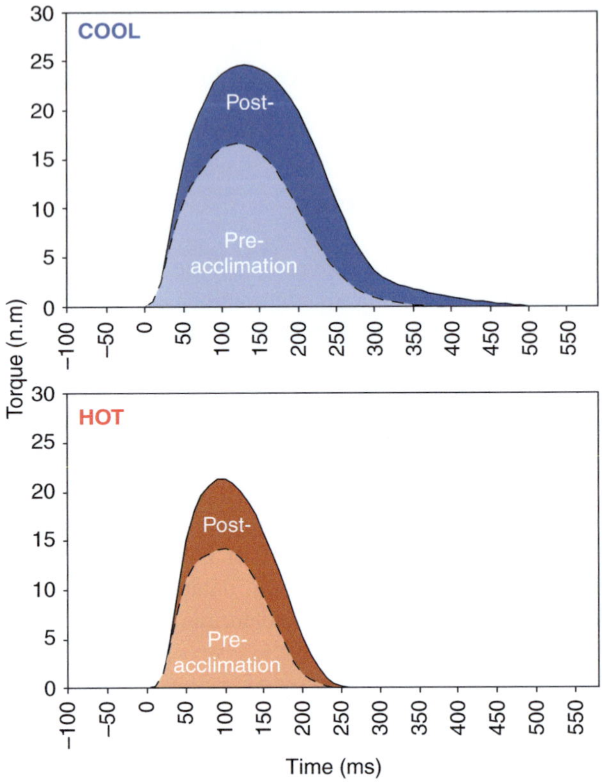

Fig. 4.6 Muscle twitch at resting basal (COOL, blue) and elevated (HOT, red) muscle temperature. An increase in muscle temperature increases the rate of force development and relaxation without modifying the peak tension. Heat acclimation (dark blue and dark orange) increases peak twitch amplitude in both normothermic (COOL) and hyperthermic (HOT) state (left panel). Reproduced from Racinais et al. [1, 70]

that an increase in temperature may decrease joint and muscle viscoelastic resistance [107, 108].

The effects of muscle temperature on contractile properties have been shown to influence muscle performance during various tasks. For example, Oksa et al. [109] showed that decreasing calf temperature through 60 min of cold exposure at 27, 20, 15, or 10 °C proportionally decreased jump height. Although higher muscle temperatures have been shown to increase contraction velocity, it did not affect the isometric force of the forearm [110], indicating that dynamic movements may be more affected by temperature than isometric contractions [2]. Within dynamic movements, fast movements have been shown to be more influenced by temperature compared with slower movements [3, 111].

Of note, acute hyperthermia has been shown to acutely shorten CT and HRT, even without changes in Pt [70]. Moreover, the shorter CT and HRT were not recovered following repeated heat exposures despite an increase in Pt (Fig. 4.6) [70]. This increase has been confirmed by an increase in torque production for a given voluntary activation as well as improvement in torque/EMG linear relationship [70], and as such is likely due to an improvement in skeletal muscle contractile properties with repeated heat exposures.

4.4.2 Muscle Hypertrophy

Whilst the effects of acute heat stress on muscle contractile function have been traditionally well described, research investigating the longer term phenotypical and functional adaptations is just emerging [70, 112, 113]. Recent studies have reported an increase in knee extensor muscle strength, either following 11 days of whole-body passive heat acclimation [70] or following 10-weeks (8 h/day) of localised heat application [113]. The increase in muscle strength was characterised by an increase in Pt and MVC force production capacity, despite unchanged voluntary activation [70], as well as an increase in myofibre and whole muscle cross-sectional area [113]. These findings are supported by previous work in rodent and muscle cultures, demonstrating increased muscle protein content in vitro [114] as well as in young and aged mice in vivo [115] following 60 min of heat stress (41 °C). Heat stress has also been shown to enhance the recovery of muscle mass within atrophied muscles [116, 117], as well as attenuate the loss in muscle mass in ageing [115], disuse [118], and pharmacological [119] models of muscle atrophy. Moreover, heat exposure has been shown to preserve muscle mass and minimise protein degradation in a variety of muscle trauma models including ischemic injury [120], crush injury [121], exercise-induced muscle damage [122], and pharmacologically induced muscle toxicity [123, 124].

The molecular mechanisms underpinning heat-induced regulation of muscle mass are complex, and seem to involve a concerted mix of enhanced anabolic and cytoprotective pathways, coupled with attenuated catabolic signalling (Fig. 4.7). Specifically, heat stress has been purported to enhance anabolic signalling through the Akt-mTOR pathway. In rodents subjected to varying degrees of environmental heat stress (37–41 °C), the activation of Akt and downstream target p70S6K1 was shown to be enhanced in a temperature dependent manner, with the most pronounced activation evident at 41 °C [125]. Excessive hyperthermia though is likely to render adverse effects, as continuous (>6 h) heat stress at (>41 °C) has been shown to enhance protein degradation and impair myoblast differentiation, resulting in poorly formed myotubes [126, 127]. In parallel to anabolic signalling, the activation of Akt may regulate muscle mass through downregulating key muscle-specific ligases of the ubiquitin proteasome system (UPS) [128, 129], a major pathway modulating muscle protein degradation [130, 131].

HSPs have also been implicated in the network of mechanisms underpinning the regulation of muscle mass following heat stress. They have been suggested to confer a variety of cytoprotective functions, resulting in the attenuated loss in muscle mass during disuse atrophy, or enhanced muscle re-growth within atrophied muscles [116–118]. Additionally, HSP72 has been purported to regulate muscle mass through inhibiting the activation of nuclear factor-κB [132], a key transcription factor associated with inflammation and muscle atrophy [70].

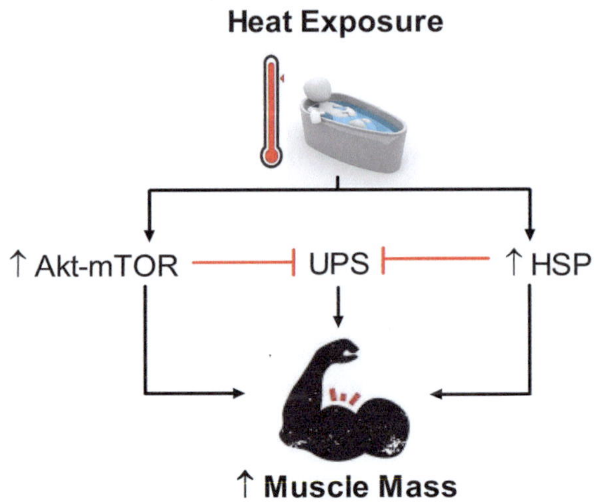

Fig. 4.7 Mechanisms by which heat exposure is suggested to maintain/increase muscle mass. Heat stress upregulates anabolic signalling through the Akt-mTOR pathway. In parallel, the activation of Akt following heat exposure may also influence muscle mass through downregulating key muscle-specific ligases of the ubiquitin proteasome system (UPS). Heat shock proteins (HSP) are suggested to preserve muscle mass through attenuating the transcription of key muscle-specific ligases of the UPP, and conferring cytoprotective effects during protein degradation. HSPs also support hypertrophy as molecular chaperones for newly formed peptides following translation

4.4.3 Muscle Oxidative Capacity

In cultured myotubes, 1 h of mild heat stress (40 °C) was shown to increase PGC-1α activity and gene expression, a key transcriptional coactivator implicated in the regulation of mitochondrial biogenesis and function [133]. The upregulation in PGC-1α was accompanied by an increase in the mRNA abundance of key mitochondrial transcription factors and mitochondrial complex-IV subunits. Moreover, when heat stress was repeatedly applied for 5 days, an increase in the protein abundance of PGC-1α along with the expression of several subunits representative of mitochondrial respiratory chain complexes I–V was evident [134]. Recent work by Tamura et al. [135] has extended on such findings within rodents in vivo, where AMPK was downregulated, whilst p38 and mTOR signalling were identified as upstream activators of PGC-1α during heat stress. Moreover, passive heat stress was reported to increase mitochondrial protein content, which in turn was additively enhanced by post-exercise heat stress [135].

Although there is presently limited research investigating mitochondrial dynamics in heat stressed humans, the available findings seem to in-part contradict that gathered from muscle culture and rodent models [136, 137]. Indeed, in humans, post-exercise recovery or exercise per se when undertaken in hot environments (~33 °C) was shown to attenuate the mRNA expression of PGC-1α, including the expression of several regulatory factors downstream, respectively [136, 137]. Interestingly, the response to localised heat exposure in humans seems to be highly diverse, as increases in PGC-1α, respiratory chain I and V complexes, as well HSP70 and HSP90 were evident following short-term localised heating of the quadriceps (2 h/day for 6 days) [138]. The mechanisms underpinning such a differential response following whole body and localised heat stress are not understood, but may be related to AMPK signalling. Indeed, whilst Hafen et al. [138] demonstrated an increase in AMPK signalling following localised heating, systemic heat stress in rodents has been shown to downregulate AMPK activation [135]. This differential AMPK response to local and systemic heat stress may in-part account for the observed disparity in PGC-1α transcriptional control [136–138]. Future research should investigate the effect of whole body and localised increases in temperature on AMPK activation and subsequent changes in the regulation of mitochondrial dynamics.

4.4.4 Substrate Utilisation

An increase in temperature can affect muscle metabolism by facilitating the release of oxygen from haemoglobin [139] and myoglobin [140]. The rate of ATP utilisation could also be increased when exercising in the heat, matched by an increase in anaerobic glycolysis and creatine phosphate hydrolysis [141, 142]. In addition, hot temperatures through modifying vasodilatation can affect substrate provision and metabolite removal [143]. Thus, exercise in the heat appears to induce a greater reliance on muscle glycogen and anaerobic metabolism [141, 142], and a greater post-exercise accumulation of ammonia, as well as muscle and blood lactate [141, 144, 145]. The greater reliance on anaerobic metabolism under heat stress may also be explained by the increased release of adrenalin and consequently enhanced β-adrenergic stimulation [141, 146, 147], which following heat acclimation has been shown to be reduced in line with reduced carbohydrate utilisation [141].

4.5 Conclusion

An increase in muscle temperature can improve physical performance in brief efforts through an increase in muscle power production. However, hyperthermia impairs both maximal neural drive during a maximal voluntary contraction and complex cognitive task performance. These impairments are due to alterations at the level of both the peripheral and central nervous systems. At the level of the peripheral nervous system, there is a decrease in the amplitude of an action potential. In humans this decrease is likely due to a faster nerve conduction velocity reducing depolarisation time, which cannot be recovered by heat acclimation. Conversely, impairments in the capacity to sustain neural drive during a prolonged contraction or to perform complex cognitive tasks are likely due to CNS alterations. The alterations may be linked to physiological (e.g., cerebral blood flow adjustments) and perceptual (e.g., suppression of arousal and disengagement to a task) perturbations, and can be partly reverted by heat acclimation. Future studies should focus on the effect of hyperthermia on integrated complex systems such as motor control. Future studies should also investigate the potential increase in myofibrillar and mitochondrial protein synthesis following heat exposure.

References

1. Racinais S, Cocking S, Périard JD. Sports and environmental temperature: from warming-up to heating-up. Temperature. 2017;4(3):227–57.
2. Bergh U, Ekblom B. Influence of muscle temperature on maximal muscle strength and power output in human skeletal muscles. Acta Physiol Scand. 1979;107(1):33–7.
3. Sargeant AJ. Effect of muscle temperature on leg extension force and short-term power output in humans. Eur J Appl Physiol Occup Physiol. 1987;56(6):693–8.
4. Racinais S, Oksa J. Temperature and neuromuscular function. Scand J Med Sci Sports. 2010;20(Suppl 3):1–18.
5. Linnane DM, Bracken RM, Brooks S, Cox VM, Ball D. Effects of hyperthermia on the metabolic responses to repeated high-intensity exercise. Eur J Appl Physiol. 2004;93(1-2):159–66.
6. Drust B, Rasmussen P, Mohr M, Nielsen B, Nybo L. Elevations in core and muscle temperature impairs repeated sprint performance. Acta Physiol Scand. 2005;183(2):181–90.
7. Bulochnik ED, Zyablov MP. Effect of hyperthermia induced by a high ambient temperature on the direct cortical response. Biull Eksp Biol Med. 1977;84:657–60.
8. Cian C, Barraud PA, Melin B, Raphel C. Effects of fluid ingestion on cognitive function after heat stress or exercise-induced dehydration. Int J Psychophysiol. 2001;42(3):243–51.
9. Hocking C, Silberstein RB, Lau WM, Stough C, Roberts W. Evaluation of cognitive performance in the heat by functional brain imaging and psychometric testing. Comp Biochem Physiol Part A Mol Integr Physiol. 2001;128(4):719–34.
10. Pilcher JJ, Nadler E, Busch C. Effects of hot and cold temperature exposure on performance: a meta-analytic review. TERG. 2002;45(10):682–98.
11. Hancock PA, Vasmatzidis I. Effects of heat stress on cognitive performance: the current state of knowledge. Int J Hyperth. 2003;19(3):355–72.
12. Gaoua N, Grantham J, Massioui El F, Girard O, Racinais S. Cognitive decrements do not follow neuromuscular alterations during passive heat exposure. Int J Hyperth. 2011;27(1):10–9.
13. Gaoua N, Grantham J, Racinais S, El Massioui F. Sensory displeasure reduces complex cognitive performance in the heat. J Environ Psychol. 2012;32:158–63.
14. Cabanac M. Physiological role of pleasure. Science. 1971;173(4002):1103–7.
15. Gaoua N, Herrera CP, Périard JD, Massioui El F, Racinais S. Effect of passive hyperthermia on working memory resources during simple and complex cognitive tasks. Front Psychol. 2018;8:2290.
16. Racinais S, Gaoua N, Grantham J. Hyperthermia impairs short-term memory and peripheral motor drive transmission. J Physiol. 2008;586(Pt 19):4751–62.
17. Gaoua N, Racinais S, Grantham J, Massioui El F. Alterations in cognitive performance during passive hyperthermia are task dependent. Int J Hyperth. 2011;27(1):1–9.
18. Radakovic SS, Maric J, Surbatovic M, Radjen S, Stefanova E, Stankovic N, et al. Effects of acclimation on cognitive performance in soldiers during exertional heat stress. Mil Med. 2007;172(2):133–6.
19. Walker SM, Dawson B, Ackland TR. Performance enhancement in rally car drivers via heat acclimation and race simulation. Comp Biochem Physiol Part A Mol Integr Physiol. 2001;128(4):701–7.
20. Racinais S, Wilson MG, Gaoua N, Périard JD. Heat acclimation has a protective effect on the central but not peripheral nervous system. J Appl Physiol. 2017;123(4):816–24.

21. Nielsen B, Savard G, Richter EA, Hargreaves M, Saltin B. Muscle blood flow and muscle metabolism during exercise and heat stress. J Appl Physiol. 1990;69(3):1040–6.
22. Nielsen B, Hales JR, Strange S, Christensen NJ, Warberg J, Saltin B. Human circulatory and thermoregulatory adaptations with heat acclimation and exercise in a hot, dry environment. J Physiol. 1993;460:467–85.
23. Flouris AD, Schlader ZJ. Human behavioral thermoregulation during exercise in the heat. Scand J Med Sci Sports. 2015;25(13):52–64.
24. Tatterson AJ, Hahn AG, Martin DT, Febbraio MA. Effects of heat stress on physiological responses and exercise performance in elite cyclists. J Sci Med Sport. 2000;3(2):186–93.
25. Ely BR, Ely MR, Cheuvront SN, Kenefick RW, Degroot DW, Montain SJ. Evidence against a 40 degrees C core temperature threshold for fatigue in humans. J Appl Physiol. 2009;107(5):1519–25.
26. Périard JD, Racinais S. Heat stress exacerbates the reduction in middle cerebral artery blood velocity during prolonged self-paced exercise. Scand J Med Sci Sports. 2015;25(Suppl 1):135–44.
27. Périard JD, Racinais S. Self-paced exercise in hot and cool conditions is associated with the maintenance of %VO2peak within a narrow range. J Appl Physiol. 2015;118(10):1258–65.
28. Périard JD, Racinais S. Performance and pacing during cycle exercise in hyperthermic and hypoxic conditions. Med Sci Sports Exerc. 2016;48(5):845–53.
29. Racinais S, Périard JD, Karlsen A, Nybo L. Effect of heat and heat acclimatization on cycling time trial performance and pacing. Med Sci Sports Exerc. 2015;47(3):601–6.
30. Morrison S, Sleivert GG, Cheung SS. Passive hyperthermia reduces voluntary activation and isometric force production. Eur J Appl Physiol. 2004;91(5-6):729–36.
31. Thomas MM, Cheung SS, Elder GC, Sleivert GG. Voluntary muscle activation is impaired by core temperature rather than local muscle temperature. J Appl Physiol. 2006;100(4):1361–9.
32. Nybo L, Nielsen B. Middle cerebral artery blood velocity is reduced with hyperthermia during prolonged exercise in humans. J Physiol. 2001;534(Pt 1):279–86.
33. Nybo L, Møller K, Volianitis S, Nielsen B, Secher NH. Effects of hyperthermia on cerebral blood flow and metabolism during prolonged exercise in humans. J Appl Physiol. 2002;93(1):58–64.
34. Rasmussen P, Stie H, Nielsen B, Nybo L. Enhanced cerebral CO2 reactivity during strenuous exercise in man. Eur J Appl Physiol. 2006;96(3):299–304.
35. Rasmussen P, Nybo L, Volianitis S, Møller K, Secher NH, Gjedde A. Cerebral oxygenation is reduced during hyperthermic exercise in humans. Acta Physiol. 2010;199(1):63–70.
36. Rowell LB. Human cardiovascular adjustments to exercise and thermal stress. Physiol Rev. 1974;54(1):75–159.
37. Thomas K, GOODALL S, Stone M, Howatson G, St Clair Gibson A, Ansley L. Central and peripheral fatigue in male cyclists after 4-, 20-, and 40-km time trials. Med Sci Sports Exerc. 2015;47(3):537–46.
38. Nybo L, Secher NH, Nielsen B. Inadequate heat release from the human brain during prolonged exercise with hyperthermia. J Physiol. 2002;545(Pt 2):697–704.
39. Nybo L, Rasmussen P. Inadequate cerebral oxygen delivery and central fatigue during strenuous exercise. Exerc Sport Sci Rev. 2007;35(3):110–8.
40. Rasmussen P, Dawson EA, Nybo L, van Lieshout JJ, Secher NH, Gjedde A. Capillary-oxygenation-level-dependent near-infrared spectrometry in frontal lobe of humans. J Cereb Blood Flow Metab. 2007;27(5):1082–93.
41. González-Alonso J, Dalsgaard MK, Osada T, Volianitis S, Dawson EA, Yoshiga CC, et al. Brain and central haemodynamics and oxygenation during maximal exercise in humans. J Physiol. 2004;557(Pt 1):331–42.
42. Trangmar SJ, Chiesa ST, Stock CG, Kalsi KK, Secher NH, González-Alonso J. Dehydration affects cerebral blood flow but not its metabolic rate for oxygen during maximal exercise in trained humans. J Physiol. 2014;592(Pt 14):3143–60.
43. Trangmar SJ, González-Alonso J. New insights into the impact of dehydration on blood flow and metabolism during exercise. Exerc Sport Sci Rev. 2017;45(3):146–53.

44. Nielsen B, Nybo L. Cerebral changes during exercise in the heat. Sports Med. 2003;33(1):1–11.
45. Thompson T, Steffert T, Ros T, Leach J, Gruzelier J. EEG applications for sport and performance. Methods. 2008;45(4):279–88.
46. Nielsen B, Hyldig T, Bidstrup F, González-Alonso J, Christoffersen GR. Brain activity and fatigue during prolonged exercise in the heat. Pflugers Arch. 2001;442(1):41–8.
47. Nybo L, Nielsen B. Perceived exertion is associated with an altered brain activity during exercise with progressive hyperthermia. J Appl Physiol. 2001;91(5):2017–23.
48. Rasmussen P, Stie H, Nybo L, Nielsen B. Heat induced fatigue and changes of the EEG is not related to reduced perfusion of the brain during prolonged exercise in humans. J Therm Biol. 2004;29:731–7.
49. Ftaiti F, Kacem A, Jaidane N, Tabka Z, Dogui M. Changes in EEG activity before and after exhaustive exercise in sedentary women in neutral and hot environments. Appl Ergon. 2010;41(6):806–11.
50. De Pauw K, Roelands B, Marušič U, Tellez HF, Knaepen K, Meeusen R. Brain mapping after prolonged cycling and during recovery in the heat. J Appl Physiol. 2013;115(9):1324–31.
51. Périard JD, De Pauw K, Zanow F, Racinais S. Cerebrocortical activity during self-paced exercise in temperate, hot and hypoxic conditions. Acta Physiol. 2018;222:1.
52. Robertson CV, Marino FE. Prefrontal and motor cortex EEG responses and their relationship to ventilatory thresholds during exhaustive incremental exercise. Eur J Appl Physiol. 2015;115(9):1939–48.
53. Kay D, Marino FE, Cannon J, St Clair Gibson A, Lambert MI, Noakes TD. Evidence for neuromuscular fatigue during high-intensity cycling in warm, humid conditions. Eur J Appl Physiol. 2001;84(1-2):115–21.
54. Tucker R, Rauch L, Harley YXR, Noakes TD. Impaired exercise performance in the heat is associated with an anticipatory reduction in skeletal muscle recruitment. Pflugers Arch. 2004;448(4):422–30.
55. Tucker R, Marle T, Lambert EV, Noakes TD. The rate of heat storage mediates an anticipatory reduction in exercise intensity during cycling at a fixed rating of perceived exertion. J Physiol. 2006;574(Pt 3):905–15.
56. Ftaiti F, Grélot L, Coudreuse JM, Nicol C. Combined effect of heat stress, dehydration and exercise on neuromuscular function in humans. Eur J Appl Physiol. 2001;84(1-2):87–94.
57. Hunter AM, St Clair Gibson A, Mbambo Z, Lambert MI, Noakes TD. The effects of heat stress on neuromuscular activity during endurance exercise. Pflugers Arch. 2002;444(6):738–43.
58. Racinais S, Bishop D, Denis R, Lattier G, Mendez-Villaneuva A, Perrey S. Muscle deoxygenation and neural drive to the muscle during repeated sprint cycling. Med Sci Sports Exerc. 2007;39(2):268–74.
59. Racinais S, Girard O. Neuromuscular failure is unlikely to explain the early exercise cessation in hot ambient conditions. Psychophysiology. 2012;49(6):853–65.
60. Nybo L, Nielsen B. Hyperthermia and central fatigue during prolonged exercise in humans. J Appl Physiol. 2001;91(3):1055–60.
61. Saboisky J, Marino FE, Kay D, Cannon J. Exercise heat stress does not reduce central activation to non-exercised human skeletal muscle. Exp Physiol. 2003;88(6):783–90.
62. Martin PG, Marino FE, Rattey J, Kay D, Cannon J. Reduced voluntary activation of human skeletal muscle during shortening and lengthening contractions in whole body hyperthermia. Exp Physiol. 2005;90(2):225–36.
63. Racinais S. Hot ambient conditions shift the force/EMG relationship. Springerplus. 2013;2(1):317.
64. Merton PA. Voluntary strength and fatigue. J Physiol. 1954;123(3):553–64.
65. Taylor JL. Point: the interpolated twitch does/does not provide a valid measure of the voluntary activation of muscle. J Appl Physiol. 2009;107(1):354–5.
66. de Haan A, Gerrits KHL, De Ruiter CJ. Counterpoint: the interpolated twitch does not provide a valid measure of the voluntary activation of muscle. J Appl Physiol. 2009;107(1):355–7.
67. Périard JD, Caillaud C, Thompson MW. Central and peripheral fatigue during passive and exercise-induced hyperthermia. Med Sci Sports Exerc. 2011;43(9):1657–65.

68. Racinais S, Girard O, Micallef JP, Perrey S. Failed excitability of spinal motoneurons induced by prolonged running exercise. J Neurophysiol. 2007;97(1):596–603.
69. Périard JD, Cramer MN, Chapman PG, Caillaud C, Thompson MW. Neuromuscular function following prolonged intense self-paced exercise in hot climatic conditions. Eur J Appl Physiol. 2011;111(8):1561–9.
70. Racinais S, Wilson MG, Periard J. Passive heat acclimation improves skeletal muscle contractility in humans. Am J Physiol Regul Integr Comp Physiol. 2017;312:R101–7.
71. Todd G, Butler JE, Taylor JL, Gandevia SC. Hyperthermia: a failure of the motor cortex and the muscle. J Physiol. 2005;563(Pt 2):621–31.
72. Périard JD, Christian RJ, Knez WL, Racinais S. Voluntary muscle and motor cortical activation during progressive exercise and passively induced hyperthermia. Exp Physiol. 2014;99(1):136–48.
73. Gandevia SC. Spinal and supraspinal factors in human muscle fatigue. Physiol Rev. 2001;81(4):1725–89.
74. Bigland-Ritchie BR, Dawson NJ, Johansson RS, Lippold OC. Reflex origin for the slowing of motoneurone firing rates in fatigue of human voluntary contractions. J Physiol. 1986;379:451–9.
75. Woods JJ, Furbush F, Bigland-Ritchie B. Evidence for a fatigue-induced reflex inhibition of motoneuron firing rates. J Neurophysiol. 1987;58(1):125–37.
76. Garland SJ, McComas AJ. Reflex inhibition of human soleus muscle during fatigue. J Physiol. 1990;429:17–27.
77. Garland SJ. Role of small diameter afferents in reflex inhibition during human muscle fatigue. J Physiol. 1991;435:547–58.
78. Duchateau J, Balestra C, Carpentier A, Hainaut K. Reflex regulation during sustained and intermittent submaximal contractions in humans. J Physiol. 2002;541(Pt 3):959–67.
79. Racinais S, Cresswell AG. Temperature affects maximum H-reflex amplitude but not homosynaptic postactivation depression. Physiol Rep. 2013;1(2):e00019.
80. Taylor JL, Todd G, Gandevia SC. Evidence for a supraspinal contribution to human muscle fatigue. Clin Exp Pharmacol Physiol. 2006;33(4):400–5.
81. Schoepfle GM, Erlanger J. The action of temperature on the excitability, spike height and configuration, and the refractory period observed in the responses of single medullated nerve fibers. Am J Phys. 1941;134:694–704.
82. Hodgkin AL, Katz B. The effect of temperature on the electrical activity of the giant axon of the squid. J Physiol. 1949;109(1-2):240–9.
83. Huxley AF. Ion movements during nerve activity. Ann N Y Acad Sci. 1959;81:221–46.
84. Dewhurst S, Riches PE, Nimmo MA, De Vito G. Temperature dependence of soleus H-reflex and M wave in young and older women. Eur J Appl Physiol. 2005;94(5-6):491–9.
85. Bell DG. The influence of air temperature on the EMG/force relationship of the quadriceps. Eur J Appl Physiol Occup Physiol. 1993;67(3):256–60.
86. Kelty JD, Noseworthy PA, Feder ME, Robertson RM, Ramirez J-M. Thermal preconditioning and heat-shock protein 72 preserve synaptic transmission during thermal stress. J Neurosci. 2002;22(1):RC193.
87. Karunanithi S, Barclay JW, Robertson RM, Brown IR, Atwood HL. Neuroprotection at Drosophila synapses conferred by prior heat shock. J Neurosci. 1999;19(11):4360–9.
88. Karunanithi S, Barclay JW, Brown IR, Robertson RM, Atwood HL. Enhancement of presynaptic performance in transgenic Drosophila overexpressing heat shock protein Hsp70. Synapse. 2002;44(1):8–14.
89. Rutkove SB. Effects of temperature on neuromuscular electrophysiology. Muscle Nerve. 2001;24(7):867–82.
90. Bolton CF, Sawa GM, Carter K. The effects of temperature on human compound action potentials. J Neurol Neurosurg Psychiatry. 1981;44(5):407–13.
91. Rutkove SB, Kothari MJ, Shefner JM. Nerve, muscle, and neuromuscular junction electrophysiology at high temperature. Muscle Nerve. 1997;20(4):431–6.

92. Mtibaa K, Thomson A, Nichols D, Hautier C, Racinais S. Hyperthermia-induced Neural Alterations Impair Proprioception and Balance. Med Sci Sports Exerc. 2018;50(1):46–53.
93. Ingersoll CD, Knight KL, Merrick MA. Sensory perception of the foot and ankle following therapeutic applications of heat and cold. J Athl Train. 1992;27(3):231–4.
94. Dewhurst S, Riches PE, De Vito G. Moderate alterations in lower limbs muscle temperature do not affect postural stability during quiet standing in both young and older women. J Electromyogr Kinesiol. 2007;17(3):292–8.
95. Close R, Hoh JF. Influence of temperature on isometric contractions of rat skeletal muscles. Nature. 1968;217(5134):1179–80.
96. Segal SS, Faulkner JA, White TP. Skeletal muscle fatigue in vitro is temperature dependent. J Appl Physiol. 1986;61(2):660–5.
97. Stephenson DG, Williams DA. Calcium-activated force responses in fast- and slow-twitch skinned muscle fibres of the rat at different temperatures. J Physiol. 1981;317:281–302.
98. Bennett AF. Thermal dependence of muscle function. Am J Phys. 1984;247(2 Pt 2):R217–29.
99. Place N, Yamada T, Zhang S-J, Westerblad H, Bruton JD. High temperature does not alter fatigability in intact mouse skeletal muscle fibres. J Physiol. 2009;587(19):4717–24.
100. Périard JD, Racinais S, Thompson MW. Adjustments in the force-frequency relationship during passive and exercise-induced hyperthermia. Muscle Nerve. 2014;50(5):822–9.
101. Oksa J, Rintamäki H, Mäkinen T, Martikkala V, Rusko H. EMG-activity and muscular performance of lower leg during stretch-shortening cycle after cooling. Acta Physiol Scand. 1996;157(1):1–8.
102. Asmussen E, Bonde-Petersen F, Jorgensen K. Mechano-elastic properties of human muscles at different temperatures. Acta Physiol Scand. 1976;96(1):83–93.
103. Hartshorne DJ, Barns EM, Parker L, Fuchs F. The effect of temperature on actomyosin. Biochim Biophys Acta. 1972;267(1):190–202.
104. Ranatunga KW. Temperature-dependence of shortening velocity and rate of isometric tension development in rat skeletal muscle. J Physiol. 1982;329:465–83.
105. Bárány M. ATPase activity of myosin correlated with speed of muscle shortening. J Gen Physiol. 1967;50(6):197–218.
106. Stein RB, Gordon T, Shriver J. Temperature dependence of mammalian muscle contractions and ATPase activities. Biophys J. 1982;40(2):97–107.
107. Buchthal F, Kaiser E, Knappeis GG. Elasticity, viscosity and plasticity in the cross striated muscle fibre. Acta Physiol. 1944;8:16–37.
108. Wright V, Johns RJ. Quantitative and qualitative analysis of joint stiffness in normal subjects and in patients with connective tissue diseases. Ann Rheum Dis. 1961;20:36–46.
109. Oksa J, Rintamäki H, Rissanen S. Muscle performance and electromyogram activity of the lower leg muscles with different levels of cold exposure. Eur J Appl Physiol Occup Physiol. 1997;75(6):484–90.
110. Binkhorst RA, Hoofd L, Vissers AC. Temperature and force-velocity relationship of human muscles. J Appl Physiol. 1977;42(4):471–5.
111. Oksa J, Rintamäki H, Mäkinen T, Hassi J, Rusko H. Cooling-induced changes in muscular performance and EMG activity of agonist and antagonist muscles. Aviat Space Environ Med. 1995;66(1):26–31.
112. Ihsan M, Watson G, Abbiss C. PGC-1α mediated muscle aerobic adaptations to exercise, heat and cold exposure. Cell Mol Exerc Physiol. 2014;3(1):e7.
113. Goto K, Oda H, Kondo H, Igaki M, Suzuki A, Tsuchiya S, et al. Responses of muscle mass, strength and gene transcripts to long-term heat stress in healthy human subjects. Eur J Appl Physiol. 2011;111:17–27.
114. Goto K, Okuyama R, Sugiyama H, Honda M, Kobayashi T, Uehara K, et al. Effects of heat stress and mechanical stretch on protein expression in cultured skeletal muscle cells. Pflugers Arch. 2003;447:247–53.
115. Ohno Y, Yamada S, Goto A, Ikuta A, Sugiura T, Ohira Y, et al. Effects of heat stress on muscle mass and the expression levels of heat shock proteins and lysosomal cathepsin L in soleus muscle of young and aged mice. Mol Cell Biochem. 2012;369(1-2):45–53.

116. Selsby JT, Rother S, Tsuda S, Pracash O, Quindry J, Dodd SL. Intermittent hyperthermia enhances skeletal muscle regrowth and attenuates oxidative damage following reloading. J Appl Physiol. 2007;102(4):1702–7.
117. Goto K, Honda M, Kobayashi T, Uehara K, Kojima A, Akema T, et al. Heat stress facilitates the recovery of atrophied soleus muscle in rat. Jpn J Physiol. 2004;54(3):285–93.
118. Selsby JT, Dodd SL. Heat treatment reduces oxidative stress and protects muscle mass during immobilization. Am J Physiol Regul Integr Comp Physiol. 2005;289(1):R134–9.
119. Tsuchida W, Iwata M, Akimoto T, Matsuo S, Asai Y, Suzuki S. Heat stress modulates both anabolic and catabolic signaling pathways preventing dexamethasone-induced muscle atrophy in vitro. J Cell Physiol. 2017;232(3):650–64.
120. Garramone RR, Winters RM, Das DK, Deckers PJ. Reduction of skeletal muscle injury through stress conditioning using the heat-shock response. Plast Reconstr Surg. 1994;93(6):1242–7.
121. Takeuchi K, Hatade T, Wakamiya S, Fujita N, Arakawa T, Miki A. Heat stress promotes skeletal muscle regeneration after crush injury in rats. Acta Histochem. 2014;116(2):327–34.
122. Touchberry CD, Gupte AA, Bomhoff GL, Graham ZA, Geiger PC, Gallagher PM. Acute heat stress prior to downhill running may enhance skeletal muscle remodeling. Cell Stress Chaperones. 2012;17(6):693–705.
123. Kojima A, Goto K, Morioka S, Naito T, Akema T, Fujiya H, et al. Heat stress facilitates the regeneration of injured skeletal muscle in rats. J Orthop Sci. 2007;12(1):74–82.
124. Shibaguchi T, Sugiura T, Fujitsu T, Nomura T, Yoshihara T, Naito H, et al. Effects of icing or heat stress on the induction of fibrosis and/or regeneration of injured rat soleus muscle. J Physiol Sci. 2016;66(4):345–57.
125. Yoshihara T, Naito H, Kakigi R, Ichinoseki-Sekine N, Ogura Y, Sugiura T, et al. Heat stress activates the Akt/mTOR signalling pathway in rat skeletal muscle. Acta Physiol. 2013;207(2):416–26.
126. Yamaguchi T, Suzuki T, Arai H, Tanabe S, Atomi Y. Continuous mild heat stress induces differentiation of mammalian myoblasts, shifting fiber type from fast to slow. Am J Physiol Cell Physiol. 2010;298(1):C140–8.
127. Luo GJ, Sun X, Hasselgren PO. Hyperthermia stimulates energy-proteasome-dependent protein degradation in cultured myotubes. Am J Physiol Regul Integr Comp Physiol. 2000;278(3):R749–56.
128. Sandri M, Sandri C, Gilbert A, Skurk C, Calabria E, Picard A, et al. Foxo transcription factors induce the atrophy-related ubiquitin ligase atrogin-1 and cause skeletal muscle atrophy. Cell. 2004;117:399–412.
129. Stitt TN, Drujan D, Clarke BA, Panaro F, Timofeyva Y, Kline WO, et al. The IGF-1/PI3K/Akt pathway prevents expression of muscle atrophy-induced ubiquitin ligases by inhibiting FOXO transcription factors. Mol Cell. 2004;14:395–403.
130. Csibi A, Leibovitch MP, Cornille K, Tintignac LA, Leibovitch SA. MAFbx/Atrogin-1 controls the activity of the initiation factor eIF3-f in skeletal muscle atrophy by targeting multiple C-terminal lysines. J Biol Chem. 2009;284(7):4413–21.
131. Lagirand-Cantaloube J, Cornille K, Csibi A, Batonnet-Pichon S, Leibovitch MP, Leibovitch SA. Inhibition of atrogin-1/MAFbx mediated MyoD proteolysis prevents skeletal muscle atrophy in vivo. PLoS ONE. 2009;4(3):e4973.
132. Ohno Y, Yamada S, Sugiura T, Ohira Y. A possible role of NF-κB and HSP72 in skeletal muscle hypertrophy induced by heat stress in rats. Gen Physiol Biophys. 2010;29(3):234–42.
133. Wu Z, Puigserver P, Andersson U, Zhang C, Adelmant G, Mootha V, et al. Mechanisms controlling mitochondrial biogenesis and respiration through the thermogenic coactivator PGC-1. Cell. 1999;98(1):115–24.
134. Liu C-T, Brooks GA. Mild heat stress induces mitochondrial biogenesis in C2C12 myotubes. J Appl Physiol. 2012;112(3):354–61.
135. Tamura Y, Matsunaga Y, Masuda H, Takahashi Y, Takahashi Y, Terada S, et al. Postexercise whole body heat stress additively enhances endurance training-induced mitochondrial adaptations in mouse skeletal muscle. Am J Physiol Regul Integr Comp Physiol. 2014;307(7):R931–43.

136. Heesch MW, Shute RJ, Kreiling JL, Slivka DR. Transcriptional control, but not subcellular location, of PGC-1α is altered following exercise in a hot environment. J Appl Physiol. 2016;121(3):741–9.
137. Slivka DR, Dumke CL, Tucker TJ, Cuddy JS, Ruby B. Human mRNA response to exercise and temperature. Int J Sports Med. 2012;33(2):94–100.
138. Hafen PS, Preece CN, Sorensen JR, Hancock CR, Hyldahl RD. Repeated exposure to heat stress induces mitochondrial adaptation in human skeletal muscle. J Appl Physiol. 2018;221(14):427.
139. McCutcheon LJ, Geor RJ, Hinchcliff KW. Effects of prior exercise on muscle metabolism during sprint exercise in horses. J Appl Physiol. 1999;87(5):1914–22.
140. Theorell H. The effect of temperature on myoglobin. Biochem Z. 1934;73:268.
141. Febbraio MA, Snow RJ, Hargreaves M, Stathis CG, Martin IK, Carey MF. Muscle metabolism during exercise and heat stress in trained men: effect of acclimation. J Appl Physiol. 1994;76(2):589–97.
142. Febbraio MA, Carey MF, Snow RJ, Stathis CG, Hargreaves M. Influence of elevated muscle temperature on metabolism during intense, dynamic exercise. Am J Phys. 1996;271(5 Pt 2):R1251–5.
143. Karvonen J. Warming up and its physiological effects. Acta Univ Ouluensis. 1978;31:6.
144. Dolny DG, Lemon PW. Effect of ambient temperature on protein breakdown during prolonged exercise. J Appl Physiol. 1988;64(2):550–5.
145. Febbraio MA, Snow RJ, Stathis CG, Hargreaves M, Carey MF. Effect of heat stress on muscle energy metabolism during exercise. J Appl Physiol. 1994;77(6):2827–31.
146. Richter EA, Ruderman NB, Gavras H, Belur ER, Galbo H. Muscle glycogenolysis during exercise: dual control by epinephrine and contractions. Am J Physiol Endocrinol Metab. 1982;242(5):E25–32.
147. Febbraio MA, Lambert DL, Starkie RL, Proietto J, Hargreaves M. Effect of epinephrine on muscle glycogenolysis during exercise in trained men. J Appl Physiol. 1998;84(2):465–70.

Part II

Heat Illnesses and Countermeasures

Common Misconceptions in Classic and Exertional Heat Stroke

5

Orlando Laitano, Michelle A. King, and Lisa R. Leon

5.1 A Long History of Heat Stroke

Heat stroke is a complex physiological condition that has plagued humankind for centuries. It is considered one of the oldest known medical conditions with descriptions of human death from heat exposure dating back to the earliest writings of man [1]. Sunstroke was an early term that was mentioned several times in the Bible in reference to the death of farmers as well as armored fighting forces ~1000 B.C. [2]. Fever in men and madness of dogs later became correlated with the summer appearance of Sirius, the Dog Star of the constellation Canis Major, from which stems the phrase "dog days" of summer (~3000 B.C.). The term "siriasis" was later introduced into the medical literature in reference to all heat illnesses and remained the preferred term well into the twentieth century [3]. Despite this long history of the debilitating effects of heat stroke, there still remain gaps in our understanding of the complex etiological factors (e.g., host immunity, environment, physical injury) that predispose to collapse, as well as the mechanisms mediating the wide array of organ responses. Within the last decade, considerable effort has been put forth attempting to identify mechanisms of multi-organ injury that may lead to the identification of novel targeted clinical therapies to mitigate morbidity and hasten recovery. While much progress has been realized in this realm, much also remains to be discovered. The purpose of this chapter is to identify misconceptions that have permeated the scientific literature and clouded our judgement as to the pathophysiologic basis for the devastation that wreaks havoc on multiple organ systems following heat stroke collapse.

O. Laitano
Department of Applied Physiology and Kinesiology, University of Florida, Gainesville, FL, USA

M. A. King · L. R. Leon (✉)
Thermal and Mountain Medicine Division, US Army Research Institute of Environmental Medicine, Natick, MA, USA
e-mail: lisa.r.leon.civ@mail.mil

J. D. Périard, S. Racinais (eds.), *Heat Stress in Sport and Exercise*,
https://doi.org/10.1007/978-3-319-93515-7_5

5.2 Heat Illness Defined

The responses to exercise-heat stress can be categorized as minor "heat-related" conditions versus exertional heat "illnesses." An abbreviated list of minor heat-related conditions includes physiological hyperthermia of exercise, exercise-associated muscle cramps (EAMCs, also known as heat cramps), and heat syncope. Note that these conditions are not categorized as heat illness per se, that is, they occur typically, but not always, during environmental heat exposure, but are usually not severe enough to place an individual at increased risk for true heat illness. Physiological hyperthermia of exercise may occur in hot or temperate environments as a natural response to the increase in metabolic heat production that occurs during exercise. That is because only ~20% of energy during exercise is used for skeletal muscle contractions, while the remaining ~80% is released as heat. As long as the rate of heat loss remains in balance with heat production, an elevated, although steady-state body core temperature can be sustained for a relatively long period until extreme dehydration or energy depletion occurs. Body core temperatures in excess of 40.6 °C have been reported in conditioned athletes and these individuals will cool naturally upon the cessation of exercise with no adverse effects [4]. EAMCs or heat cramps occur following strenuous exercise and appear as brief, recurrent, and often agonizing skeletal muscle cramps of the limbs and trunk, although smooth, cardiac, and diaphragm muscles are not involved. Cramps may be precipitated by vigorous use of affected skeletal muscles and may recur in the same individual, but are not associated with significant complications and do not predispose to exertional heat illness. EAMCs often occur in salt-depleted persons during a period of recovery (up to many hours) after prolonged, intense sweating [5]. The mechanism of EAMC is not fully understood. It has been suggested that EAMCs are due to electrolyte depletion, but this is based mainly on anecdotal and observational studies rather than sound experimental evidence [6, 7]. Malignant hyperthermia (MH) is a genetic condition characterized by a mutation in the ryanodine I receptor (RyR1) that normally regulates Ca^{2+} flux in skeletal muscle [8]. Whether there is a connection between EAMCs, MH, and complications associated with exertional heat illness or exertional rhabdomyolysis is a hypothesis that has been recently introduced and will be discussed further in this chapter [9–12]. Heat syncope is a temporary circulatory insufficiency due to pooling of blood in the peripheral veins (typically the lower extremities), which reduces diastolic filling of the heart resulting in inadequate cerebral perfusion. Heat syncope often, but not always, occurs after prolonged standing or cessation of exercise in hot weather with symptomology ranging from lightheadedness to a loss of consciousness. Dehydration may be a contributing factor to heat syncope, but body core temperature is typically not elevated, and victims will re-establish cerebral perfusion and recover rapidly once seated or supine.

Heat "illnesses" are best described as a spectrum of conditions with overlapping features that exist on a continuum and often lead to severe organ damage or death if not rapidly recognized and effectively treated. In World War II, attempts were made to provide distinct categories, with associated symptoms, for each of the various

Table 5.1 Definition of exertional heat illness

Condition	Definition	Symptoms
Heat exhaustion (HE)	Inability to sustain cardiac output to meet combined demands of increased skin blood flow for heat dissipation and blood flow for the metabolic requirements of exercising skeletal muscle and vital organs	Undue fatigue, transient ataxia, dizziness, headache, nausea, vomiting, malaise, tachycardia, hyperventilation, and transient mildly impaired cognition
Exertional heat injury (EHI)	A condition intermediate between HE and EHS with clinical evidence of end-organ injury	More sustained mild confusion and disorientation than HE with elevated core body temperature
Exertional heat stroke (EHS)	Serious, life-threatening condition characterized by central nervous system (CNS) dysfunction in the presence of severe hyperthermia	Delirium, agitation, inappropriate aggressiveness, convulsions, or coma. Core body temperature (i.e., rectal) >40 °C, although reliance on a specific rectal temperature value is not advised

severities of heat illness, but we recognize today that it is difficult to regard them as discrete disorders with their own distinct pathogenesis. For the purposes of this chapter, discussion will be focused primarily on exertional heat illnesses, which typically occur in young, healthy, fit individuals engaging in physical activity in hot or temperate environments. Exertional heat illnesses are categorized as heat exhaustion (HE), exertional heat injury (EHI), and exertional heat stroke (EHS) as highlighted in Table 5.1 [13].

HE is a mild to moderate condition that occurs when cardiac output can no longer be sustained due to competing demands for increased skin blood flow for heat dissipation vs. blood flow to support skeletal muscle contractions and vital organ function. Hypovolemia due to dehydration may contribute to the development of HE. Symptoms typically consist of fatigue, transient ataxia, dizziness, headache, and nausea with most individuals recovering following the cessation of physical activity, removal from the heat, and adequate hydration. EHI is a condition that is intermediate in severity between HE and EHS with patients exhibiting more sustained mild confusion and disorientation. It may be difficult to distinguish EHI from HE during the first few hours of illness without a clinical evaluation for end-organ dysfunction (e.g., acute kidney injury), which is not present with HE. EHI patients will maintain thermoregulatory control, but active cooling is recommended to accelerate organ function recovery. EHS is a serious, life-threatening condition characterized by central nervous system (CNS) dysfunction (e.g., delirium, agitation, inappropriate aggressiveness, convulsions, or coma) that occurs in the presence of severe hyperthermia. The co-occurrence of CNS dysfunction with hyperthermia is necessary to distinguish EHS from other conditions, such as exercise-associated hyponatremia (EAH), that may not present with increased body core temperature but has similar symptomology as EHS at the time of presentation. EAH is a consequence of hyper-hydration and high stress in which antidiuretic hormone levels are often significantly elevated. Symptomology similar to EHS includes acute mental status changes, seizures, coma, headache, confusion, visual disturbances/changes, nausea, and recurrent vomiting. Due to these overlapping

Table 5.2 Comparison of classic and exertional heat stroke

Patient characteristics	Classic	Exertional
Age	Young children or elderly	15–45 years
Health	Chronic illness common	Usually healthy
Recent febrile illness or immunization	Uncommon	Not uncommon
Prevailing weather	Recent, prolonged heat waves	Variable
Activity	Sedentary	Strenuous exercise
Drug use	Diuretics, antidepressants, anticholinergics, phenothiazines	Usually none, sometimes ergogenic stimulants or cocaine
Sweating	Usually absent	Typically present
Lactic acidosis	Uncommon	Common
Acute kidney injury	Fairly rare (<5%)	Common (~30%)
Rhabdomyolysis	Unusual	Common, may be severe
Hyperuricemia	Mild	Variable
Potassium	Usually normal	Hypo- or hyperkalemia may be present
Calcium	Usually normal	Hypocalcemia not uncommon
Disseminated intravascular coagulation (DIC)	Mild	May be marked
Glucose	Hypoglycemia common	Variable; hyper- or hypoglycemia may be present

features, the presence of hyperthermia with CNS dysfunction is the important distinguishing characteristic of EHS.

Although many of the clinical manifestations of classic heat stroke (CHS) will overlap with EHS, CHS is a condition most commonly experienced by vulnerable individuals, such as the elderly or very young (Table 5.2).

These individuals either have pre-existing conditions (i.e., illness, medication use, cardiovascular insufficiency) or lack the behavioral mechanisms to escape prolonged heat exposure during annual heat waves (e.g., children or pets in vehicles). CHS is a condition that occurs under resting conditions, thus eliminating the influence of skeletal muscle heat production and/or overuse injury from the clinical pathology (i.e., rhabdomyolysis). Perhaps one of the most intriguing aspects of heat stroke, whether exertional or classic in nature, is frequent reports of the victim collapsing under conditions that he/she had been exposed to many times before, or while others were concurrently exposed to the same condition without incident [14, 15]. This suggests that these victims were inherently more vulnerable on that particular day or some unique circumstance triggered the heat stroke event. There are individual and environmental factors that have been recognized to increase heat stroke risk, but it is difficult to "score" or assign cumulative risk since any combination of factors may be present on a particular day and will vary among individuals (Table 5.3).

The main distinction between CHS and EHS is the presence of physical exertion and the participation of skeletal muscles in the etiology of the latter condition. The role that skeletal muscles play in EHS etiology remains debatable, but a consensus is the occurrence of extensive skeletal muscle damage with EHS, a condition that is

Table 5.3 Individual and environmental risk factors for exertional heat illness

Congenital	Functional	Acquired
Individual		
Anhidrosis Ectodermal dysplasia	Low physical fitness Low work efficiency for task High body mass index (BMI) Lack of heat acclimatization	Dehydration Sleep deprivation Infection Prior heat illness Febrile illness Burn graft Miliaria rubra/sunburn Alcohol use Supplement use Medication
Environmental		
High ambient temperature/wet bulb globe temperature (WBGT)		
Previous day heat stress		
High humidity		
Lack of air movement		
Lack of access to air conditioning/cool respite		
Lack of shelter or shade		
Clothing		

normally absent in CHS. Clinically relevant muscle damage is also known as rhabdomyolysis which is defined by rapid (*rhabdo*) skeletal muscle (*myo*) breakdown (*lysis*) resulting in the leakage of intramuscular content such as electrolytes, purines, enzymes, and myoglobin into the circulation [16]. As renal failure can be a consequence of rhabdomyolysis, it is considered part of the multi-organ dysfunction that often characterizes EHS. For the remainder of this chapter we will discuss five misconceptions associated with EHS and CHS that have permeated the scientific literature and seem to have clouded our judgement as to the pathophysiologic basis for the devastation that wreaks havoc on multiple organ systems following heat stroke collapse. We present the most current literature on the topic and highlight the evidence, or lack thereof, behind each misconception.

5.2.1 Misconception #1: The Severity of Heat Stroke Can Be Defined by a Critical Body Core Temperature Value at the Time of Collapse

Medical textbooks [13], research studies [17–19], athletic position statements [20, 21], and military regulations [22, 23] are almost universal in defining the body core temperature of heat stroke (whether CHS or EHS) as >40 °C. Interestingly, fever during an infectious illness can also cause body core temperature to exceed 40 °C, but does not always invoke the same warnings as heat stroke. Perhaps this is due to regulated (fever) vs. unregulated (hyperthermia) increases in body core temperature being mediated by opposing thermoregulatory and cardiovascular adjustments such that fever is tolerated (in most cases) whereas severe hyperthermia is not (Fig. 5.1).

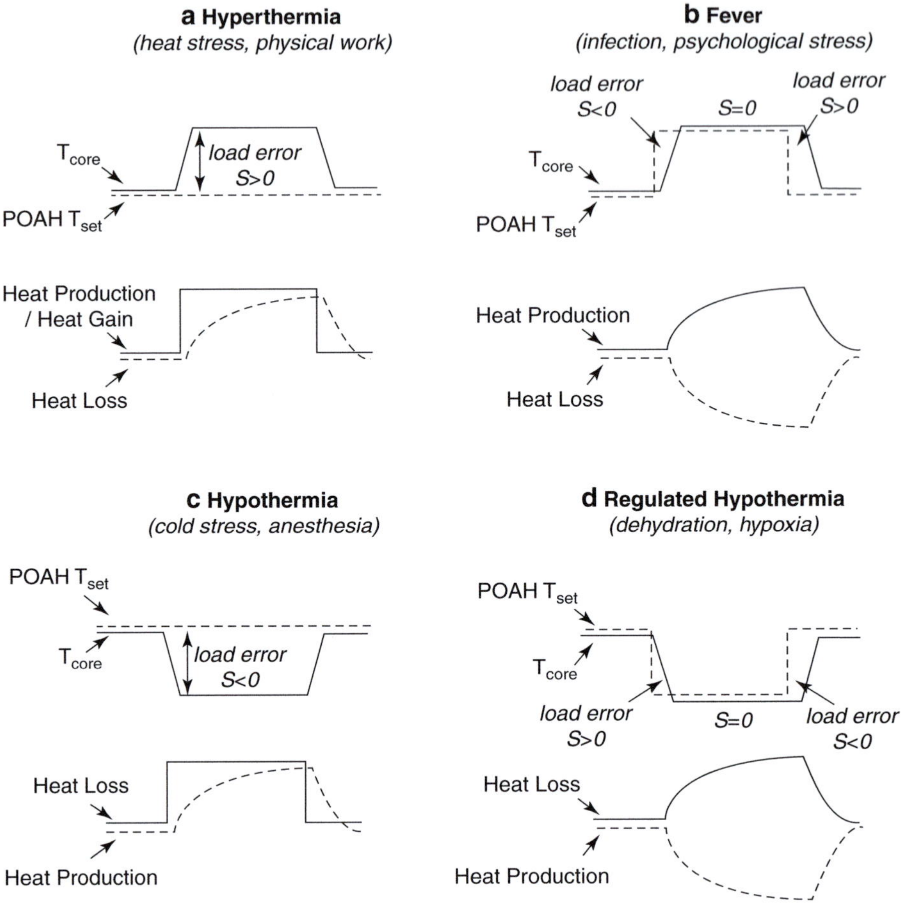

Fig. 5.1 Diagrammatic representation of unregulated (left side) and regulated (right side) body core temperature (T_{core}) and set point temperature (T_{set}) changes in response to environmental stimuli. (**a**) Hyperthermia is an increase in T_{core} that occurs independent of a change in T_{set} as heat gain (heat stress) and heat production (physical exertion) are increased. (**b**) Fever is a regulated increase in T_{core} in response to an upward setting of T_{set}. (**c**) Hypothermia is the mirror image of hyperthermia, whereas (**d**) regulated hypothermia is the mirror image of fever

Because it has been reported that heat stroke victims collapse with a wide range of body core temperatures, it is now widely recognized that the absolute value is not a sensitive predictor of severity, and what could be considered extreme temperatures are survivable if active cooling is implemented rapidly and effectively. Furthermore, there are reports of highly trained athletes experiencing temperatures as high as ~42 °C [4] with no adverse consequences, which is thought to be due to the protective effect of training and acclimation. These examples highlight the complicated nature of heat stroke and indicate that the pathology is not just a consequence of high body core temperature but is due to multiple interacting factors that include tissue ischemia/reperfusion injury, a systemic inflammatory response syndrome (SIRS), coagulation, and other physiological disturbances.

There are a multitude of factors that may be responsible for the wide range of body core temperature values associated with heat stroke. While it is tempting to rank risk factors according to the magnitude of impact on the heat stress response, the difficulty is that many exist concurrently and in unique combinations that are not always predictable, recognizable (i.e., subclinical infection), or universally present in a population under observation. Physical effort unmatched to physical fitness was identified as a significant risk factor for EHS [24]. In many instances, EHS occurs within the first 2 h of exertion and not necessarily at high ambient temperatures. Additional risk factors may include, but are not limited to, circadian rhythmicity, site of body temperature measurement, different times of clinical presentation, presence of pre-existing or concurrent illness, heat acclimatization state, physical fitness, type of physical activity, clothing ensemble, as well as a number of environmental factors (e.g., solar load, relative humidity; for a thorough review on the topic, see Leon and Bouchama [19]).

The ideal site for the measurement of body core temperature should be convenient, unbiased by the environmental conditions and reflective of small changes in arterial blood temperature. Temperature within a deep body region will vary due to different metabolic rates of surrounding tissues, local blood flow changes, and temperature gradients between adjacent tissues. Skin temperature is the most accessible measurement, but is influenced by changes in blood flow, sweat secretion, environmental effects on heat exchange mechanisms (i.e., evaporative cooling and radiation), or inaccurate measurements such as occurs with a loss of contact with the skin surface [25]. Esophageal temperature is considered the most accurate noninvasive measure of body core temperature in humans because it is rapidly responsive to changes in blood temperature. Unfortunately, this method is not feasible in unresponsive or combative heat stroke patients so other methods must be relied upon under these conditions; because of this limitation, rectal temperature is considered the gold standard for the measurement of body core temperature in heat stroke patients. The only considerations are that rectal temperature will be slightly lower than esophageal temperature because it responds more slowly to transient changes in core temperature, which could result in an underestimation of the response. Interestingly, rectal temperature tends to be ~0.8 °C lower than esophageal temperature even during cooling interventions [26]. There is also the procedural caution that perforation of the rectum can occur in those individuals who are not trained in the use of rectal probes. For this reason, many rely on tympanic auditory canal temperatures, which are simple to obtain, but are biased by head skin temperature and provide values that may be higher or lower than steady-state rectal and esophageal temperatures [27, 28].

One of the more intriguing aspects of heat stroke is the remarkable inter-individual (i.e., within a species or strain) variability of the hyperthermic response that is observed during heat exposure. Large inter-individual variability is readily apparent in animal models, despite the absence of pre-existing risk factors (e.g., specific pathogen free or disease free) or their genetic similarity as many are several generations inbred [29–36]. Despite controlling for immune health, genetic diversity, environmental temperature, humidity and lighting, feed and water intake, body weight and age, there still exists quite wide variability in the maximum body core temperature that is associated with CHS or EHS collapse [20–22, 28]. This has been shown for a number of species including cats, guinea pigs, mice, rats, and salamanders (for a review, see Leon [38]). These findings suggest an inherent variability in how organisms respond to environmental heat stress on any given day, which is difficult if not nearly impossible to predict or define. Basically, something is just different for that organism, whether animal or human, on that particular day relative to other individuals with the same exposure or the same physiological and genetic make-up. It is conceivable that we will never be able to precisely elucidate what those factors are, or when they might arise, but can be sure that reliance on a specific body core temperature value is not the panacea for heat stroke prevention that we wish it to be.

5.2.2 Misconception #2: Heat Stroke Causes a "Failure" of Thermoregulatory Control Due to Hypothalamic Injury

Hyperthermia is the body core temperature disturbance that occurs *during heat exposure*, whereas hypothermia and/or recurrent hyperthermia (also referred to as rebound fever) are often observed *during recovery*. In humans, hypothermia is typically a direct response to active cooling that represents a rapidly correctable undershoot of body core temperature (e.g., <37 °C). This rebound hypothermia is typically regarded as an adverse response, but there is speculation that a mild hypothermia (~32–34 °C) could provide protection against tissue injury as has been shown with cardiopulmonary bypass and other conditions [4, 39]. This rationale has resulted in the suggested practice of inducing hypothermia in the most serious EHS cases as a treatment strategy to protect against liver failure (personal communication to LR Leon). A protective effect of hypothermia has been demonstrated by small rodents that develop protracted (several hours) *regulated* hypothermia (<35 °C and often as low as 30 °C) when left to passively cool in a temperate environment. If rodents are prevented from developing this response, CHS mortality is significantly increased [29]. During the recovery period, a fever-like response (≥24 h after collapse) may also be observed, which is typically transient in nature, but can be associated with poor outcome in humans [29, 40].

All of the body core temperature disturbances described above have been traditionally regarded, and are still mentioned today, as a reflection of heat-induced injury to the preoptic anterior hypothalamic (POAH) region of the brain, which many consider the main CNS site for temperature regulation. In particular, this hypothesis suggests that prolonged heat exposure causes POAH *injury* that results in a "failure" of thermoregulatory control that manifests as severe hyperthermia during heat exposure as well as hypothermia and/or recurrent fever during heat stroke recovery. We contend that severe hyperthermia during heat exposure is a consequence of failure of the cardiovascular system to support competing blood flow demands to the skin, organs, and skeletal muscle such that heat dissipation cannot keep up with heat gain and heat production. On the other hand, a recent report in rats showed glutamine and monoamine imbalances that correlated with neuroinflammation in the hypothalamus at the point of collapse with severe CHS [41]. These imbalances may account for transient POAH "dysfunction" that explains the development of hypothermia during the immediate recovery period (e.g., during active or passive cooling). For example, behavioral analysis of severe CHS mice in a temperature gradient indicated that the thermoregulatory feedback loop was altered in some way following collapse—either due to a breakdown of the afferent (sensory) relay of information or inability of the POAH to integrate the sensory information to make corrective effector actions as highlighted in Fig. 5.2 [42].

We are not aware of any permanent disturbances in thermoregulation of severe CHS or EHS patients or animal models that would suggest irreversible thermal injury to the POAH. Furthermore, there are no clinical or experimental data that show injury to the POAH to account for more protracted hypothermia or rebound fever episodes during the days or weeks of recovery. Malamud et al. [43] provided the most striking clinical data from 125 fatal military EHS cases that exhibited no damage to the POAH despite extensive damage to other CNS regions. In fact, there were several noteworthy observations from this study: First, EHS in these patients was defined as CNS dysfunction with body core temperature >41.1 °C, rather than the current definition of >40 °C; second, body core temperature at the time of hospital admission ranged from 36.1 °C, presumably due to active cooling, to 43.9 °C; third, shock rather than the degree of hyperthermia was the best prognostic index; fourth, rises in body core temperature following active cooling (ice packs, etc.) were interpreted as a persistent disturbance in thermoregulation; yet, the lack of demonstrable damage (in the hypothalamus) in these patients differed from injury in other portions of the brain. The emergence of sensitive imaging technologies has provided more detailed examination of the CNS abnormalities experienced by EHS patients but has also failed to detect damage to the hypothalamus despite detectable, and in some cases, quite severe injury to the cortex, thalami, paraventricular nucleus, cerebellum, and third ventricle [44, 45].

While the hypothalamus may not be grossly damaged, investigators have shown increases in pyknotic neurons along with significant elevation in inflammatory mediators such as NF-κB, interleukin (IL)-1β, cyclooxygenase (COX)2, and glial fibrillary acidic protein (GFAP) following severe heat stress [41]. Similarly, there is a body of literature that contends heat stroke is due to damage caused by brain ischemia, inflammation, and neuronal damage [36, 37]. While inflammatory markers are increased following heat stress, it is still unknown if this is the cause of temperature dysregulation at the time of collapse. Obtaining these data in human populations will prove to be challenging, as heat stroke is unpredictable and collecting data at the point of collapse especially in an area such as the hypothalamus is difficult, if not impossible.

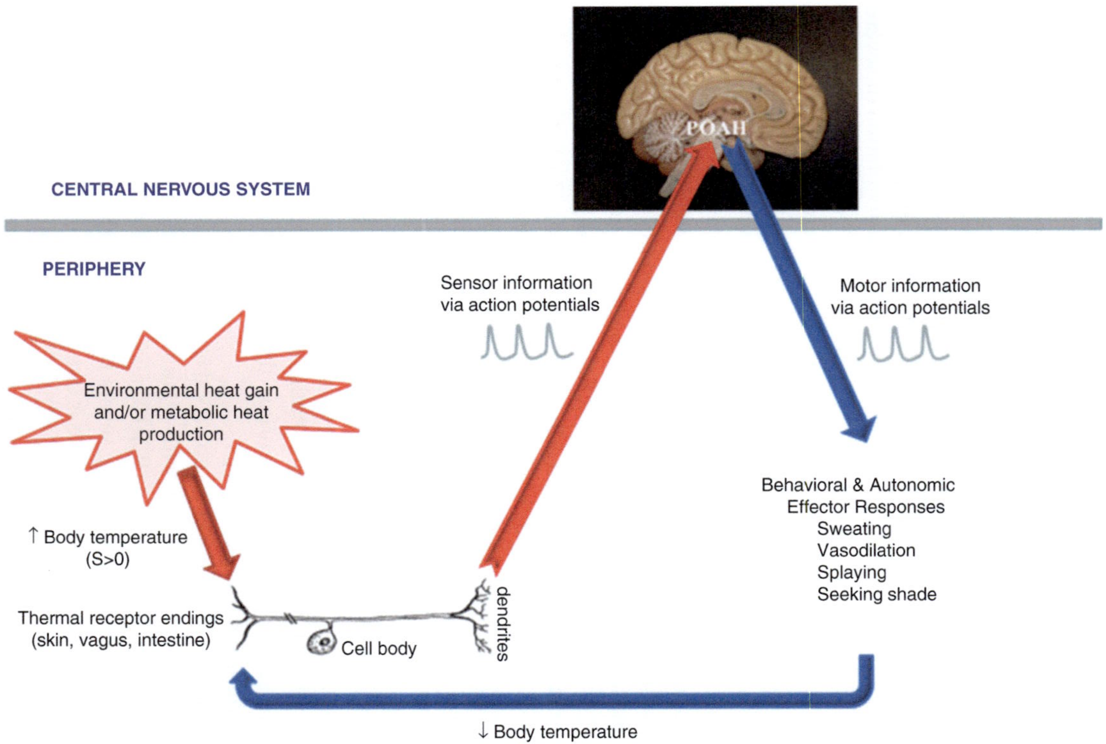

Fig. 5.2 Negative feedback loop of body core temperature regulation. Thermal receptor endings in the skin and other anatomical sites detect increases in skin and body core temperature during environmental heat stress and/or metabolic heat production (exercise). Sensory information is sent via action potentials through ganglions to the POAH. The POAH acts as a thermoregulatory integration center and compares peripheral temperature changes to the set point temperature. A load error is generated that evokes graded behavioral and autonomic effector responses to aid in the dissipation of heat for the return of body core temperature toward homeostasis

5.2.3 Misconception #3: Hyperthermia During Heat Exposure Is a Fever Response to Endotoxin Leakage from the Gut and/or Leakage of Intramuscular Content

One of the major misconceptions regarding CHS or EHS is that endotoxemia, or the presence of endotoxin in the circulation, is a universal phenomenon observed in all animal models and patients. This misconception was based on early reports that showed reductions in splanchnic blood flow at temperatures as low as 40 °C [46], which falls in line with the definition of heat stroke. As blood flow is diverted from the splanchnic circulation to the skin for heat dissipation, gut epithelial membranes experience nitrosative and oxidative stress that degrades tight junction integrity to facilitate endotoxin leakage into the portal circulation. Whereas the liver reticuloendothelial system normally clears endotoxin, under severe CHS or EHS conditions, heat-induced dysfunction or injury to the liver may compromise the ability to perform this function. It is under these severe conditions, often associated with liver failure, that endotoxemia has been observed. For example, a young football player who collapsed and died from EHS presented with significant endotoxemia as well as hemorrhagic liver necrosis and body core temperature of 40.6 °C [47]. Similarly, circulating endotoxin showed a precipitous increase in primates under CHS conditions once body core temperatures exceeded the fatal level of 43.0 °C [48]. Liver injury was not assessed in this study, but is typically detectable at body core temperatures ranging from ~ 42 to 43 °C [49–51]. Endotoxin neutralization studies in dogs, rabbits, and rats have shown a protective effect of antibiotics and endotoxin tolerance against heat stroke mortality, but once again these were lethal models with temperatures at or above the threshold where liver injury would be expected [48, 52, 53]. In contrast to these studies, a mouse CHS model that induced body core temperature as high as 42.7 °C did not correlate with detectable levels of circulating endotoxin despite significant gut histological injury [29, 37]. This was mostly likely due to a lack of liver injury, once again supporting the hypothesis that dysfunction of this organ is a primary mechanism for endotoxemia. Similarly, Chung et al. [54] failed to show increased endotoxin in former heat stroke patients who were subsequently exposed to a 60-min heat stress that only caused body core temperature to increase to <39.5 °C.

We propose three alternative interpretations for the role of endotoxin in the heat stroke response. First, endotoxemia is not a universal phenomenon and is manifest under the most severe, typically chronic exposure conditions that lead to organ (primarily liver) failure. As mentioned above, this has been shown in several animal models and patients, although it does conflict with several recent reports showing a correlation between indirect biomarkers of gut leakage (namely, intestinal fatty-acid binding protein, or I-FABP) with endotoxin-like symptoms with exercise heat-stress or EHS [30]. However, caution is warranted in the interpretation of these findings due to the potential lack of sensitivity and specificity of I-FABP (or other biomarkers) as measures of endotoxin leakage and reliance on correlation studies that do not equate to causality. Second, we do not believe that endotoxin leakage is a major contributor to hyperthermia *during heat exposure* under most conditions (i.e., lack of pre-existing conditions or significant organ injury). That is, there is little evidence to support the contention that endotoxin leakage during heat exposure is inducing an increase in the POAH temperature set point for fever development. This is based on a number of studies that have failed to show a consistent effect of non-steroidal anti-inflammatory drugs (NSAIDs, i.e., fever inhibitors) on the hyperthermic response during heat exposure. NSAIDs are a class of over-the-counter and prescription drugs that are the most widely prescribed worldwide. NSAIDs efficacy for improved heat tolerance (*during heat exposure*) has been examined in animal models and men working in hot environments with contradictory results. In animal models, indomethacin (a prescription NSAID) or sodium salicylate (aspirin) improved heat tolerance by inducing a lower rate of body core temperature rise, suggesting this was at least partially a true fever response [55, 56]. Prostaglandins (PGs), which are the main mediators of endotoxin-induced fever, have been shown to contribute to active cutaneous vasodilation in humans suggesting their release via endotoxin stimulation in the heat could affect hyperthermia and heat tolerance [57, 58]. However, Jacobsen and Bass [59] failed to show an effect of a high chronic dose of sodium salicylate on skin temperature or pulse rate of men walking in a hot environment, yet the hyperthermic response was *potentiated* (rather than inhibited) suggesting an alternative mechanism for this response [59]. More recently, Audet et al. [32] showed that indomethacin had no effect on the body core temperature response of mice during the development of CHS, but significantly increased mortality due to toxic effects on the gut mucosa that caused excessive hemorrhaging. Given that many athletes use NSAIDs as potential performance enhancing drugs, the advantage of this strategy for heat mitigation is questionable and the probability of gut injury is concerning. One can suppose that under conditions of NSAIDs usage with heat exposure, endotoxemia, fever, and gut and/or liver injury are likely consequences.

The third intriguing possibility is that extensive skeletal muscle damage (i.e. rhabdomyolysis) during exercise triggers a pyrogenic response similar to fever [60, 61] leading to further increases in body core temperature and EHS—note that in this instance, exertional rhabdomyolysis is inducing a regulated increase in body core temperature that is distinct from the hyperthermia induced by heat stress or endotoxemia per se. Fever is caused by the release of endogenous pyrogens (e.g., IL-1β and IL-6, among others; reviewed in [62]) that leads to a subsequent rise in PGs. PGs, whether produced systemically or locally within the POAH, raise the thermoregulatory set point triggering multiple behavioral and autonomic responses to induce fever. Endogenous antipyretics such as arginine vasopressin, glucocorticoids, IL-10, and TNF-α are also released to modulate fever and prevent it from reaching dangerous levels [62, 63]. Conversely, hyperthermia due to exertion and heat stress occurs because the metabolic heat production of active skeletal muscles and the environmental source of heat surpasses the body's capacity to exchange heat to the surroundings, thus driving an increase in body core temperature (Fig. 5.3).

Patients with exertional rhabdomyolysis often exhibit fever symptoms, which may be a consequence of skeletal muscle damage that induces a fever in the absence of environmental heat stress. That is, the pyrogenic response may be elicited directly

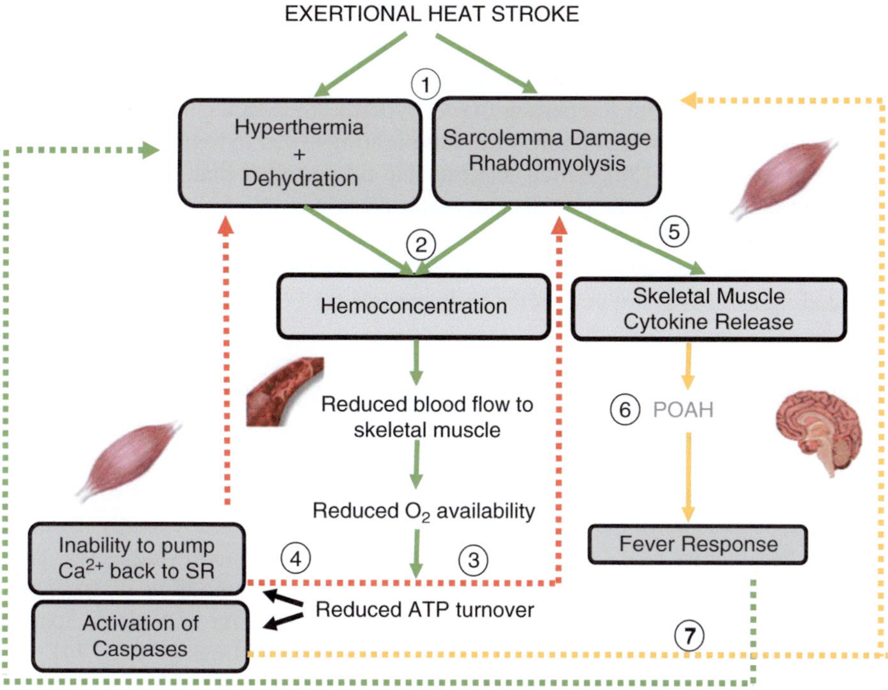

Fig. 5.3 Integrative model of EHS. (1) EHS is associated with whole-body hyperthermia and (sometimes) dehydration and sarcolemma damage; (2) reductions in plasma volume and leak of intracellular content lead to hemoconcentration; (3) reduced ATP turnover due to reduced blood flow and oxygen delivery to skeletal muscle leads to (4) inability to pump calcium back to the sarcoplasmic reticulum and activation of caspases; (5) this results in sustained muscle contractions and further sarcolemma damage which leads to (6) cytokine release into the circulation leading to (7) fever response via POAH in the CNS which further enhances hyperthermia and dehydration. The arrows stand for a "traffic light" system where red = poor evidence; yellow = some evidence; and green = strong evidence

by cytokines released from inflammatory cells as well as skeletal muscle [64–66]. The stimulus for cytokine release by skeletal muscle may be an increased epinephrine and norepinephrine response [67, 68], triggered by exercise of sufficient intensity, and/or via direct sarcolemma damage [69]. An alternative pathway for cytokine release from skeletal muscle may arise from the presence of endotoxin in the circulation as mentioned earlier in this section. Endotoxemia has been observed in more chronic CHS, although not universally (see above) [70], and has been shown to lead to cytokine release (i.e., IL-6, IL-1β, IL-8) from inflammatory cells and other tissues including skeletal muscle. Cytokines released into the circulation can trigger PG production via COX enzyme with PGE_2 considered the main mediator of fever [61]. Evidence for exercise-induced muscle damage that induces a pyrogenic response is provided in studies that administered NSAIDs in exercising humans and rodents [61, 71, 72]. NSAIDs block COX activity and have been shown to halt the increase in body core temperature during exercise in temperate environments suggesting that a component of the elevation in body core temperature may partly have a fever-like origin [61, 72]. On the other hand, the use of NSAIDs during exposure to heat stress has been shown to enhance organ damage while having no effect on the rise in body core temperature (i.e., hyperthermia rather than fever) during heat exposure [32]. Another point worth highlighting is that pyrogenic cytokines are also observed in CHS [36], which suggests that tissues other than the skeletal muscle, presumably inflammatory cells, are major players in this form of elevated body core temperature. Nevertheless, the extent by which the exercise-induced pyrogenic response is enhanced by skeletal muscle damage and thus contributes to hyperthermia leading to EHS is not yet known and warrants further investigation.

5.2.4 Misconception #4: Cytokines Are Adverse Mediators of the Systemic Inflammatory Response Syndrome

The severity of multi-organ dysfunction with EHS is thought to be a consequence of cytokines, such as IL-6, that induce a SIRS. IL-6 and TNF-α are typically regarded as pro-inflammatory (e.g., adverse) mediators of the SIRS due to studies that have demonstrated endotoxemia- or sepsis-like symptoms and initiation of the acute phase response following injection, as well as the correlation of high circulating levels with heat stroke severity in clinical and laboratory settings [47, 73, 74]. The misconception that cytokines are solely adverse mediators of the SIRS, and do not provide any protective advantage, is based almost exclusively on correlation studies that have shown high circulating concentrations (e.g., IL-1β, IL-6, TNF-α, IFN-γ) with morbidity and mortality, which is more evident in CHS than EHS [17, 70, 75–80]. The most consistent observation has been the correlation of high circulating IL-6 levels with heat stroke death in primates and patients. IL-6 is typically elevated in 100% of CHS patients [75], levels tend to be highest in those who do not survive [75, 77, 79], and sustainment of high IL-6 levels during cooling in primates correlated with heat stroke severity [17, 75]. Yet, there are two aspects of these studies that warrant caution in our interpretation of cytokines (IL-6 or others) in the heat stroke syndrome: First, the majority of the early CHS studies showed a (typically weak) correlation of high circulating cytokine levels with heat stroke symptoms, but failed to show a causal relationship between these responses. Whereas more recent animal studies have closed this gap and shown, with the use of exogenous cytokine treatment or gene knockout technology, that IL-6 and TNF have protective functions that enhance performance in the heat and improve survival during recovery, respectively [81, 82]. As such, IL-6 and TNF-α, as well as other cytokines, appear to have both pro- and anti-inflammatory actions that depend on the cytokine milieu in which they function as well as the concentration that is achieved (likely) at the local, tissue level [38]. Second, the contribution of circulating soluble cytokine receptors on the actions of these proteins has virtually been ignored, despite early evidence of their potential protection for CHS survival. For example, CHS survivors had higher levels of soluble TNF receptors (sTNFRs; natural antagonists of TNF actions) than non-survivors, suggesting a detrimental effect of high TNF levels in these patients [77]. Note that the binding of TNF to the sTNFR occurs only at high concentrations suggesting that the pro-inflammatory effects are concentration-dependent. The soluble IL-6R (sIL-6R) plays a different role in that it will integrate into cell membranes that do not contain the receptor to make them responsive to IL-6 via ubiquitously expressed glycoprotein 130 (GP130) receptor. The role of the sIL-6R in CHS pathophysiology has not been clearly delineated, but increased circulating sIL-6R correlated with effective cooling in patients supporting a protective effect on outcome.

Clearly, additional research is needed to understand the complex nature of cytokine actions in CHS and EHS. Not only are current misconceptions based primarily on correlation studies, but the differences in the etiology and time course of cytokine responses between CHS and EHS suggest that these immune modulators may function differently with these two conditions. Perhaps the misunderstanding of cytokine actions in heat stroke is based on their known role with endotoxemia, which has always been considered the initiating stimulus for the SIRS. As new research unfolds to further delineate the role of endotoxin and novel approaches continue to be employed to study cytokine actions, the interplay of these responses will become clearer.

5.2.5 Misconception #5: EHS, Exertional Rhabdomyolysis, and Malignant Hyperthermia Are Triggered by Similar Mechanisms of Calcium Dysregulation in the Skeletal Muscle

Controversy exists regarding the molecular events resulting in extensive sarcolemma damage leading to clinically relevant exertional rhabdomyolysis, mainly when it is associated with EHS, but these events seem to involve intracellular calcium (Ca^{2+}) dysregulation of some sort. For instance, exertional rhabdomyolysis is believed to share similarities with another myopathy known as malignant hyperthermia (MH) [8, 83, 84]. MH is a reaction triggered by anesthetic drugs (e.g., halothane, enflurane, isoflurane, sevoflurane, desflurane) that causes sustained skeletal muscle contractions (in a positive feedback loop) that drive body core temperature toward heat stroke levels if not rapidly treated. The syndrome is associated with a mutation in the ryanodine 1 receptor (RyR1), which causes dysregulation in Ca^{2+} homeostasis, namely, increased Ca^{2+} influx in skeletal muscle. This influx of Ca^{2+} triggers skeletal muscle contractions resulting in metabolic heat production that can lead to severe whole-body hyperthermia. Presumably, this Ca^{2+} dysregulation due to mutation of the RyR1 receptor in the sarcoplasmic reticulum would be present in those at increased risk of EHS and exertional rhabdomyolysis and this would conveniently separate responders and non-responders in terms of susceptibility to these conditions [85, 86]. The majority of evidence for this relationship among MH, exertional rhabdomyolysis, and EHS comes from case reports [87, 88]. For instance, Hopkins et al. reported a case where two men in the military service, who previously experienced EHS episodes, tested as MH susceptible after their skeletal muscle samples were submitted to an in vitro contracture test (IVCT) in response to caffeine and halothane challenges. The authors concluded that abnormalities seen in both EHS and MH were similar, but not identical [88]. Tobin et al. [87] suggested an association between MH and EHS with a case report of a 12-year-old boy who displayed signs of MH during surgery for a humerus fracture. Eight months later, the boy displayed signs and symptoms of EHS after a football match-play at an ambient temperature of 26 °C. Body core temperature at the time of arrival at the hospital was >42.2 °C. Postmortem analysis of the DNA from the patient, and their parents, revealed mutations of the RyR1 gene [87]. Wrappler et al. observed an association between exertional rhabdomyolysis and MH by performing IVCT in patients with previous history of exertional rhabdomyolysis. Ten out of 12 subjects with previous episodes of exertional rhabdomyolysis were classified as MH-susceptible [84]. Even though these studies indicate a possible association among EHS, exertional rhabdomyolysis, and MH, the sample size was too small to establish causality. More recently, a study explored the link between EHS and MH in a larger cohort of subjects [12]. Skeletal muscle samples were collected from military personnel who previously experienced EHS and submitted the skeletal muscle samples to an IVCT. Authors observed a high prevalence of MH susceptibility in those who experienced EHS previously, presumably indicating that Ca^{2+} dysregulation is part of the EHS syndrome. Indeed, there is no direct evidence of Ca^{2+} dysregulation in EHS, at least when exertional rhabdomyolysis is absent. Even though exertional rhabdomyolysis could be part of the multi-organ dysfunction that often characterizes EHS, and perhaps those with a genetic mutation are at elevated risk of developing EHS and exertional rhabdomyolysis, there are several aspects of exertional rhabdomyolysis and EHS that are incongruent with the conclusion that these syndromes share similar mechanisms related to RyR1 myopathy such as MH. In summary, MH, exertional rhabdomyolysis, and EHS may display a common feature of loss of intracellular Ca^{2+} control (although yet to be demonstrated with EHS), it is unlikely that they share a single common genetic predisposition. As pointed out later in this chapter, there are other features of both EHS and exertional rhabdomyolysis that can trigger intracellular Ca^{2+} dysregulation and are independent of a MH genotype (Fig. 5.4).

EHS involves a physiological crisis characterized by increased energy demand, whole-body hyperthermia, and often sweat-induced dehydration, all of which can

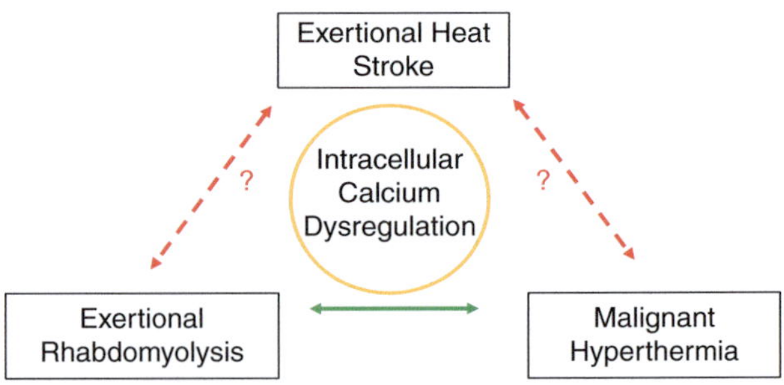

Fig. 5.4 Intracellular calcium dysregulation is a common feature of both exertional rhabdomyolysis and malignant hyperthermia. But whether it is present in exertional heat stroke requires further evidence

potentially result in exertional rhabdomyolysis, as highlighted in Fig. 5.3, and other stress responses such as reactive oxygen species formation as reviewed by King et al. [89]. The misconceptions concerning EHS-related rhabdomyolysis revolve around the extent by which each of these factors contributes to the exertional rhabdomyolysis syndrome in those experiencing EHS. A potential alternative mechanism involved in exertional rhabdomyolysis and EHS is the fact that hyperthermia decreases muscle blood flow due to a competition for the available cardiac output between skin, for thermoregulation, and active muscles, which can make skeletal muscles ischemic during exercise [90]. A reduction in O_2 delivery decreases ATP turnover in muscle—despite a reported compensatory increase in O_2 extraction in hyperthermia [91]—which can presumably increase intramuscular Ca^{2+} content due to an inability to actively transport Ca^{2+} back to the endoplasmic reticulum. Increased intramuscular Ca^{2+} leads to the activation of proteolytic enzymes (i.e., caspases) that contribute to skeletal muscle breakdown seen in exertional rhabdomyolysis.

5.3 Conclusion

We have focused on the five most common misconceptions regarding heat stroke that arise from a combination of lack of evidence, the misinterpretation of inconclusive studies, and reliance on other conditions with similar pathophysiology as CHS or EHS. There are extensive data in the literature indicating that the body core temperature response to heat exposure varies widely and is not a sensitive predictor of CHS or EHS. Due to a variety of different thermoregulatory disturbances at the time of collapse, or during the early periods of recovery, heat stroke has been thought to cause injury to the POAH, which is considered the main thermoregulatory center of the brain. This misconception persists despite autopsy of 125 military EHS patients who showed that the POAH was the one region of the brain that did not experience overt injury. It is possible that the technique used to examine injury at that time was not advanced to detect subtle changes, yet more sophisticated imaging techniques continue to support those original findings from over 70 years ago. Why the misconception that the POAH is damaged continues to persist today remains a mystery. Cytokine release in response to endotoxemia has been purported by many authors (including those of this chapter) to be key stimulators of the SIRS with heat stroke. However, perhaps the error on our part has been to present that hypothesis as a universal finding, whereas data are showing that endotoxemia is not a universal finding and cytokines can be both pro- and anti-inflammatory in their actions. Finally, even though extensive rhabdomyolysis is part of the EHS syndrome, there is no evidence that intramuscular content leakage leads to a pyrogenic response that contributes to EHS. Furthermore, EHS, exertional rhabdomyolysis, and MH do not share a single mutation (i.e., RyR1 receptor), even though calcium dysregulation may be present in each syndrome. The take home message from this summary of misconceptions is that heat stroke is a complex condition with multiple, overlapping etiologies and mechanisms that are not readily understood and continue to perplex physicians and researchers in terms of the most effective prevention (beyond cooling) and treatment strategies. While initially disappointing to realize, we view this as an exciting opportunity to explore new methodologies and approaches to solve this mystery once and for all!

Disclaimers The opinions or assertions contained herein are the private views of the authors and are not to be construed as official or as reflecting the views of the Army or the Department of Defense. Any citations of commercial organizations and trade names in this report do not constitute an official Department of the Army endorsement of approval of the products or services of these organizations.

References

1. Osler SW. The principles and practice of medicine: designed for the use of practitioners and students of medicine. D. Appleton; 1892. 1116 p.
2. Pandolf KB, Burr RE. Medical aspects of harsh environments, vol 2. [Internet]. Washington: Walter Reed Army Medical Center; 2002. http://www.dtic.mil/docs/citations/ADA433963. Cited 19 Mar 2018.
3. Goldman R. Introduction to heat related problems in military operations. In: Pandolf KB, Burr RE, editors. Medical aspects of harsh environments. Washington: Department of the Army; 2002. p. 3–42.
4. Maron MB, Wagner JA, Horvath SM. Thermoregulatory responses during competitive marathon running. J Appl Physiol. 1977;42(6):909–14.
5. Jung AP, Bishop PA, Al-Nawwas A, Dale RB. Influence of hydration and electrolyte supplementation on incidence and time to onset of exercise-associated muscle cramps. J Athl Train. 2005;40(2):71–5.
6. Miller KC, Stone MS, Huxel KC, Edwards JE. Exercise-associated muscle cramps: causes, treatment, and prevention. Sports Health. 2010;2(4):279–83.
7. Schwellnus MP. Cause of exercise associated muscle cramps (EAMC)—altered neuromuscular control, dehydration or electrolyte depletion? Br J Sports Med. 2009;43(6):401–8.
8. Muldoon S, Deuster P, Brandom B, Bunger R. Is there a link between malignant hyperthermia and exertional heat illness? Exerc Sport Sci Rev. 2004;32(4):174–9.
9. Poussel M, Guerci P, Kaminsky P, Heymonet M, Roux-Buisson N, Faure J, et al. Exertional heat stroke and susceptibility to malignant hyperthermia in an athlete: evidence for a link? J Athl Train. 2015;50(11):1212–4.
10. Hosokawa Y, Casa DJ, Rosenberg H, Capacchione JF, Sagui E, Riazi S, et al. Round table on malignant hyperthermia in physically active populations: meeting proceedings. J Athl Train. 2017;52(4):377–83.
11. Bendahan D, Kozak-Ribbens G, Confort-Gouny S, Ghattas B, Figarella-Branger D, Aubert M, et al. A noninvasive investigation of muscle energetics supports similarities between exertional heat stroke and malignant hyperthermia. Anesth Analg. 2001;93(3):683.
12. Sagui E, Montigon C, Abriat A, Jouvion A, Duron-Martinaud S, Canini F, et al. Is there a link between exertional heat stroke and susceptibility to malignant hyperthermia? PLoS One. 2015;10(8):e0135496.
13. Winkenwerder M, Sawka MN. Disorders due to heat and cold. In: Goldman L, Ausiello D, editors. Cecil medicine. Philadelphia: Elsevier Science; 2007. p. 763–7.
14. Carter R, Cheuvront SN, Williams JO, Kolka MA, Stephenson LA, Sawka MN, et al. Epidemiology of hospitalizations and deaths from heat illness in soldiers. Med Sci Sports Exerc. 2005;37(8):1338–44.
15. Nelson NG, Collins CL, Comstock RD, McKenzie LB. Exertional heat-related injuries treated in emergency departments in the U.S., 1997–2006. Am J Prev Med. 2011;40(1):54–60.

16. Alpers JP, Jones LK. Natural history of exertional rhabdomyolysis: a population-based analysis. Muscle Nerve. 2010;42(4):487–91.
17. Bouchama A, Roberts G, Al Mohanna F, El-Sayed R, Lach B, Chollet-Martin S, et al. Inflammatory, hemostatic, and clinical changes in a baboon experimental model for heatstroke. J Appl Physiol. 2005;98(2):697–705.
18. Bouchama A, Mohanna FA, El-sayed R, Eldali A, Saussereau E, Chollet-martin S, et al. Experimental heatstroke in baboon: analysis of the systemic inflammatory response. Shock. 2005;24(4):332–5.
19. Leon LR, Bouchama A. Heat stroke. Compr Physiol. 2015;5(2):611–47.
20. American College of Sports Medicine, Armstrong LE, Casa DJ, Millard-Stafford M, Moran DS, Pyne SW, et al. American College of Sports Medicine position stand. Exertional heat illness during training and competition. Med Sci Sports Exerc. 2007;39(3):556–72.
21. Casa DJ, DeMartini JK, Bergeron MF, Csillan D, Eichner ER, Lopez RM, et al. National Athletic Trainers' Association position statement: exertional heat illnesses. J Athl Train. 2015;50(9):986–1000.
22. Department of the Army TMB. Heat stress control and heat casualty management. 2003.
23. Department of the Army. Army regulation 40-501 standards of medical fitness. Department of the Army; 2017.
24. Rav-Acha M, Hadad E, Epstein Y, Heled Y, Moran DS. Fatal exertional heat stroke: a case series. Am J Med Sci. 2004;328(2):84–7.
25. Smith ADH, Crabtree DR, Bilzon JLJ, Walsh NP. The validity of wireless iButtons and thermistors for human skin temperature measurement. Physiol Meas. 2010;31(1):95–114.
26. Sarkar S, Donn SM, Bhagat I, Dechert RE, Barks JD. Esophageal and rectal temperatures as estimates of core temperature during therapeutic whole-body hypothermia. J Pediatr. 2013;162(1):208–10.
27. Greenleaf JE, Castle BL. External auditory canal temperature as an estimate of core temperature. J Appl Physiol. 1972;32(2):194–8.
28. McCaffrey TV, McCook RD, Wurster RD. Effect of head skin temperature on tympanic and oral temperature in man. J Appl Physiol. 1975;39(1):114–8.
29. Leon LR, DuBose DA, Mason CW. Heat stress induces a biphasic thermoregulatory response in mice. Am J Physiol Regul Integr Comp Physiol. 2005;288(1):R197–204.
30. King MA, Leon LR, Mustico DL, Haines JM, Clanton TL. Biomarkers of multiorgan injury in a preclinical model of exertional heat stroke. J Appl Physiol Bethesda Md. 2015;118(10):1207–20.
31. King MA, Leon LR, Morse DA, Clanton TL. Unique cytokine and chemokine responses to exertional heat stroke in mice. J Appl Physiol Bethesda Md. 2017;122(2):296–306.
32. Audet GN, Dineen SM, Stewart DA, Plamper ML, Pathmasiri WW, McRitchie SL, et al. Pretreatment with indomethacin results in increased heat stroke severity during recovery in a rodent model of heat stroke. J Appl Physiol. 2017;123(3):544–57.
33. Quinn CM, Duran RM, Audet GN, Charkoudian N, Leon LR. Cardiovascular and thermoregulatory biomarkers of heat stroke severity in a conscious rat model. J Appl Physiol Bethesda Md. 2014;117(9):971–8.
34. Quinn CM, Audet GN, Charkoudian N, Leon LR. Cardiovascular and thermoregulatory dysregulation over 24 h following acute heat stress in rats. Am J Physiol Heart Circ Physiol. 2015;309(4):H557–64.
35. Audet GN, Quinn CM, Leon LR. Point-of-care cardiac troponin test accurately predicts heat stroke severity in rats. Am J Physiol Regul Integr Comp Physiol. 2015;309(10):R1264–72.
36. Welc SS, Clanton TL, Dineen SM, Leon LR. Heat stroke activates a stress-induced cytokine response in skeletal muscle. J Appl Physiol Bethesda Md. 2013;115(8):1126–37.
37. Leon LR, Blaha MD, DuBose DA. Time course of cytokine, corticosterone, and tissue injury responses in mice during heat strain recovery. J Appl Physiol. 2006;100(4):1400–9.
38. Leon LR. Heat stroke and cytokines. Prog Brain Res. 2007;162:481–524.
39. Dietrich WD, Kuluz JW. New research in the field of stroke: therapeutic hypothermia after cardiac arrest. Stroke. 2003;34(4):1051–3.

40. Mendeloff A, Smith D. Extreme physical effort in summer heat followed by collapse, stupor, purpura, jaundice and azotemia. Am J Med. 1955;18:659–70.
41. Chauhan NR, Kapoor M, Prabha Singh L, Gupta RK, Chand Meena R, Tulsawani R, et al. Heat stress-induced neuroinflammation and aberration in monoamine levels in hypothalamus are associated with temperature dysregulation. Neuroscience. 2017;358:79–92.
42. Leon LR, Gordon CJ, Helwig BG, Rufolo DM, Blaha MD. Thermoregulatory, behavioral, and metabolic responses to heatstroke in a conscious mouse model. Am J Physiol-Regul Integr Comp Physiol. 2010;299(1):R241–8.
43. Malamud N, Haymaker W, Custer RP. Heat stroke. A clinico-pathologic study of 125 fatal cases. Mil Surg. 1946;99(5):397–449.
44. Albukrek D, Bakon M, Moran DS, Faibel M, Epstein Y, Moran D. Heat-stroke-induced cerebellar atrophy: clinical course, CT and MRI findings. Neuroradiology. 1997;39(3):195–7.
45. Szold O, Reider-Groswasser II, Abraham RB, Aviram G, Segev Y, Biderman P, et al. Gray–white matter discrimination—a possible marker for brain damage in heat stroke? Eur J Radiol. 2002;43(1):1–5.
46. Hall DM, Buettner GR, Oberley LW, Xu L, Matthes RD, Gisolfi CV. Mechanisms of circulatory and intestinal barrier dysfunction during whole body hyperthermia. Am J Physiol-Heart Circ Physiol. 2001;280(2):H509–21.
47. Graber CD, Reinhold RB, Breman JG, Harley RA, Hennigar GR. Fatal heat stroke: circulating endotoxin and gram-negative sepsis as complications. JAMA. 1971;216(7):1195–6.
48. Gathiram P, Gaffin SL, Brock-Utne JG, Wells MT. Time course of endotoxemia and cardiovascular changes in heat-stressed primates. Aviat Space Environ Med. 1987;58(11):1071–4.
49. Kew M, Bersohn I, Seftel H, Kent G. Liver damage in heatstroke. Am J Med. 1970;49(2):192–202.
50. Bowers WD, Hubbard RW, Leav I, Daum R, Conlon M, Hamlet MP, et al. Alterations of rat liver subsequent to heat overload. Arch Pathol Lab Med. 1978;102(3):154–7.
51. Chao TC, Sinniah R, Pakiam JE. Acute heat stroke deaths. Pathology. 1981;13(1):145–56.
52. Bynum G, Brown J, Dubose D, Marsili M, Leav I, Pistole TG, et al. Increased survival in experimental dog heatstroke after reduction of gut flora. Aviat Space Environ Med. 1979;50(8):816–9.
53. DuBose DA, Basamania K, Maglione L, Rowlands J. Role of bacterial endotoxins of intestinal origin in rat heat stress mortality. J Appl Physiol. 1983;54(1):31–6.
54. Chung NK, Shabbir M, Lim CL. Cytokine levels in patients with previous heatstroke under heat stress. Mil Med. 1999;164(4):306–10.
55. Lin MT, Chai CY. Effects of sodium acetylsalicylate on body temperature of monkeys under heat exposure. J Pharmacol Exp Ther. 1975;194(1):165–70.
56. Sakurada S, Hales JRS. A role for gastrointestinal endotoxins in enhancement of heat tolerance by physical fitness. J Appl Physiol. 1998;84(1):207–14.
57. McCord GR, Cracowski J-L, Minson CT. Prostanoids contribute to cutaneous active vasodilation in humans. Am J Physiol-Regul Integr Comp Physiol. 2006;291(3):R596–602.
58. Holowatz LA, Thompson-Torgerson CS, Kenney WL. The human cutaneous circulation as a model of generalized microvascular function. J Appl Physiol. 2008;105(1):370–2.
59. Jacobson ED, Bass DE. Effects of sodium salicylate on physiological responses to work in heat. J Appl Physiol. 1964;19(1):33–6.
60. Rowsey PJ, Metzger BL, Carlson J, Gordon CJ. Long-term exercise training selectively alters serum cytokines involved in fever. Biol Res Nurs. 2009;10(4):374–80.
61. Bradford CD, Cotter JD, Thorburn MS, Walker RJ, Gerrard DF. Exercise can be pyrogenic in humans. Am J Physiol Regul Integr Comp Physiol. 2007;292(1):R143–9.
62. Kluger MJ, Kozak W, Conn CA, Leon LR, Soszynski D. Role of fever in disease. Ann N Y Acad Sci. 1998;856:224–33.
63. Leon LR, Kozak W, Rudolph K, Kluger MJ. An antipyretic role for interleukin-10 in LPS fever in mice. Am J Physiol. 1999;276(1 Pt 2):R81–9.
64. Cartmell T, Poole S, Turnbull AV, Rothwell NJ, Luheshi GN. Circulating interleukin-6 mediates the febrile response to localised inflammation in rats. J Physiol. 2000;526(3):653–61.

65. Rummel C, Sachot C, Poole S, Luheshi GN. Circulating interleukin-6 induces fever through a STAT3-linked activation of COX-2 in the brain. Am J Physiol-Regul Integr Comp Physiol. 2006;291(5):R1316–26.
66. Chiu WT, Kao TY, Lin MT. Increased survival in experimental rat heatstroke by continuous perfusion of interleukin-1 receptor antagonist. Neurosci Res. 1996;24(2):159–63.
67. Welc SS, Morse DA, Mattingly AJ, Laitano O, King MA, Clanton TL. The impact of hyperthermia on receptor-mediated interleukin-6 regulation in mouse skeletal muscle. PLoS One. 2016;11(2):e0148927.
68. Mattingly AJ, Laitano O, Clanton TL. Epinephrine stimulates CXCL1 IL-1α, IL-6 secretion in isolated mouse limb muscle. Physiol Rep. 2017;5(23):e13519.
69. Peake J, Della Gatta P, Suzuki K, Nieman D. Cytokine expression and secretion by skeletal muscle cells: regulatory mechanisms and exercise effects. Exerc Immunol Rev. 2015;21:8–25.
70. Bouchama A, Parhar RS, el-Yazigi A, Sheth K, al-Sedairy S. Endotoxemia and release of tumor necrosis factor and interleukin 1 alpha in acute heatstroke. J Appl Physiol. 1991;70(6):2640–4.
71. Foster J, Mauger AR, Chrismas BCR, Thomasson K, Taylor L. Is prostaglandin E2 (PGE2) involved in the thermogenic response to environmental cooling in healthy humans? Med Hypotheses. 2015;85(5):607–11.
72. Veltmeijer MTW, Veeneman D, Bongers CCCW, Netea MG, van der Meer JW, Eijsvogels TMH, et al. The impact of central and peripheral cyclooxygenase enzyme inhibition on exercise-induced elevations in core body temperature. Int J Sports Physiol Perform. 2016;12(5):662–7.
73. Epstein Y, Roberts WO, Golan R, Heled Y, Sorkine P, Halpern P. Sepsis, septic shock, and fatal exertional heat stroke. Curr Sports Med Rep. 2015;14(1):64–9.
74. Leon LR. The thermoregulatory consequences of heat stroke: are cytokines involved? J Therm Biol. 2006;31(1):67–81.
75. Bouchama A, al-Sedairy S, Siddiqui S, Shail E, Rezeig M. Elevated pyrogenic cytokines in heatstroke. Chest. 1993;104(5):1498–502.
76. Bouchama A, Hammami MM, Shail EA, DeVol E. Differential effects of in vitro and in vivo hyperthermia on the production of interleukin-10. Intensive Care Med. 2000;26(11):1646–51.
77. Hammami MM, Bouchama A, Al-Sedairy S, Shail E, AlOhaly Y, Mohamed GED. Concentrations of soluble tumor necrosis factor and interleukin-6 receptors in heatstroke and heatstress. Crit Care Med. 1997;25(8):1314.
78. Hammami MM, Bouchama A, Shail E, Al-Sedairy S. Elevated serum level of soluble interleukin-2 receptor in heatstroke. Intensive Care Med. 1998;24(9):988.
79. Hashim IA, Al-Zeer A, Al-Shohaib S, Al-Ahwal M, Shenkin A. Cytokine changes in patients with heatstroke during pilgrimage to Makkah [Internet]. Mediators Inflamm. 1997. https://www.hindawi.com/journals/mi/1997/213017/abs/. Cited 18 Mar 2018.
80. Lu K-C, Wang J-Y, Lin S-H, Chu P, Lin Y-F. Role of circulating cytokines and chemokines in exertional heatstroke. Crit Care Med. 2004;32(2):399–403.
81. Phillips NA, Welc SS, Wallet SM, King MA, Clanton TL. Protection of intestinal injury during heat stroke in mice by interleukin-6 pretreatment. J Physiol. 2015;593(3):739–52, discussion 753
82. Leon LR, Dineen S, Blaha MD, Rodriguez-Fernandez M, Clarke DC. Attenuated thermoregulatory, metabolic, and liver acute phase protein response to heat stroke in TNF receptor knockout mice. Am J Physiol Regul Integr Comp Physiol. 2013;305(12):R1421–32.
83. Carsana A. Exercise-induced rhabdomyolysis and stress-induced malignant hyperthermia events, association with malignant hyperthermia susceptibility, and RYR1 gene sequence variations. ScientificWorldJournal. 2013;2013:531465.
84. Wappler F, Fiege M, Steinfath M, Agarwal K, Scholz J, Singh S, et al. Evidence for susceptibility to malignant hyperthermia in patients with exercise-induced rhabdomyolysis. Anesthesiology. 2001;94(1):95–100.
85. Capacchione JF, Muldoon SM. The relationship between exertional heat illness, exertional rhabdomyolysis, and malignant hyperthermia. Anesth Analg. 2009;109(4):1065–9.
86. Bourdon L, Canini F. On the nature of the link between malignant hyperthermia and exertional heatstroke. Med Hypotheses. 1995;45(3):268–70.

87. Tobin JR. Malignant Hyperthermia and Apparent Heat Stroke. JAMA. 2001;286(2):168.
88. Hopkins PM, Ellis FR, Halsall PJ. Evidence for related myopathies in exertional heat stroke and malignant hyperthermia. Lancet Lond Engl. 1991;338(8781):1491–2.
89. King MA, Clanton TL, Laitano O. Hyperthermia, dehydration, and osmotic stress: unconventional sources of exercise-induced reactive oxygen species. Am J Physiol Regul Integr Comp Physiol. 2016;310(2):R105–14.
90. González-Alonso J, Crandall CG, Johnson JM. The cardiovascular challenge of exercising in the heat. J Physiol. 2008;586(1):45–53.
91. González-Alonso J, Calbet JA, Nielsen B. Muscle blood flow is reduced with dehydration during prolonged exercise in humans. J Physiol. 1998;513(Pt 3):895–905.

Hydration in Sport and Exercise

6

Louise M. Burke

6.1 Introduction

During most types of exercise, the evaporation of sweat plays a substantial role in the dissipation of excess heat produced by the working muscle or absorbed from the environment (see Chap. 2). Although this activity is a foundation of the thermoregulatory response to exercise, its side-effect is to reduce body water stores. Athletes and other people who exercise can counter sweat losses by consuming fluid during the session or, in a lesser contribution, by hyperhydrating prior to the exercise bout. However, in the majority of situations, the opportunity or desire to drink cannot keep pace with sweat losses, leading to a body fluid deficit. Furthermore, some athletes may commence the session with pre-existing hypohydration, due to the failure to replace daily environmental sweat losses and/or reverse the fluid deficit from a prior exercise bout. The goal of this chapter is to summarise the current status of our knowledge around hydration and exercise, with a focus on practical strategies to assess and address fluid deficits associated with *exercise*. Consideration of hydration and *sport* requires additional context, since it must address the real-world significance of the effects of hypohydration on performance as well as the effect of the rules and logistics of organised events/activities on the practicality of fluid intake.

L. M. Burke (✉)
Australian Institute of Sport, Belconnen, ACT, Australia

Mary MacKillop Institute for Health Research, Australian Catholic University, Melbourne, VIC, Australia
e-mail: Louise.burke@ausport.gov.au

J. D. Périard, S. Racinais (eds.), *Heat Stress in Sport and Exercise*,
https://doi.org/10.1007/978-3-319-93515-7_6

6.2 Terminology and Physiology of Hypohydration/Dehydration

In brief, the fluid content of the human body (total body water; TBW) typically fluctuates by ~1% of body mass (BM) each day, according to changes in the intake of water (food and fluids), metabolic water production and fluid loss (sweat, urine, faeces and respiratory fluid losses) [1, 2]. Figure 6.1 summarises the components involved in daily fluid balance in a sedentary individual, noting that the higher water losses of athletes can create a large imbalance unless water intake from fluids and foods is increased appropriately. Although exercise changes TBW balance by increasing metabolic water production and insensible losses via respiration, its principle effect is to increase total sweat loss. The reduction of body fluids by 2% BM (~3% TBW) from levels of euhydration, to clear the threshold of daily biological variation [2], is termed "hypohydration", and although the strict definition of "dehydration" is the process of reaching a fluid deficit, these terms are often used interchangeably.

Consumption of fluid to (temporarily) increase TBW above the normal daily variation is termed hyperhydration. In the context of exercise, this strategy may be deliberately undertaken by athletes prior to events in which a large fluid deficit is expected to occur due to the inability to drink sufficient volumes of fluid in relation to large sweat losses. Here, the athlete may drink large amounts of fluids in the hours prior to the event, often in conjunction with an osmotically active agent such as glycerol or sodium, with the goal of producing a temporary increase in total body water to offset sweat losses and delay the progression of absolute hypohydration from becoming physiologically significant [3]. However, it may also occur

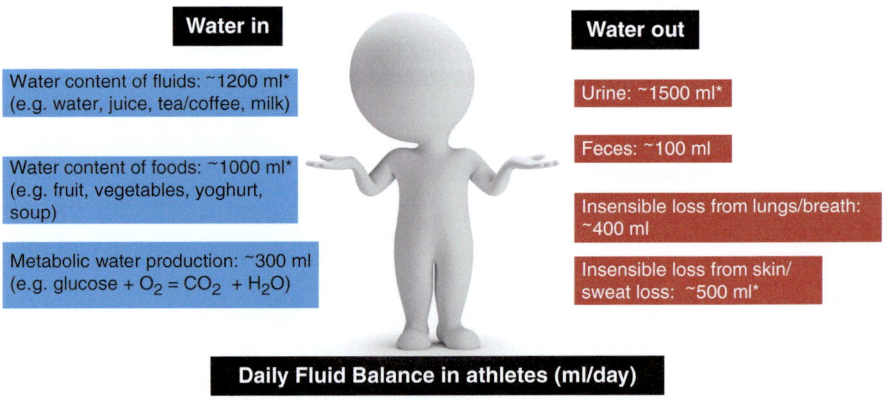

Fig. 6.1 Daily water balance in an athlete. Figures are approximate and represent baseline values for a sedentary person. * An athlete's sweat losses are likely to be substantially increased and must be balanced by greater intake of fluids, as well as greater intake of water from foods

unintentionally during exercise if an athlete consumes fluid in excess of their actual sweat losses: this practice has been identified as an outcome of flawed fluid intake guidelines for sport (or, flawed understanding of these guidelines) and when carried out to extremes may lead to the serious consequences associated with water intoxication/hyponatremia [4].

6.2.1 Sweat Losses in Exercise and Sport

Sweat rates during exercise vary markedly according to the type and intensity of the activity, the environmental conditions in which it occurs (temperature, humidity, wind speed, altitude, etc.), the size and body composition of the athlete, their clothing or other protective gear and their degree of training status and acclimatisation [5–7]. Prediction equations have been established to estimate sweat losses associated with many of these variables [6]. Although these estimates are valuable in providing general advice around education guidelines or fluid provision to populations who exercise purposefully or within their daily occupation (e.g. the military or manual labourers) [6], it is acknowledged that they are generally unable to account for the complexity of the interaction of these factors in real life [7]. Measurements of sweat losses associated with various types of exercise and sporting pursuits show that the typical sweat rates of athletes can vary from 300 to 2000 L/h [8, 9]. Tables 6.1 and 6.2 summarise some of the published observations of fluid balance during competitive sports events, noting sweat rates, typical fluid intakes and the degree of BM loss (i.e. proxy for fluid mismatch) encountered during a variety of different types of activities.

6.2.2 Brief Summary of the Physiological Effects of Hypohydration

Dehydration causes loss of intracellular and extracellular (plasma and interstitial) fluid in proportion to the loss of water and solutes. Typically, sweat is hypotonic (i.e. lower in electrolyte concentration) with respect to plasma; therefore, exercise-associated sweat losses lead to a reduction in plasma volume and increase in its electrolyte concentration (principally, sodium) known as hypertonic hypovolemia. Meanwhile, hypohydration due to the use of some diuretics, or the diuresis associated with altitude and cold environments, produces an isotonic hypovolemia (for review, see [5]). The physiological effects of hypohydration on exercise have been best characterised during aerobic/submaximal exercise, particularly when performed in the heat. It is noted that exercise in the heat creates a "relative hypohydration" of its own, due to the redistribution of blood flow to the skin to assist with heat dissipation [40, 41]. In combination, there is a significant increase in cardiovascular strain, with an elevation in heart rate to accommodate the reduction in stroke

Table 6.1 Fluid balance characteristics of sporting competitions of continuous duration

Study	Subjects	Event	Duration (min)	Environment (°C, %)	Sweat rate (L/h)	Δ Body mass (%)	Fluid intake (L/h)
Single-day endurance and ultra-endurance events							
Beis et al. [10]	10 M Elite (race winners)	13 different Olympic/Big City Marathons	126 ± 1	Air: 0–30 Humidity: 39–89	NR	N/A	0.55 ± 0.34
Mettler et al. [11]	128 M, 39 F Mixed-calibre athletes	Zurich Marathon	220 ± 32 (M) 245 ± 23 (F)	Air: ~10 Humidity: raining	NR	−0.8 ± 0.8 (M) −0.2 ± 0.8 (F)	0.47 (M) 0.36 (F)
Hew [12]	63 M, 54 F Mixed-calibre athletes	Houston Marathon	269 ± 45 (M) 303 ± 54 (F)	NR	NR	−2.1 (M) [−1.7 ± 1.8 kg] −1.0 (F) [−0.6 ± 1.1 kg]	0.74 (M) 0.68 (F)
Pahnke et al. [13]	26 M, 20 F Mixed-calibre athletes	Hawaii Ironman triathlon	NR	Air: 27.6 Humidity: NR	NR	−2.1 ± 2.1	0.85 ± 0.30 (M) 1.05 ± 0.30 (F)
Speedy et al. [14]	292 M, 38 F Mixed-calibre athletes	New Zealand Ironman triathlon	734	Air: 21, Water: 20.7 Humidity: 91	NR	−4.3 ± 2.3 (M) −2.7 ± 3.1 (F)	NR
Speedy et al. [15]	46 M + 2 F Mixed-calibre athletes	Coast to Coast New Zealand (paddle/ride/run)	879 ± 83	Air: 7.5–19.6 Humidity: 56–94	NR	−3.1 ± 2.1	NR
Kao et al. [16]	19 M, 4 F Mixed-calibre athletes	Soochow University International 24-h running event	1440 (199.4 ± 37 km)	Air: 11.5–14.6 Humidity: 55–60	NR	−5.1 ± 2.3	NR
Glace et al. [17]	13 M + F Mixed-calibre athletes	160-km trail run Start time 0430	1572 ± 216	Air: 21–38 Humidity: NR	NR	−0.5 (−0.5 ± 1.5 kg)	0.74
Fallon et al. [18]	7 M Mixed-calibre athletes	100-km road run	629 ± 113	Air: 2–17 Humidity: 45	0.86 ± 0.15	−3.3 ± 1.1	0.54 ± 0.21

Armstrong et al. [19]	42 M, 6 F Mixed-calibre athletes	164-km cycle event, USA	546 ± 72 (M) 540 ± 12 (F)	Air: 34.5 ± 5.0 Humidity: 53	1.13	N/A	0.65 (M) 0.52 (F)
Knechtle et al. [20]	37 M Mixed-calibre athletes	Swiss MTB Bike Masters 120 km	540 ± 80	Air: 11 (at start) Humidity: NR	NR	−1.9 ± 1.6	0.7 ± 0.2
Brearley et al. [21]	4 M Professional drivers	V8 Supercar Championship Race	31 ± 7	Air: 29, Cabin: 49 Humidity: NR	1.06 ± 0.12	−0.6 ± 0.6	NR
Multi-day Stage race (cycling or running)							
Ross et al. [22]	5 M Elite cycling team	Tour of Gippsland (NRS stage race) 9 stages over 5 days		Air: 15.8 ± 1.4 Humidity: 54 ± 12	1.1 ± 0.3	− 1.5 ± 0.3 (road race) − 1.1 ± 0.2 (criterium)	0.41 ± 0.19 (road race) 0.24 ± 0.19 (criterium)
Ebert et al. [23]	8 M Elite Professional team	Tour down under 719 km in 6 days	NR	Air: 20.2–32.9 Humidity: 14–69	1.60 ± 0.10	−2.8	1.00 ± 0.10
Ebert et al. [23]	6 F Elite cycling team	Tour De L'Aude 788 km in 10 days	NR	Air: 7.7–27.8 Humidity: 29–76	0.90	−2.6	0.40 ± 0.06
Garcia-Roves et al. [24]	10 M Elite Professional team	3 × 24-h periods during the 3-week Tour of Spain	NR	NR	NR	NR	1.26 ± 0.55 L
Knechtle et al. [25]	25 M Mixed-calibre athletes	Swiss Jura Marathon 350 km in 7 stages	373 ± 50	NR	NR	−1.4 ± 2.0	0.54–0.75 (range)

Data are mean ± SD or [range] or not reported (NR) for male (M) and female (F) athletes

Table 6.2 Fluid balance characteristics of sporting events of intermittent nature or conducted in water environment

Study	Subjects	Event	Duration (min)	Environment (°C, %)	Sweat rate (L/h)	Δ Body mass (%)	Fluid intake (L/h)
Field sports: football and cricket							
Da Silva et al. [26]	10 M Elite youth players	Soccer match—Brazil	110	Air: 31.2 ± 2.0 Humidity: 48	~1.22	− 1.6 ± 0.8	~0.61
Aragon-Vargus et al. [27]	17 M Elite professional players	Soccer Match—Costa Rice Premier division	180	Air: 34.9 ± 1.2 Humidity: 35	1.48 ± 0.36	−3.4 ± 1.1	0.65
Maughan et al. [28]	20 M Elite professional players	Soccer Match—English Premier League	96	Air 6–8 Humidity: 50–60	1.09 (Team A) 1.01 (Team B)	− 0.9 ± 0.7 (Team A) − 1.3 ± 0.6 (Team B)	0.68 (Team A) 0.43 (Team B)
O'Hara et al. [29]	14 M Elite professional players	UK Super League Rugby League match	NR	Air :12.1 ± 5.3 Humidity: 70	NR	−1.2 ± 0.6 (Team A) −1.4 ± 0.7 (Team B)	1.12 L (Team A) 1.56 L (Team B)
Meir et al. [30]	28 M Elite development squad	<21 Rugby Union Championship (4 games; G1-4)	NR	Air: 18.5 ± 1.6 Humidity: 40	0.54 ± 0.55 (G1) 0.49 ± 0.65 (G2) 0.89 ± 0.65 (G3) 0.92 ± 0.88 (G4)	−0.8 ± 0.8 (G1) −0.7 ± 0.9 (G2) −1.3 ± 0.9 (G3) −1.3 ± 1.2 (G4)	NR
Gore et al. [31]	3 M First-grade bowlers	Cricket match	360[c]	Air: 32.8 ± 0.5 Humidity: 29	1.37 ± 0.06	−4.3 ± 0.7	0.46
Court sports—tennis, basketball and ice hockey							
Osterberg et al. [32]	29 M Professional basketball players	NBA Match	40 (playing time = 21 ± 8)	Air: 20–22 °C Humidity: 18–22	3.30	−1.4 ± 0.6	1.50

Tippet et al. [33]	7 F Professional tennis players	WTA match Hard court	119.9 ± 40.1	Air: 30.3 ± 2.3 Humidity: NR	2.00 ± 0.50	−1.2 ± 1.0	1.5 ± 0.50
Bergeron et al. [34]	12 M, 8 F Sub-elite players	US Division 1 Collegiate tournament: Hard court	90	Air: 32.2 ± 1.5 Humidity: 54 ± 2	1.8 (M) 1.1 (F)	−1.3 ± 0.8 (M) −0.7 ± 0.8 (F)	1.13 (M) 0.87 (F)
Logan-Sprenger et al. [35]	24 M Elite junior ice hockey players	Ontario Hockey League	210[c]	Air: 10.8 ± 0.2 Humidity: 30 ± 2	0.90	−1.3 ± 0.3	0.68
Water-based sports							
Cox et al. [36]	23 M Elite Australian squad	Tournament	47	Air: 24.1 Water: 27.3 Humidity: 54	0.79	−0.4	0.38
Wagner et al. [37]	25 M, 11 F Mixed-calibre athletes	26.4-km swim Switzerland	528 (M) 599 (F)	Air: 18.5–28.1 Water: 22.9–24.1 Humidity: 42–93	NR	− 0.5 ± 1.1 (M) − 0.1 ± 1.6 (F)	0.56 ± 0.22 (M) 0.44 ± 0.17 (F)
Neville et al. [38]	32 M Professional crew	Lead up race to America's Cup	150	Air: 32 ± 1 Humidity: 52 ± 5	0.96 ± 0.38	−0.7 ± 0.8	0.64
Slater et al. [39]	26 M, 9 F Club level dinghy crew	Club regatta Singapore	300	Air: 29–33 Humidity: 62–81	0.47 (M) 0.23 (F)	−2.1 (M) −0.9 (F)	0.24 (M) 0.16 (F)

Data are mean ± SD or [range] or not reported (NR) for male (M) and female (F) athletes volume [42]. Competition between peripheral and central circulation for the reduced plasma volume can lead to a decrease in muscle blood flow and aerobic reserve, and particularly as plasma osmolality increases, a reduced sweat rate for any given core temperature and an increase in heat storage (for review, see [40, 41, 43]). Increased rates of muscle glycogen usage, motor unit recruitment and afferent feedback, as well as elevations in skin temperature, discomfort and thirst-derived distraction are among the many factors associated with hypohydration, particularly when exercise is undertaken in a warm-hot environment [40, 41, 43]. Studies that have monitored the effect of hypohydration on physiological responses to exercise in the heat have noted that the magnitude of impairment is linearly related to the magnitude of the fluid deficit [42]. For a more detailed account of the effects of hypohydration on physiological characteristics during exercise, the reader is referred to recent comprehensive reviews [8, 40, 41, 43].

6.3 Assessment of Hydration Status

In the laboratory, various techniques such as the use of tracer-labelled water, bioelectrical impedance and blood characteristics allow the tracking of TBW. In the real world of both research and athletic practice, however, there is a need for techniques that are less expensive, invasive or resource/time heavy to gain information about an individual's current hydration status as well as to monitor the fluid deficit that might occur during exercise and require replacement. The accuracy of any assessment technique involves the concept of sensitivity (identifying correctly when a state is present) and specificity (recording a negative outcome when the state is absent).

6.3.1 Assessing Current Hydration Status

In research, clinical practice or athlete servicing, there are various situations in which it is desirable to know an individual's current hydration status. The requirement here is to identify changes in body fluid status that are greater than biological variability. The gold standard for measurement of TBW involves the isotope dilution method (usually deuterium oxide) which can measure changes in TBW of ~1% [44]. However, this is an expensive method requiring analytical resources and expertise; furthermore, it requires a baseline measurement in the individual to monitor change rather than to identify an absolute cut-off deemed to represent hypohydration. The measurement of plasma osmolality, which is controlled around a euhydration set-point of ~285 mOsm/kg, can also be used as a marker of hydration status, with a change of ~5 mOsm/kg equating to a fluid loss of ~2% BM [45]. However, this is an invasive technique, which again requires expense and laboratory resources, and may not always reflect changes in TBW in some scenarios faced by athletes (e.g. altitude). Bioelectrical impedance analysis (BIA) is a non-invasive technique that can be used to estimate TBW, on the basis that the resistance to low amperage current (single or multiple frequencies) passed between skin electrodes varies inversely with tissue water and electrolyte content. Although it may provide a reliable measure of TBW in euhydrated individuals, it lacks sensitivity in measuring loss of TBW and shifts between body fluid compartments [46].

The simplest and most practical measurements of hydration status, particularly for field work, involve measurements of urine characteristics and alterations in BM. In the absence of energy deficits/surpluses and changes in diet that alter the mass of gut contents, acute monitoring of morning BM (after voiding) provides a simple measurement of success in maintaining TBW [47]. Changes in urinary concentration can be readily assessed via comparison to colour charts (Fig. 6.2) or by the use of portable refractometers and osmometers to measure specific gravity and osmolality, respectively [48]. Spot checks of urine characteristics over the day are liable to false negatives for various reasons. For example, if a dehydrated individual rapidly rehydrates or consumes large amounts of fluid without replacement of the solutes lost in sweat, diuresis of pale and dilute urine will occur until there is full replacement of electrolytes, with sufficient time for requilibration of plasma osmolality and volume, and distribution of fluid across the various compartments [49]. Therefore, it is recommended that first void morning urine samples be used for assessment of hydration status [1, 48], particularly for serial assessments in the same individual over time, where an individual "normal" can be established. Investigation of the utility of monitoring other body fluids such as saliva or tears has so far failed to produce alternatives that are accurate and free of artefacts [1, 2].

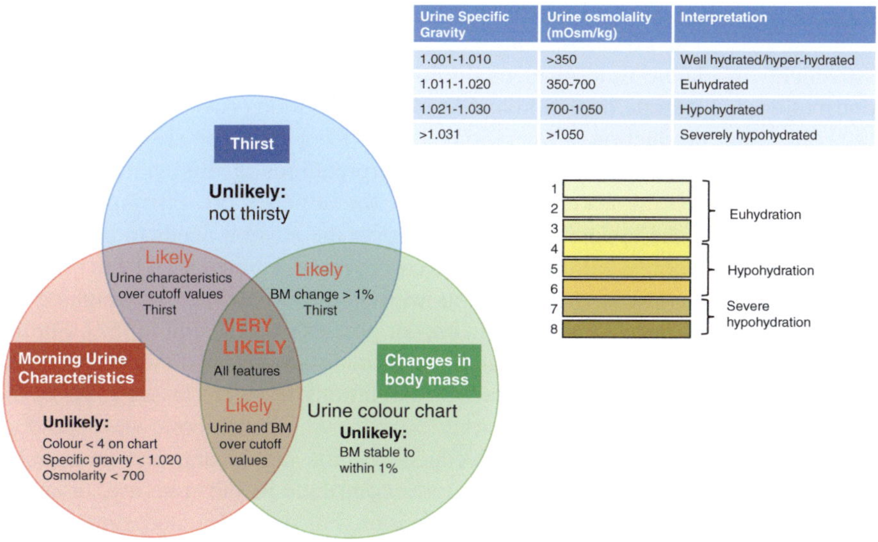

Urine Specific Gravity	Urine osmolality (mOsm/kg)	Interpretation
1.001-1.010	>350	Well hydrated/hyper-hydrated
1.011-1.020	350-700	Euhydrated
1.021-1.030	700-1050	Hypohydrated
>1.031	>1050	Severely hypohydrated

Fig. 6.2 An assessment of fluid balance can be made by examining thirst, characteristics of the first voided morning urine sample and changes in morning body mass (BM). If two or more of the characteristics fall outside the levels associated with euhydration, there is a likely risk of hypohydration. Adapted from [50, 51]. Note that cut-offs in the reference ranges for euhydration and hypohydration may differ slightly between different expert groups

Finally, given the opportunity for each assessment technique to be confounded by artefacts or lack of precision around the individual demarcation of euhydration vs hypohydration, it has been suggested that athletes develop their personal metrics around cut-offs or use a combination of techniques to provide a more robust diagnosis (see Fig. 6.2). Note that these strategies may affirm the presence of hypohydration, but can provide only a qualitative assessment of the magnitude of the fluid deficit (e.g. what might be considered moderate/tolerable vs severe/problematic by the individual athlete or specific scenario) rather than precise determination (e.g. the fluid deficit is x mL). Indeed, although there have been many attempts to provide specific metrics to describe these situations (e.g. a fluid deficit equivalent to 2–3% BM might be considered "mild", while severe equates to a fluid deficit of >5% BM), the context around these values requires greater consideration and interpretation. Further information on methods of hydration assessment can be found in recent publications [1, 2, 51].

6.3.2 Assessing Fluid Balance Across an Exercise Session

Athletes often want to assess fluid balance across an exercise session to gauge their typical sweat rates, the success of their usual fluid intake practices in addressing these and/or the residual fluid deficit that needs to be replaced during the recovery period. Typically, this is undertaken by monitoring changes in BM, adjusted for intake over the session, in the belief that, during short-medium duration exercise (2–3 h) at least, BM changes mirror changes in TBW. In such circumstances, calculations of sweat and TBW loss summarised in Table 6.3 can provide a reasonable assessment of relative fluid changes during the workout or event, since both the contribution of substrate depletion to BM changes and the provision of TBW from metabolic water production or liberation of water from glycogen are minor in comparison with sweat losses. It is noted that interpretations of fluid status from this information requires an acknowledgement of the athlete's pre-existing hydration level (i.e. whether the apparent fluid change over the session needs to be adjusted for pre-exercise hyperhydration tactics or a carryover of a fluid deficit from previous exercise or weight-making strategies), as well as errors involved in weighing the trapped fluid in clothes and hair during pre- or post-exercise weigh in [52]. Studies show that the adjustment of calculations for urine losses, trapped sweat and intake during the exercise session can improve calculations of sweat losses [53]. In the case of very prolonged exercise sessions (>3–4 h) where metabolic water production and liberation of water from glycogen become significant, and measurable amounts of body carbohydrate and fat stores contribute to substrate use, there is a need to adjust sweat rate and TBW calculations. For example, modelling of fluid characteristics of an ultra-endurance run lasting 24–30 h suggested that ~2–5% BM needed to be subtracted from BM changes to assess true change in TBW [54]. Despite the challenges, assessing fluid balance over typical exercise sessions with careful strategies and insights can help athletes to develop or monitor drinking practices that neither over- or under-hydrate inappropriately.

Table 6.3 Strategies to estimate fluid balance across a session of exercise

Steps	1. Weigh athlete's body mass (BM) before session, using reliable digital scales (ideally measuring to 0.01 kg). This should be done wearing minimal clothing and after the athlete has gone to the toilet 2. Weigh athlete again after session in the same clothing, and after towelling dry 3. Weigh athlete's drink bottle before and after the session (ideally measuring to 1 g using kitchen scales) to calculate the volume (g/mL) of fluid consumed 4. Note the mass (g) of any foods or sports products (e.g. gels) consumed during the session
Extra steps for further accuracy	5. If the athlete has to go to the toilet during the session, weigh in before and after, or collect urine in a beaker to measure the volume/mass 6. Estimate total urine losses during the session by having athlete weigh in post-session, go to the toilet, and reweigh (alternatively, collect urine in beaker and measure the volume). Add this to the volume/mass of urine produced at any mid-session toilet stops
Calculations	*Fluid intake (mL)* = drink bottle before – drink bottle after (g) *705 g – 104 g* = 601 *g or 601 mL* *Urine losses (mL)* = change in BM due to toilet stops during and/or after the session: kg × 1000 or g *e.g. weight change: 60.25 – 60.00 = 0.25 kg = 250 mL or 251 g urine in beaker* *Fluid deficit (mL)* = Pre-session BM – Post-session BM (kg) × 1000. (Note: to measure total fluid deficit which includes sweat and urine losses, use post-session value taken after the toilet visit) *60.50 – 59.05 = 1.45 kg = 1450 mL* *Fluid deficit (% BM)* = (Fluid deficit [in kg] × 100)/pre-session BM (kg) *(1.45 × 100)/(60.50) = 2.4%* *Total sweat losses over the session* = Fluid deficit (g) + fluid intake (g) + food intake (g) – urine losses (g) *1450 + 601 + 40 g (sports gel) – 250 = 1841 mL* *Sweat rate over the session* = sweat losses converted to mL per h *Session lasted for 90 min: sweat rate = 1841 × 60/90 = 1227 mL or 1.23 L/h*
Interpretation issues	While this activity can provide an estimate of the net fluid deficit incurred across a session of exercise, some issues should be considered • Pre-session hydration status needs to be taken into account to distinguish between relative and absolute fluid deficits. If the athlete has pre-existing hypohydration, the total fluid deficit is underestimated by these calculations. By contrast, if the athlete has hyper-hydrated prior to the session, the net fluid deficit will be overestimated • During prolonged exercise (>2–3 h) at high workloads, the contribution of other factors to TBW changes may no longer be insignificant. For example: failure to adjust for BM loss from depletion of fuels (e.g. glycogen), liberation of water from glycogen breakdown and metabolic water production may contribute to an overestimation of the true fluid loss. For ultra-endurance events of 4–24 h, estimates of the true fluid deficit might require an adjustment of 2–5% BM • Sources of error due to fluids trapped in clothes and hair should also be considered

6.4 Effect of Dehydration on Performance

Despite more than five decades of research, the impact of hypohydration on the performance of exercise, or more specifically, sport, is a contentious topic [55, 56]. Although the effects of hypohydration on physiological parameters are well established, there are a number of factors related both to the published literature on hydration and performance and observations of real-life outcomes of sporting events that create debate about the real impact of hypohydration on sports performance. Clearly, the translation of physiological and psychological effects of hypohydration into changes in sports performance will be mediated by characteristics such as the individual athlete, the type of event, the environmental conditions in which it is conducted, the importance of absolute vs relative performance (i.e. does the athlete need to perform optimally or just better than other competitors?) and whether the event was commenced with pre-existing hypohydration or whether it accumulated over the session. Several reviews have identified flaws in the conduct of laboratory-based studies of hydration and performance, including their lack of integration of the environmental conditions, motivational incentives and success determinants of real-life sport as well as their failure to capture the timing and magnitude of hypohydration to which athletes are commonly exposed [56, 57]. Some of the issues underpinning the debate around the application of the literature on hypohydration to sporting success are identified in Fig. 6.3. These include factors that support the hypothesis that performance is impaired by hypohydration, as well as factors that suggest that the effects of hypohydration are overstated.

Table 6.4 summarises the conclusions from a number of review papers which have conducted narrative and systematic reviews, including some meta-analyses, of the studies of hypohydration and exercise/sports performance. The summation of this literature shows some discrepancies, but also the likelihood that hypohydration impairs performance to a degree that is meaningful to competitive success in certain scenarios. This risk is greatest for aerobic exercise undertaken in hot conditions and when skin temperature is raised, when the fluid deficit exceeds 3% BM and perhaps, when there is an overlay of skill and cognitive performance [59–63]. Of course, it should be noted that amalgamation of data may mask the frailties of individual studies, particularly those conducted without insight into the specific conditions of real-life sport, and that athletes should be more interested in understanding the risks associated with their individual scenarios rather than seeking a universal truth. Data on fluid balance characteristics measured across a range of competitive sporting events (Tables 6.1 and 6.2) show that both groups and individual athletes commonly achieve a BM deficit of >2% within their activities, and although measurements of elite competitors are relatively scarce, it is likely that they are at greater risk of exposure to levels of hypohydration that have been associated with a performance decrement.

There is a need for new research on hypohydration and sports performance with protocols that reflect the conditions under which hypohydration occurs and measure the impact on performance in ways that are meaningful to competition outcomes.

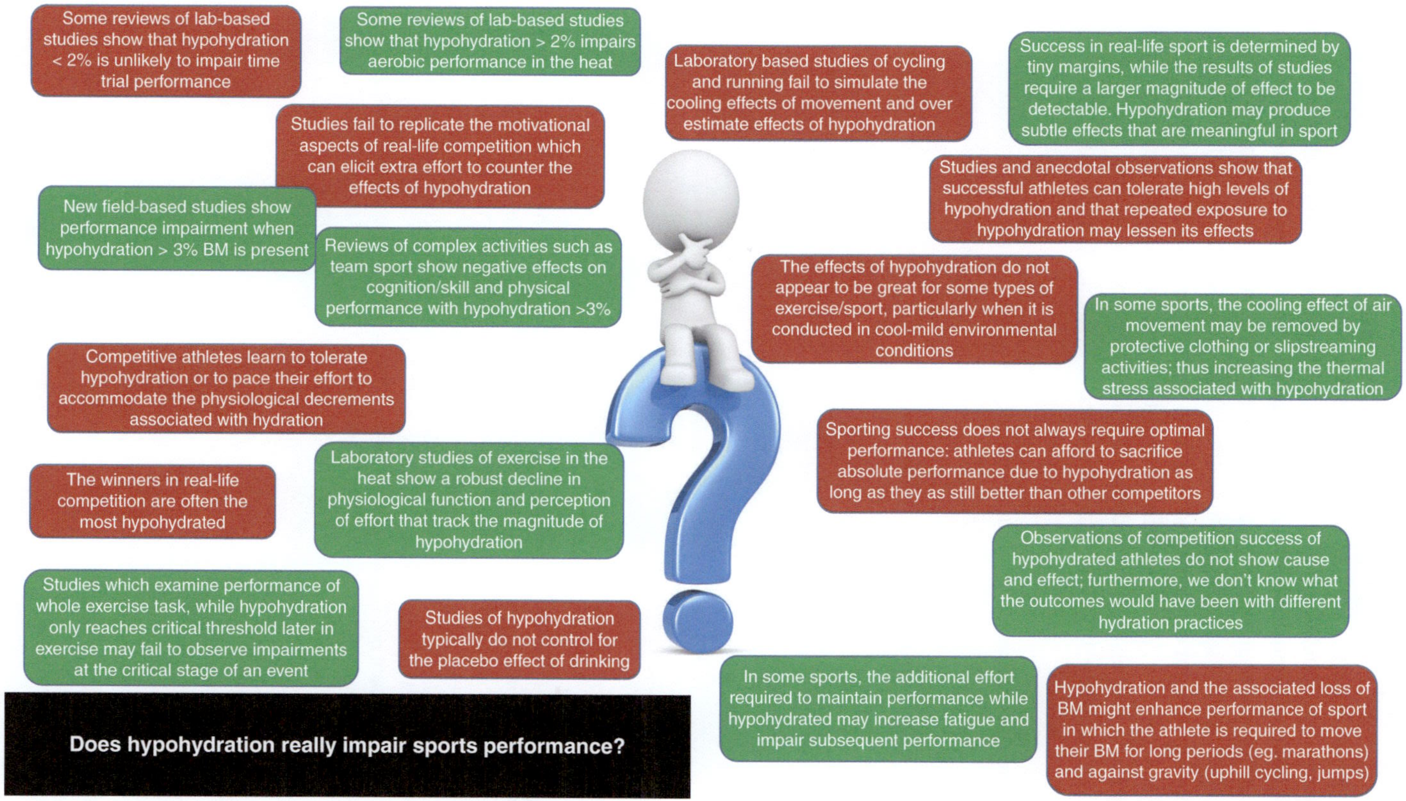

Fig. 6.3 Considerations underpinning the debate over the effect of hypohydration on sports performance; red boxes denote concepts that downplay the effect, while green boxes denote concepts that support the negative effect of hypohydration on performance. Adapted from [58]

One theme of interest is the removal of an important confounder of the hydration literature: removal of the placebo effect of fluid intake during exercise and/or the blinding of study participants to alterations in their hydration status during performance trials. This challenge was tackled in a study of recreational athletes who cycled in hot conditions [64], by utilising a protocol involving oral intake of small amounts of fluid to provide identical sensory exposure in each trial while manipulating hydration via the infusion of different volumes of fluid intake into the stomach with a nasogastric tube. The hypohydration trial in this investigation achieved a mean BM loss of 2.4% at the end of 155 min of intermittent cycling (8 × 15 min + 5 min rest) compared with 0.1% in the euhydration trial. This protocol achieved an increase in heart rate, perceived effort and thirst, and a hypertonic

Table 6.4 Reviews of hypohydration and exercise/sports performance from the recent decade

Review	Topic	Method	Key conclusions
Nuccio et al. [59]	Effect of hypohydration on cognitive, technical and physical performance in team sports	Systematic narrative review of lab/field studies of high-intensity intermittent exercise involving team athletes as participants, examining effects of hypohydration on cognitive performance and technical skill (n = 17) and/or physical performance (sprinting, jumping, etc.) [n = 15] with defined criteria for acceptance into analysis	• The effect of hypohydration on team sport performance is mixed • Hypohydration is more likely to impair cognition, technical skill and physical performance at higher levels of BM loss (>3% BM), especially when heat stress is involved • Increased ratings of perceived effort and fatigue consistently accompany hypohydration and could explain some of the performance impairment
Savoie et al. [60]	Effect of hypohydration on muscle endurance, strength, anaerobic power and capacity, and vertical jumping ability	Systematic review and meta-analysis of lab studies (n = 28) of controlled and single efforts involving muscle strength (upper body = 14 effects, and lower body = 25 effects), endurance (n = 6 and 10), anaerobic power and capacity (n = 9), and vertical jumping ability (n = 12) with defined criteria for acceptance into analysis	• Hypohydration significantly impairs muscular endurance in both upper and lower body by ~8% • Hypohydration significantly impairs muscular strength by ~5% with similar effects on upper and lower body • Anaerobic power is significantly reduced (~6%) by hypohydration but effects on anaerobic capacity (-3%) and vertical jump (+1%) are not significant • Trained individuals demonstrate a 3% lower decrease in effects on performance than untrained subjects

Table 6.4 (continued)

Review	Topic	Method	Key conclusions
Cheuvront and Kenefick [61]	Effect of hypohydration on endurance and strength/ power exercise	Narrative review of lab/field studies of hypohydration on endurance ($n = 34$ with 60 effects) and strength ($n = 43$ studies with 267 effects) with defined criteria for acceptance into analysis Review counted how many effects recorded an impairment of performance/ exercise capacity with different levels of hypohydration (see Fig. 6.4)	• Among studies of endurance exercise, 41/60 (68%) effects were significantly impaired by hypohydration of ≥2% BM, rising to 53/60 (88%) if absolute impairment (including non-significant changes) was considered • The effects are increased in hot environments, and when skin temperature is increased • Among studies of strength/ power exercise, 54/276 (20%) reported significant impairment of effects, showing that the effect of hypohydration on strength or power are marginal • Effects on cognition appear to be small and related to distraction or discomfort • The effect of hypohydration on a particular sport related to the makeup of the task involved
Goulet [62]	Effect of hypohydration on endurance performance	Systematic review and meta-analysis of lab studies ($n = 15$, with 28 effects) with defined criteria for acceptance into analysis	• Hypohydration of >2% BM did not alter time trial performance (mean change of +0.1%); it is unlikely that hypohydration <4% BM impairs performance under real-world conditions • Hypohydration of >2% significantly impaired performance of fixed intensity protocols by 1.9%; hypohydration is associated with an impairment of endurance capacity
Judelson et al. [63]	Effect of hypohydration on strength, power and high-intensity exercise	Narrative review of lab studies of hypohydration on muscle power, strength and high-intensity exercise (>2 min and <10 min) noting those with exacerbating and masking factors, and emphasising 11 studies determined as being accurate in isolating effect of hypohydration on exercise	• Hypohydration appears to consistently attenuate strength by ~2%, power by ~3% and high-intensity endurance by ~10%

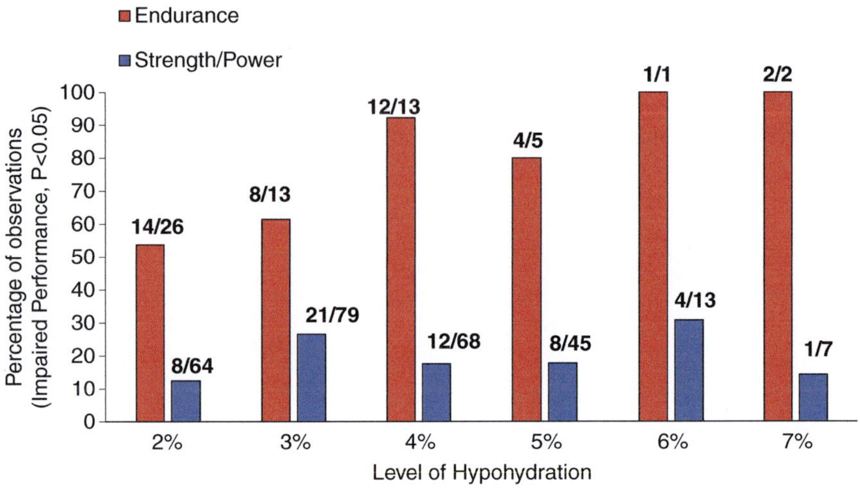

Fig. 6.4 Summary of a narrative review of studies of the effect of hypohydration on endurance (34 studies) and strength/power performance (43 studies) [61]. The studies are arranged according to the level of hypohydration achieved in the study, noting the percentage of the studies which found a significant impairment of performance. Redrawn with permission

hypovolemia in the hypohydration trial, as would be expected from a real-life trial. Performance (work achieved during a 15-min time trial at the end of this pre-load) was reduced by 8% in the hypohydration trial. The authors acknowledge that two previous studies [65, 66], using similar protocols around gastric infusion of fluids to blind hydration status, recorded contradictory outcomes, failing to detect an impairment of performance in the hypohydration trials. However, the protocols used in these studies (no oral intake and delivery of isotonic fluid to the gut) failed to mimic the differential physiological and perceptual effects consistent with real-life hypohydration and adequate fluid intake.

A further investigation from a different laboratory implemented the same protocol in a more highly trained population, who cycled for 2 h before undertaking a 5-km time trial on a 4% grade on a laboratory ergometer in hot conditions [67]. Conditions just before the start of the time trial were a mean BM loss of −0.2 and −2.2% in the euhydrated and hypohydrated trials, respectively. Heart rates were elevated during the steady state pre-load with the hypohydrated condition but were identical during the time trial ride although performance was improved in the euhydration trial (32.9 s faster, 6% increase in power output). Rectal temperatures were elevated during the hypohydration time trial above the control condition, although thirst sensation was identical between trials suggesting that impairment of performance via hypohydration can occur independently of thirst. Further research of this type which eliminates the effects of expectation on performance outcomes associated with hypohydration, and differentiates thirst perception from physiological effects such as thermoregulatory and cardiac strain is encouraged.

6.5 Guidelines for Fluid Intake During Sport and Exercise

The past five decades have also seen a marked evolution of guidelines for fluid intake during sport and exercise and the range of issues address within them. Several critical events and position stands are described in brief summary. The first guidelines were issued in the 1970s and were directed to distance running events, addressing both the restrictions on fluid intake within official race rules (fluid intake was restricted prior to 15 km and then 11 km in road races, under the direction of the International Amateur Athletic Federation) as well as the culture of runners to consider fluid intake during marathons and ultra-marathons to be detrimental and a sign of weakness/lack of fitness [68]. These guidelines, within a larger paper on prevention of heat injuries during distance running, recommended that water or dilute glucose/electrolyte solutions be provided regularly in events longer than 16 km [69]. The 1996 publication of specific guidelines for fluid intake during exercise by the American College of Sports Medicine (ACSM) also largely focused on running and promoted pro-active and formulaic intake of fluid with the goal of minimising the mismatch between fluid intake and sweat losses [70].

During the following decade, there was increased recognition of the syndrome of hyponatremia/water intoxication caused by the excessive intake of fluids by some (usually recreational) participants in endurance/ultra-endurance sports activities resulting in a small but concerning number of cases of morbidity and mortality from cerebral oedema [15, 71, 72]. Hyponatremia, which is diagnosed by measurement of plasma sodium concentrations, ranges from a mild drop below the normal range (e.g. <135 mmol/L) which may be largely asymptomatic to the severe drop (<120 mmol/L) which is associated with major problems [4]. The 2007 update of the ACSM guidelines [8] was undertaken to address this issue and reduce any misunderstanding that its previous iteration encouraged athletes/recreational exercisers to "drink as much as possible" during exercise. The revised position promoted the development of individualised fluid intake plans ("programmed drinking") across a range of sporting activities, with the goal of defending (where possible) a "gold standard" of hydration (suggested as loss of <2% BM over the event) but warning against over-drinking (shown by a gain in BM) [8].

Despite this clarification, the ACSM position stand has attracted harsh criticism and counterarguments that humans need only drink fluids "to thirst" or "ad libitum" during exercise/sporting activities. Furthermore, it has been postulated that opinions to the contrary represent flawed research supported by biased sports scientists and commercial interest in the sale of sports drinks [73–75]. Other more recent expert statements on fluid intake during exercise, underpinned by an interest in preventing hyponatremia rather than focusing on performance effects and the cardiovascular/thermoregulatory stress associated with hypohydration, have championed this latter approach of "drinking to thirst" [4]. Curiously, case histories of hyponatremia have included athletes who claimed to be following thirst-driven drinking practices [4] and scenarios involving sodium depletion in concert with hypohydration [76], showcasing the complexity of fluid and electrolyte balance during prolonged exercise. Unfortunately, the current situation in relation to exercise fluid guidelines presents a challenge to contemporary sports scientists and athletes to choose between two camps promoting apparently opposite approaches ("programmed drinking" vs "drinking to thirst"/ad libitum drinking).

This author and others [77] propose that there is a middle ground for fluid guidelines for sport and exercise, noting that both viewpoints support the benefits of drinking during exercise and warn against drinking in excess of sweat losses or the TBW deficit. However, the route to achieving this is multi-factorial because of the differences in the conditions and context of an exercise session. These differences include the priorities of the exerciser (performance vs health/safety), and whether the conditions under which they exercise make it easy or difficult to drink in relation to their sweat rates. Just as there are a variety of influences on sweat rates and losses during exercise, there are a range of issues that influence fluid intake; these include access to fluid, ability to drink, gastrointestinal comfort, thirst, and cultural and personal beliefs [58]. Furthermore, fluid may be consumed not only to meet needs for water replacement but as a vehicle for intake of other performance-enhancing nutrients (carbohydrate and caffeine), to contribute to thermoregulation via the ingestion of cold or icy drinks or to promote a sensation of well-being via the oropharyngeal sensing of fluid [58]. Some of these issues are unique to the event and within the control of the athlete/exercising person, while others are beyond their control. This is especially the case for competitive sports where fluid intake can be determined by event rules, logistics or the need to balance the time spent consuming fluids within a race with the performance benefits [58]. Indeed, in many competitive sports, it is impossible for athletes to drink to thirst or to drink ad libitum, since opportunities to drink sufficient volumes of fluid do not align with the accrual of a fluid deficit or development of thirst. For example, the rules of soccer (football) prevent the intake of fluid during match play (45 min halves plus overtime). Meanwhile, practical restrictions around the volume of fluid can be obtained, consumed and tolerated at any single aid station by an elite marathon runner make it unlikely that thirst could be sufficiently subdued if allowed to fully develop.

Although drinking during exercise is considered an innate behaviour, controlled by biological cues and fine-tuned by evolution [78], Tables 6.1 and 6.2 illustrate that the patterns and drivers of fluid intake by athletes during competitive sporting events are extremely complex. These data and those from more comprehensive reviews [9, 59] report that across a range of sports, non-elite athletes typically appear to consume fluids at rates of 300–1000 mL/h, limiting the mean change in BM across the session to ~2%. From the sparse information on elite competitors, however, it seems likely that their fluid intakes during events are less able to keep pace with their greater sweat rates. Indeed, weight losses of ≥5% BM have been observed in some individuals, particularly during events undertaken in hot and/or humid conditions [10]. By contrast, some athletes (especially non-elite competitors) may have drinking behaviours that result in an undesirable gain in body mass over an event, with a small proportion of these progressing to symptomatic hyponatremia [4]. This apparent range in practices and outcomes confirms the multi-factorial nature of drinking behaviours.

Figure 6.5 identifies a range of factors that can exist within sport or exercise that can promote under-drinking or over-hydration, even within the same event or

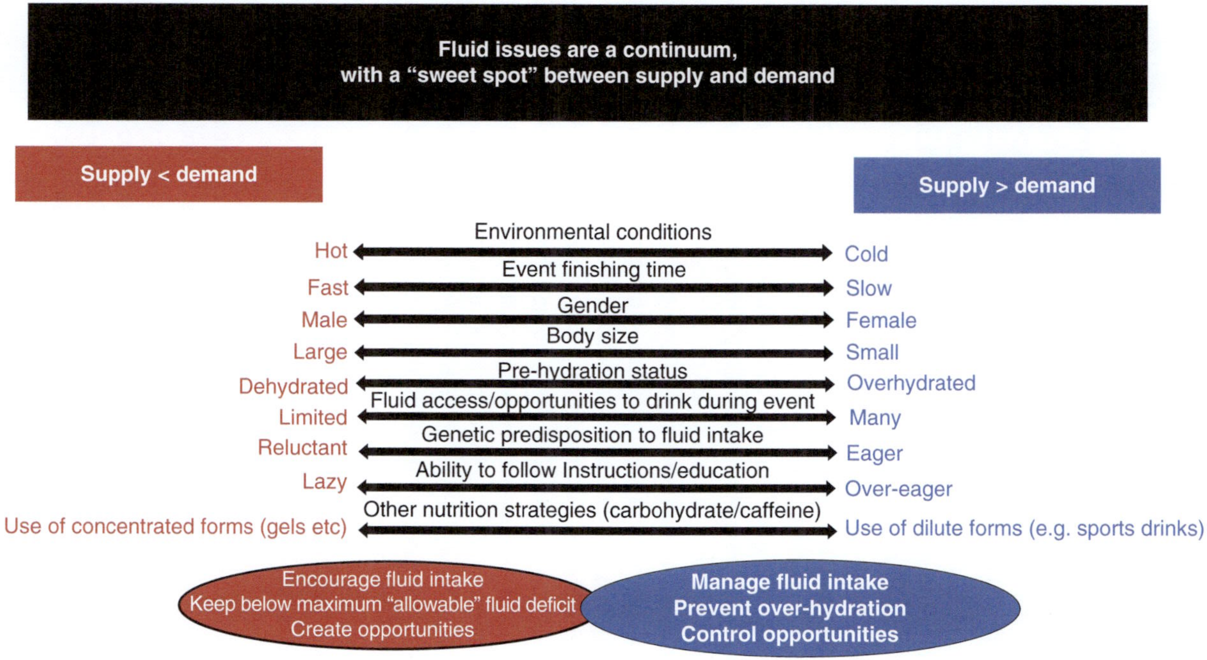

Fig. 6.5 A paradigm to illustrate that guidelines and practices around fluid intake in sport need to recognise that scenarios exist in which athletes may either overdrink or underdrink in comparison to their fluid losses, and that different strategies are needed to address these

individual. This suggests that guidance can be provided to athletes based on the likelihood that their sweat rates will be greatly higher than their opportunities/desire to drink, or the reverse. When factors align in the latter direction (e.g. it is cool, exercise is of moderate intensity and there are plenty of opportunities to consume fluids), it may be sufficient to drink to thirst, although some over-eager drinkers may also benefit from programmed drinking that limits total fluid intake. In contrast, when factors align to make likely sweat losses well in excess of fluid intake (e.g. hot weather, higher intensity exercise, few opportunities to access fluid and difficulty drinking large volumes on a single occasion), athletes can be guided to exploit the available opportunities with a fluid intake plan that limits the accrual of the TBW deficit.

In finishing this discussion, it is worth noting that over the last decade, as the debate between "drinking to thirst" and "programmed drinking" has developed, several studies have attempted to investigate the benefits of the "drinking to thirst" model, or in some cases, to discredit the benefit of a planned approach to hydration for athletic performance [79–82]. As in the general literature on hydration and sports performance, there are methodological issues arising from different practices being used to achieve "drinking to thirst/ad libitum" fluid intake as well as different practices of "programmed drinking"(ranging from complete replacement of BM losses to allowance of an accrual of 2% BM loss). It is interesting that the majority of studies have not found any differences in performance following these different approaches to managing hydration during running and cycling protocols [79–82]; this suggests that a range of practices can be tolerated or are associated with successful outcomes. Furthermore, a study involving 30 km of ergometer cycling in the heat [83] reported that programmed fluid intake which preserved BM to a mean loss <0.5% was associated with a faster performance in the last of three 5-km cycling time trials, than ad libitum drinking (mean BM loss of 1.8%). Clearly, more research is needed, but the available studies support a common sense and flexible attitude to the range of drinking practices that might be suitable for different athletes or exercise scenarios.

6.6 Conclusion

Hydration status adds an overlay to the physiological and perceptual responses to exercise, especially when it is carried out in the heat. Sweat losses vary according to features of the exercise activity, the environment and the athlete, while self-chosen fluid intakes are influenced by a similarly large range of factors. This is especially true during sporting competitions which contribute an additional layer of rules, logistical concerns and performance trade-offs. The real-life picture of hydration practices during exercise/sport is a complicated mix, where some athletes/events experience a large (>5% BM) loss of TBW, while others maintain TBW within a range that is likely well tolerated in terms of physiological strain and performance (<2% BM loss), some drink in excess leading to a gain in TBW and/or BM and a few develop problematic symptoms of hyponatremia/water intoxication. The difficulty of detecting small but meaningful performance effects associated with hypohydration as well as absolute and relative changes in hydration status adds further complexity to the topic. Therefore, it is unlikely that a simple and single guideline for fluid intake during sport and exercise activities will suit all needs and all scenarios. Further research on hypohydration, fluid intake and sports performance is needed, especially involving field scenarios or lab protocols that can better mimic real-life conditions and measure meaningful changes in physiology and performance. Both "ad libitum drinking" or "drinking to thirst" and "programmed fluid intake" protocols have a role in guiding hydration practices for exercise and sport; the best approach will vary according to the factors that influence the relative magnitude of sweat rates and opportunities to drink. Finding some unifying themes to guide education messages around fluid intake to athletes and other people who exercise will help to solve the current debates and confusion.

References

1. Cheuvront SN, Ely BR, Kenefick RW, Sawka MN. Biological variation and diagnostic accuracy of dehydration assessment markers. Am J Clin Nutr. 2010;92(3):565–73.
2. Cheuvront SN, Kenefick RW. Dehydration: physiology, assessment, and performance effects. Comp Physiol. 2014;4(1):257–85.
3. van Rosendal SP, Coombes JS. Glycerol use in hyperhydration and rehydration: scientific update. Med Sport Sci. 2012;59:104–12.
4. Hew-Butler T, Rosner MH, Fowkes-Godek S, Dugas JP, Hoffman MD, Lewis DP, et al. Statement of the 3rd international exercise-associated hyponatremia consensus development conference, Carlsbad, California, 2015. Br J Sports Med. 2015;49(22):1432–46.
5. Sawka MN, Cheuvront SN, Kenefick RW. Hypohydration and human performance: impact of environment and physiological mechanisms. Sports Med. 2015;45(Suppl 1):S51–60.
6. Gonzalez RR, Cheuvront SN, Montain SJ, Goodman DA, Blanchard LA, Berglund LG, et al. Expanded prediction equations of human sweat loss and water needs. J Appl Physiol. 2009;107(2):379–88.
7. Jay O, Webb P. Improving the prediction of sweat losses during exercise. J Appl Physiol. 2009;107(2):375–6.
8. American College of Sports Medicine, Sawka MN, Burke LM, Eichner ER, Maughan RJ, Montain SJ, et al. American College of Sports Medicine position stand. Exercise and fluid replacement. Med Sci Sports Exerc. 2007;39(2):377–90.
9. Garth A, Burke L. What do athletes drink during competitive sporting activities? Sports Med. 2013;43(7):539–64.
10. Beis LY, Wright-Whyte M, Fudge B, Noakes T, Pitsiladis YP. Drinking behaviors of elite male runners during marathon competition. Clin J Sport Med. 2012;22:254–61.
11. Mettler S, Rusch C, Frey WO, Bestmann L, Wenk C, Colombani PC. Hyponatremia among runners in the Zurich Marathon. Clin J Sport Med. 2008;18:344–9.
12. Hew TD. Women hydrate more than men during a marathon race. Hyponatremia in the houston marathon: a report on 60 cases. Clin J Sport Med. 2005;15:148–53.
13. Pahnke MD, Trinity JD, Zachwieja JJ, Stofan JR, Hiller WD, Coyle EE. Serum sodium concentration changes are related to fluid balance and sweat sodium loss. Med Sci Sports Exerc. 2010;42:1669–74.
14. Speedy DB, Noakes TD, Rogers IR, Thompson JM, Campbell RG, Kuttner JA, et al. Hyponatremia in ultradistance triathletes. Med Sci Sports Exerc. 1999;31:809–15.

15. Speedy DB, Campbell R, Mulligan G, Robinson DJ, Walker C, Gallagher P, et al. Weight changes and serum sodium concentrations after an ultradistance multisport triathlon. Clin J Sport Med. 1997;7:100–3.
16. Kao WF, Shyu CL, Yang XW, Hsu TF, Chen JJ, Kao WC, et al. Athletic performance and serial weight changes during 12- and 24-hour ultra-marathons. Clin J Sport Med. 2008;18:155–8.
17. Glace BW, Murphy CA, McHugh MP. Food intake and electrolyte status of ultramarathoners competing in extreme heat. J Am Coll Nutr. 2002;21:553–9.
18. Fallon KE, Broad E, Thompson MW, Reull PA. Nutritional and fluid intake in a 100-km ultramarathon. Int J Sport Nutr. 1998;8:24–35.
19. Armstrong LE, Casa DJ, Emmanuel H, Ganio MS, Klau JF, Lee EC, et al. Nutritional, physiological, and perceptual responses during a summer ultraendurance cycling event. J Strength Cond Res. 2012;26:307–18.
20. Knechtle B, Knechtle P, Rosemann T, Senn O. No dehydration in mountain bike ultramarathoners. Clin J Sport Med. 2009;19:415–20.
21. Brearley MB, Finn JP. Responses of motor-sport athletes to V8 supercar racing in hot conditions. Int J Sports Physiol Perform. 2007;2:182–91.
22. Ross ML, Stephens B, Abbiss CR, Martin DT, Laursen PB, Burke LM. Fluid balance, carbohydrate ingestion, and body temperature during men's stage-race cycling in temperate environmental conditions. Int J Sports Physiol Perform. 2014;9(3):575–82.
23. Ebert TR, Martin DT, Stephens B, McDonald W, Withers RT. Fluid and food intake during professional men's and women's road-cycling tours. Int J Sports Physiol Perform. 2007;2: 58–71.
24. Garcia-Roves PM, Terrados N, Fernandez SF, Patterson AM. Macronutrients intake of top level cyclists during continuous competition—change in the feeding pattern. Int J Sports Med. 1998;19:61–7.
25. Knechtle B, Gnadinger M, Knechtle P, Imoberdorf R, Kohler G, Ballmer P, et al. Prevalence of exercise-associated hyponatremia in male ultraendurance athletes. Clin J Sport Med. 2011;21:226–32.
26. Da Silva RP, Mundel T, Natali AJ, Bara Filho MG, Alfenas RC, Lima JR, et al. Pre-game hydration status, sweat loss, and fluid intake in elite Brazilian young male soccer players during competition. J Sports Sci. 2012;30:37–42.
27. Aragon-Vargas LF, Moncada-Jimenez J, Hernandez-Elizondo J, Barrenechea A, Monge-Alvarado M. Evaluation of pre-game hydration status, heat stress, and fluid balance during professional soccer competition in the heat. Eur J Sport Sci. 2009;9:269–76.
28. Maughan RJ, Watson P, Evans GH, Broad N, Shirreffs SM. Water balance and salt losses in competitive football. Int. J. Sport Nutr Exerc Metab. 2007;17:583–94.
29. O'Hara JP, Jones BL, Tsakirides C, Carroll S, Cooke CB, King RF. Hydration status of rugby league players during home match play throughout the 2008 Super League season. Appl Physiol Nutr Metab. 2010;35:790–6.
30. Meir RA, Halliday AJ. Pre- and post-game body mass changes during an international rugby tournament. A Practical Perspective. J Strength Cond Res. 2005;19:713–6.
31. Gore CJ, Bourdon PC, Woolford SM, Pederson DG. Involuntary dehydration during cricket. Int J Sports Med. 1993;14:387–95.
32. Osterberg KL, Horswill CA, Baker LB. Pregame urine specific gravity and fluid intake by National Basketball Association players during competition. J Ath Train. 2009;44:53–7.
33. Tippet ML, Stofan JR, Lacambra M, Horswill CA. Core temperature and sweat responses in professional women's tennis players during tournament play in the heat. J Ath Train. 2011;46:55–60.
34. Bergeron MF, Maresh CM, Armstrong LE, Signorile JF, Castellani JW, Kenefick RW, et al. Fluid-electrolyte balance associated with tennis match play in a hot environment. Int J Sport Nutr. 1995;5:180–93.
35. Logan-Sprenger HM, Palmer MS, Spriet LL. Estimated fluid and sodium balance and drink preferences in elite male junior players during an ice hockey game. Appl Physiol Nutr Metab. 2011;36:145–52.

36. Cox GR, Broad EM, Riley MD, Burke LM. Body mass changes and voluntary fluid intakes of elite level water polo players and swimmers. J Sci Med Sport. 2002;5:183–93.
37. Wagner S, Knechtle B, Knechtle P, Rüst CA, Rosemann T. Higher prevalence of exercise-associated hyponatremia in female than in male open-water ultra-endurance swimmers: the 'Marathon-Swim' in Lake Zurich. Eur J Appl Physiol. 2012;112:1095–106.
38. Neville V, Gant N, Folland JP. Thermoregulatory demands of elite professional America's Cup yacht racing. Scand J Med Sci Sports. 2010;20:475–84.
39. Slater G, Tan B. Body mass changes and nutrient intake of dinghy sailors while racing. J Sports Sci. 2007;25:1129–35.
40. Sawka MN, Leon LR, Montain SJ, Sonna LA. Integrated physiological mechanisms of exercise performance, adaptation, and maladaptation to heat stress. Compr Physiol. 2011;1(4):1883–928.
41. Cheuvront SN, Kenefick RW, Montain SJ, Sawka MN. Mechanisms of aerobic performance impairment with heat stress and dehydration. J Appl Physiol. 2010;109(6):1989–95.
42. Montain SJ, Coyle EF. Influence of graded dehydration on hyperthermia and cardiovascular drift during exercise. J Appl Physiol. 1992;73(4):1340–50.
43. Sawka MN, Cheuvront SN, Kenefick RW. High skin temperature and hypohydration impair aerobic performance. Exp Physiol. 2012;97(3):327–32.
44. Ritz P. Methods of assessing body water and body composition. In: Arnaud MJ, editor. Hydration throughout life. Vittel: Perrier Vittel Water Institute; 1998. p. 63–74.
45. Popowski LA, Oppliger RA, Patrick Lambert G, Johnson RF, Kim Johnson A, Gisolf CV. Blood and urinary measures of hydration status during progressive acute dehydration. Med Sci Sports Exerc. 2001;33(5):747–53.
46. O'Brien C, Young AJ, Sawka MN. Bioelectrical impedance to estimate changes in hydration status. Int J Sports Med. 2002;23:361–6.
47. Cheuvront SN, Carter R, Montain SJ, Sawka MN. Daily body mass variability and stability in active men undergoing exercise-heat stress. Int J Sport Nutr Exerc Metab. 2004;14:532–40.
48. Armstrong LE, Maresh CM, Castellani JW, Bergeron MF, Kenefick RW, LaGasse KE, et al. Urinary indices of hydration status. Int J Sport Nutr. 1994;4:265–79.
49. Kovacs EM, Senden JM, Brouns F. Urine color, osmolality and specific electrical conductance are not accurate measures of hydration status during postexercise rehydration. J Sports Med Phys Fitness. 1999;39(1):47–53.
50. Cheuvron SN, Sawka MN. SSE#97: hydration assessment of athletes. Gatorade Sports Sci Exchange 2005;18(2).
51. Armstrong LE, Pumerantz AC, Fiala KA, Roti MW, Kavouras SA, Casa DJ, et al. Human hydration indices: acute and longitudinal reference values. Int J Sport Nutr Exerc Metab. 2010;20(2):145–53.
52. Baker LB, Reimel AJ, Sopeña BC, Barnes KA, Nuccio RP, De Chavez PJD, Stofan JR, et al. Trapped sweat in basketball uniforms and the effect on sweat loss estimates. Physiol Rep. 2017 5(18). pii: e13463. https://doi.org/10.14814/phy2.13463.
53. Cheuvront SN, Haymes EM, Sawka MN. Comparison of sweat loss estimates for women during prolonged high-intensity running. Med Sci Sports Exerc. 2002;34(8):1344–50.
54. Hoffman MD, Goulet EDB, Maughan RJ. Considerations in the use of body mass change to estimate change in hydration status during a 161-kilometer ultramarathon running competition. Sports Med. 2018;48(2):243–50.
55. Sawka MN, Noakes TD. Does dehydration impair exercise performance? Med Sci Sports Exerc. 2007;39(8):1209–17.
56. Cotter JD, Thornton SN, Lee JK, Laursen PB. Are we being drowned in hydration advice? Thirsty for more? Extrem Physiol Med. 2014;3:18.
57. Maughan RJ. Investigating the associations between hydration and exercise performance: methodology and limitations. Nutr Rev. 2012;70(Suppl 2):S128–31.
58. Burke LM, Patterson AM, Cox GR, Armstrong LE. Concepts in developing better guidelines for fluid intake during sporting events: is drinking to thirst the answer? In review.

59. Nuccio RP, Barnes KA, Carter JM, Baker LB. Fluid balance in team sport athletes and the effect of hypohydration on cognitive, technical, and physical performance. Sports Med. 2017;47(10):1951–82.
60. Savoie FA, Kenefick RW, Ely BR, Cheuvront SN, Goulet ED. Effect of hypohydration on muscle endurance, strength, anaerobic power and capacity and vertical jumping ability: a meta-analysis. Sports Med. 2015;45(8):1207–27.
61. Cheuvront SN, Kenefick RW. Dehydration: physiology, assessment, and performance effects. Compr Physiol. 2014;4(1):257–85.
62. Goulet ED. Effect of exercise-induced dehydration on endurance performance: evaluating the impact of exercise protocols on outcomes using a meta-analytic procedure. Br J Sports Med. 2013;47(11):679–86.
63. Judelson DA, Maresh CM, Anderson JM, Armstrong LE, Casa DJ, Kraemer WJ, et al. Hydration and muscular performance: does fluid balance affect strength, power and high-intensity endurance? Sports Med. 2007;37(10):907–21.
64. James LJ, Moss J, Henry J, Papadopoulou C, Mears SA. Hypohydration impairs endurance performance: a blinded study. Physiol Rep. 2017;5(12). pii: e13315.
65. Cheung SS, McGarr GW, Mallette MM, Wallace PJ, Watson CL, Kim IM, et al. Separate and combined effects of dehydration and thirst sensation on exercise performance in the heat. Scand J Med Sci Sports. 2015;25(Suppl 1):104–11.
66. Wall BA, Watson G, Peiffer JJ, Abbiss CR, Siegel R, Laursen PB. Current hydration guidelines are erroneous: dehydration does not impair exercise performance in the heat. Br J Sports Med. 2015;49(16):1077–83.
67. Adams JD, Sekiguchi Y, Suh HG, Seal AD, Sprong CA, Kirkland TW, et al. Dehydration impairs cycling performance, independently of thirst: a blinded study. Med Sci Sports Exerc. 2018. https://doi.org/10.14814/phy2.13315
68. Noakes TD. Lore of running. 4th ed. Champaign: Human Kinetics; 2003.
69. American College of Sports Medicine. Position statement of the American College of Sports Medicine: prevention of heat injuries during distance running. Med Sci Sports Exerc. 1975;7:vii–x.
70. Convertino VA, Armstrong LE, Coyle EF, Mack GW, Sawka MN, Senay LC Jr, et al. American College of Sports Medicine position stand. Exercise and fluid replacement. Med Sci Sports Exerc. 1996;28(1):i–vii.
71. Almond CS, Shin AY, Fortescue EB, Mannix RC, Wypij D, Binstadt BA, et al. Hyponatremia among runners in the Boston Marathon. N Engl J Med. 2005;352(15):1550–6.
72. Noakes TD, Sharwood K, Speedy D, Hew T, Reid S, Dugas J, et al. Three independent biological mechanisms cause exercise-associated hyponatremia: evidence from 2,135 weighed competitive athletic performances. Proc Nat Acad Sci USA. 2005;102(51):18550–5.
73. Noakes TD, Speedy DB. Case proven: exercise associated hyponatraemia is due to overdrinking. So why did it take 20 years before the original evidence was accepted? Br J Sports Med. 2006;40(7):567–72.
74. Noakes TD. Changes in body mass alone explain almost all of the variance in the serum sodium concentrations during prolonged exercise. Has commercial influence impeded scientific endeavour? Br J Sports Med. 2011;45(6):475–7.
75. Noakes T. Waterlogged: the serious problem of overhydration in endurance sports. Champaign: Human Kinetics; 2012.
76. Lewis D, Blow A, Tye J, Hew-Butler T. Considering exercise-associated hyponatraemia as a continuum. BMJ Case Rep. 2018. pii: bcr-2017-222916.
77. Kenefick RW. Drinking strategies: planned drinking versus drinking to thirst. Sports Med. 2018;48(Suppl 1):31–7.
78. Valenzuela PL, Morales JS, de la Villa P, Lucía A. Comment on: "drinking strategies: planned drinking versus drinking to thirst". Sports Med. 2018. https://doi.org/10.1007/s40279-018-0905-5. [Epub ahead of print].
79. Daries HN, Noakes TD, Dennis SC. Effect of fluid intake volume on 2-h running performances in a 25 degrees C environment. Med Sci Sports Exerc. 2000;32:1783–9.

80. Dion T, Savoie FA, Asselin A, Gariepy C, Goulet ED. Half-marathon running performance is not improved by a rate of fluid intake above that dictated by thirst sensation in trained distance runners. Eur J Appl Physiol. 2013;113:3011–20.
81. Berkulo MA, Bol S, Levels K, Lamberts RP, Daanen HA, Noakes TD. Ad-libitum drinking and performance during a 40-km cycling time trial in the heat. Eur J Sport Sci. 2016;16:213–20.
82. Dugas JP, Oosthuizen U, Tucker R, Noakes TD. Rates of fluid ingestion alter pacing but not thermoregulatory responses during prolonged exercise in hot and humid conditions with appropriate convective cooling. Eur J Appl Physiol. 2009;105:69–80.
83. Bardis CN, Kavouras SA, Adams JD, Geladas ND, Panagiotakos DB, Sidossis LS. Prescribed drinking leads to better cycling performance than ad libitum drinking. Med Sci Sports Exerc. 2017;49:1244–51.

Heat-Stress Exercise and Cooling

7

Christopher J. Tyler

7.1 Introduction

The development of hyperthermia is central to all of the mechanisms and theories proposed to explain the impaired exercise performance and capacity observed in the heat and so interventions that reduce the actual and/or perceived level of thermal strain are sought. Cooling is one such intervention. Data regarding cooling use in elite sport are limited but Périard et al. [1] reported that 52% of athletes surveyed prior to the 2015 IAAF World Championships planned to use at least one cooling strategy during the championships and that ice-slurry ingestion was the most frequently planned for cooling invention (Fig. 7.1). This chapter will review the effect that different internal and external cooling interventions have on the physiological, perceptual, and performance responses to heat stress when administered prior to (pre-cooling), during (per-cooling), or after (post-cooling) exercise.

7.2 Mechanisms of Cooling

Athlete cooling can be internal or external. In addition to decreasing tissue temperature, external cooling activates peripheral thermoreceptors, whereas internal cooling activates central thermoreceptors and so the outcomes of the two different cooling approaches can differ but may also be complementary and additive. External cooling is probably the most common method of cooling used before, during, and after exercise; however, internal cooling was the most commonly planned for cooling strategy by athletes prior to the 2015 IAAF World Championships in Beijing [1]. A number of external and internal cooling methods

C. J. Tyler (✉)
Environmental Physiology, University of Roehampton, London, UK
e-mail: chris.tyler@roehampton.ac.uk

J. D. Périard, S. Racinais (eds.), *Heat Stress in Sport and Exercise*,
https://doi.org/10.1007/978-3-319-93515-7_7

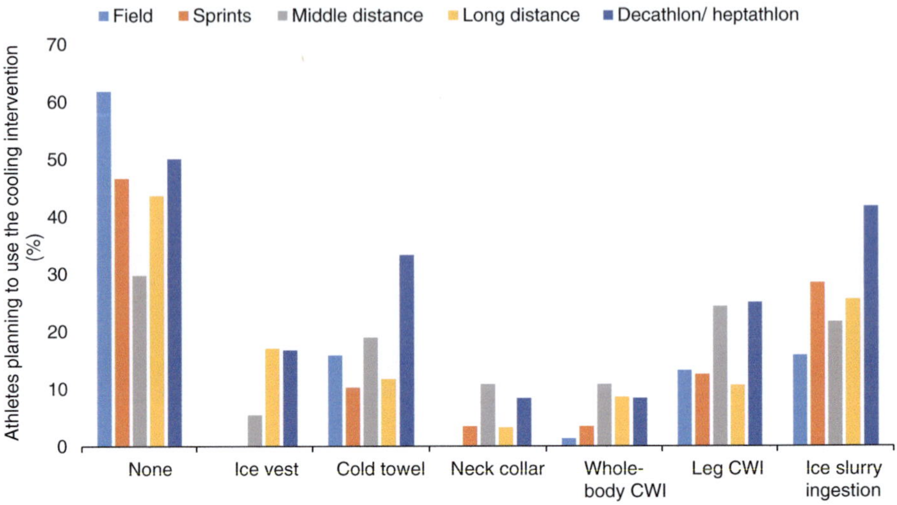

Fig. 7.1 Prearranged cooling strategies of athletes (% of responders) competing in the 2015 IAAF World Championships (redrawn from Périard et al. [1])

are used and these include interventions capable of offering an actual cooling effect (e.g. whole-body and partial-water immersion, cooling garments, cold/iced beverage ingestion) as well as those that offer a sensation of cooling (e.g. menthol application or mouth rinsing).

7.2.1 External Cooling

7.2.1.1 Cold-Water Immersion

Water immersion can be used before (pre-cooling) or after (post-cooling) exercise and involves immersion of the body (whole or partial) in water at 2–26 °C to lower core temperature through the direct removal of heat by the water and the circulation of the cooler peripheral blood to central regions of the body. Although it was previously thought that the water used should be very cold, recent data has shown comparable cooling rates with warmer water (26 °C v 14 °C) due to a reduced vasoconstrictor drive [2]. As well as reducing thermal strain, water immersion can increase central blood volume and subsequently stroke volume allowing for a lower heart rate at a given exercise intensity [3–5]. Water immersion can also elicit a change in the perceived thermal state and thermal comfort of the athlete due to rapid reductions in skin temperature [4, 6, 7]—this reduction can be pleasant or unpleasant depending on the magnitude of cooling presented. Unpleasant thermal sensations are related to lower water temperatures [8] and larger body surface areas being exposed [9]. Using very cold water for immersion post-exercise can be beneficial in reducing body temperature and treating hyperthermia [10] (see Sect. 7.3.4) but using such temperatures prior to exercise are not advisable due to the discomfort caused and the reduction in muscle temperature. A reduction in muscle temperature can reduce maximal power, force, and velocity [11] and so while cooling non-active parts can reduce cardiovascular, physiological, and perceived strain [12, 13], cooling the active muscles should be avoided.

Whole-body cold-water immersion is the most effective way of cooling the body pre- and post-exercise but it is not always practical. Partial-water immersion can cause less disruption to pre-competition routines (e.g. less time is required to dry the athlete, and some clothing can be worn) than whole-body water immersion and this may explain why more athletes planned to use this lower-limb cooling at the 2015 IAAF World Championships in Beijing (Fig. 7.1 [1]). A number of studies have looked at cooling sections of the body using water immersion and these have included cooling non-active (e.g. torso [13] or hands [12]) and active (e.g. legs [11]) body parts.

7.2.1.2 Cooling Garments

Ice vests, *cooling vests*, *ice jackets*, and *cooling jackets* are names used interchangeably in the literature to describe practical, external cooling garments designed to cool the torso through direct conductive and evaporative methods. *Ice vest* is the most commonly used description and so will be used here. Ice vests either contain cooling sections activated by water and cooled in refrigerators, freezers, or iced-water or make use of frozen or cooled inserts. Vests typically cover the entire torso (~25% of the body surface area [14]) but cool less (5–10% [14, 15]) due to the design of the vests and location of the cooling strips/packs (Fig. 7.2). Ice vests can be worn during exercise as well as before and/or after. When worn as part of a pre-cooling routine ice vests can lower core body and skin temperatures, reduce heart

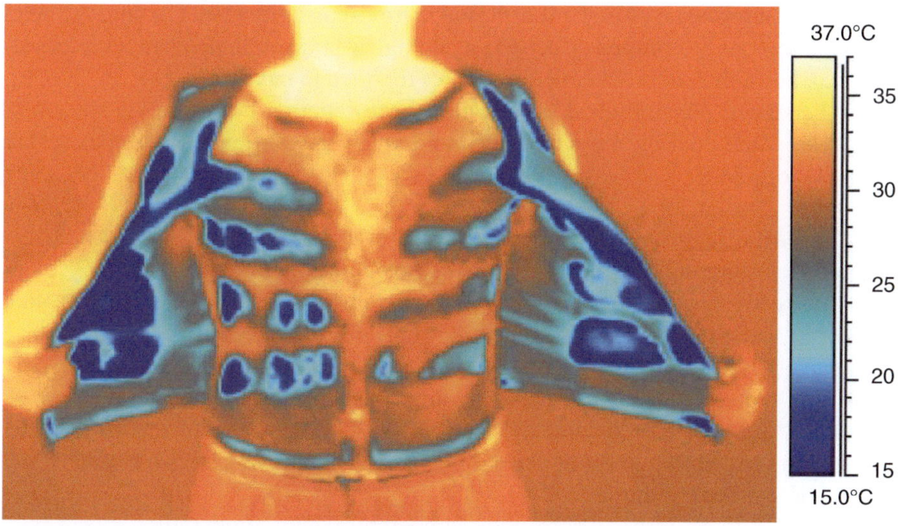

Fig. 7.2 Thermal image showing the localised cooling offered by an ice vest

rate, and improve the perception of task difficulty and thermal comfort [16–19]; however, not all vests offer a sufficiently prolonged cooling effect to induce such changes [20, 21].

Ice vests worn during exercise can prolong the cooling but there are barriers to use, e.g. the rules and regulations governing the sport vests may prohibit their use. Even if allowed, the wearing of an ice vest might be problematic during load-bearing activities or weight-sensitive events (e.g. motor racing). Commercially available cooling vests weigh 0.5–4.5 kg [16, 22] with the most commonly investigated one (Arctic Heat) weighing ~1.5 kg when activated [23]. The energy cost of walking and running increases when carrying a load [24] and so running while wearing a torso-cooling garment is likely to increase the energy demands of the activity but this has not been directly assessed (there appears to be no energy cost of cycling with a cooling vest [15]). Arngrimsson et al. [16] investigated the effects of wearing an ice vest during an active warm-up but they did not measure the effects of the extra mass on the metabolic cost of the exercise bout. Instead, they reduced the warm-up running speed by ~0.8 km h^{-1} "to compensate for the extra metabolic work done… due to the weight of the vest" (p. 1868). Ice vests worn during exercise can improve exercise capacity during heat stress [14, 15, 25]; however, it may only effectively reduce physiological strain when the magnitude of heat stress is very high [25]. When the magnitude of heat strain is low, any benefit of wearing an ice vest on exercise performance appears to be due to the sum effect of small beneficial physiological changes and reductions in perceived physiological strain [14, 15].

In addition to cooling the torso, the neck and head have been proposed as good sites for cooling due to their close proximity to the thermoregulatory control centre (i.e. hypothalamus) [26, 27] and ease of access [21]. They make up only ~1% [14] and ~8% [28] of the body surface area, respectively; however, cooling the neck and head may elicit disproportionately beneficial changes in perceived thermal strain [27] due to being a site of high alliesthesial thermosensitivity [29]. Cooling the head and neck regions has limited effect on physiological strain with no changes in physiological, hormonal, or biochemical variables reported in most cases [14, 26, 30–36]. Some beneficial physiological effects have been reported but these are generally only seen when cooling the whole area and when the participant is under severe thermal strain [37].

7.2.1.3 Fan Cooling

Frequently in thermal physiology investigations, airflow is either missing (sometimes correctly and sometimes incorrectly) or is provided at lower speeds than might be experienced when competing outside the laboratory [38]. During laboratory-based performance studies, the lack of a representative airflow may mean that the effect of other heat-mitigating interventions (e.g. cooling) may be overstated [39, 40]. For example, Morrison et al. [39] observed that cold-water immersion pre-cooling and fan per-cooling did not improve exercise capacity any more than fan per-cooling alone.

Compared to no airflow, core body temperature, skin temperature, sweat rate, and heart rate are all lower with airflow [41–44] and perceived exertion (RPE) and thermal sensation can be improved [39, 43, 44] and so it is unsurprising that increasing airflow can be ergogenic. Otani et al. [43] compared the effects of providing airspeeds of 0, 10, 20, or 30 k h^{-1} to cyclists exercising to exhaustion in hot conditions. There was a progressive reduction in capacity time as airspeed was reduced and the greatest exercise capacity was observed with an airspeed of 30 k h^{-1} (90.3 ± 16.5 min). The time to exhaustion was ~50 min shorter with no airflow at all (41.4 ± 10.1 min). Rectal temperatures were similar at the point of exhaustion in all four trials but the higher wind speeds slowed the rate at which is rose. Unsurprisingly, convective and evaporative heat loss was lower in the 0 k h^{-1} trial than in all other trials. While many sports naturally generate an airflow (e.g. cycling) the opportunities to use artificial airflow in elite sport are limited predominantly to periods of seated rest, e.g. while on the substitutes' bench or at the change of ends in tennis. Fan cooling (in combination with skin wetting to promote evaporative heat loss) increased the exercise time completed during a recent tennis simulation in the heat [44] compared to no cooling and the currently adopted ice-towel approach.

7.2.1.4 Menthol Cooling

Menthol is not a cooling method per se but it provides a sensation of cooling [45–50] due to the activation of the transient receptor potential ion channel melastatin 8 (TRPM8). The magnitude of cooling sensation appears to be dose-dependent with participants reporting a greater cooling sensation when a 0.2% menthol spay is used compared to a control or 0.05% spray [48]. Despite improving thermal sensation menthol application does not reduce thermal strain [45–47, 50] and may actually increase it [50, 51]. Menthol concentrations of 1–10% induce vasoconstriction, reduce skin blood flow, and delay the onset of sweating in some [51] but not all [52, 53] studies and a dose response relationship between the concentration of menthol and the rate in the rise of rectal temperature has been reported [51, 54]. Schlader et al. [50] reported that despite only improving perceived, rather than actual, thermal strain, the application of menthol gel (8%) to the face increased the work completed at a fixed rating of perceived exertion by 21% but other, less aggressive, applications (0.05% menthol sprayed on to clothing) have reported no improvements in exercise performance in the heat [45–47].

7.2.2 Internal Cooling

The ingestion of cold or iced beverages can lower starting core temperature and slow the rate at which it rises during exercise in the heat [55–57]. Ingesting an ice slurry can approximately double the internal sink compared to cold fluid ingestion due to harnessing the combined thermodynamic properties of the solid (specific heat capacity = 2.1 kJ kg^{-1} $°C^{-1}$) and liquid (specific heat capacity = 4.2 kJ kg^{-1} $°C^{-1}$) water as well as the enthalpy of fusion required for the change from solid into liquid form (334 kJ kg^{-1}). Due to the greater heat sink it is unsurprising that ingesting ice slurry before and during exercise can reduce core body temperature [5, 58–61]; however, a reduction is not always observed [62]. Although the ingestion of cold

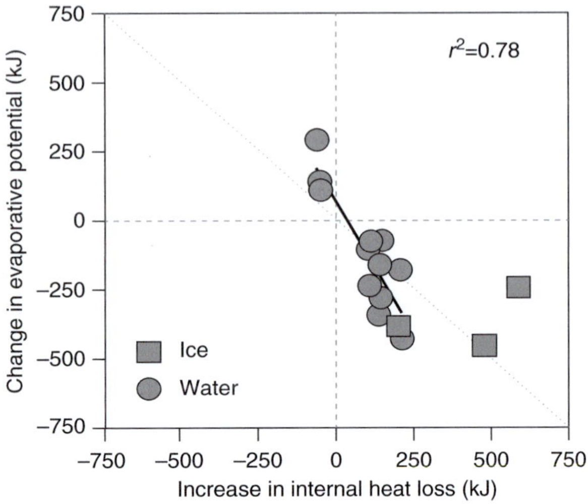

Fig. 7.3 The relationship between changes in evaporative heat loss potential and internal heat loss when ingesting ice and cold-water beverages (from Jay and Morris [66])

fluid cools the body from within and creates a heat sink, it also suppresses the sudomotor response to heat independently of changes in skin or core body temperature, resulting in lower sweat rate and sweat loss, and reductions in evaporative heat loss from the skin [5, 56, 60, 63–66]. The reduction in evaporative heat loss can result in greater heat storage during exercise in the heat [60, 65] and the magnitude of reduction is strongly related to the temperature of the ingested fluid (Fig. 7.3) [66]. Pre-cooling with cold/ice fluid ingestion prior to the initiation of a full sweating response can increase heat storage [60] and so may be beneficial; however, it may also delay the onset of sweating and/or vasodilation which may in turn increase the rate of heat storage during the early stages of exercise [66].

Regardless of changes, or not, in heat storage, the ingestion of cold/ice fluids can make athletes feel cooler and such changes can have ergogenic effects in the heat [67]. Siegel et al. [5] reported that the ingestion of an ice slurry had a disproportionate improvement in thermal perceptions. Despite greater reductions in body temperature following cold-water immersion, participants reported similar ratings of thermal sensation following the ingestion of an ice slurry [5]. As mentioned, menthol is a potent improver of thermal comfort and in addition to being used externally it can be used internally either as a mouth rinse [68–70] or ingested as a drink with cold (3 °C) and ice-slush (−1 °C) beverages [71, 72]. Menthol mouth rinsing (0.01%) or ingestion (0.05%) improves perceived thermal strain with minimal physiological change [68–72] and can improve exercise capacity and performance in the heat [68–72].

Gastrointestinal permeability is increased as body temperature increases [73] and so it has been hypothesised that the direct internal cooling offered by cold/ice fluid ingestion may reduce gastrointestinal discomfort during exercise in the heat. Snipe and Costa [61] directly tested this hypothesis and found that despite reducing thermal strain, the ingestion of cool (7 °C) or cold (0 °C) fluid before and during exertional heat stress had no effect on systemic inflammatory cytokines and only slightly reduced intestinal injury and upper-gastrointestinal symptoms in endurance-trained euhydrated athletes. In this study the comparison trial involved the ingestion of 22 °C, rather than body temperature, water and so it is possible that the lack of observed effect was due, at least in part, to the cooling offered by the "control" trial.

7.2.3 External and Internal Cooling in Combination

It seems prudent to suggest that combining internal and external cooling would be an optimal cooling strategy; however, little data exists and what does exist is equivocal. Schulze et al. [74] investigated a combined cooling approach—pre-cooling with internal (ice slurry) and external (ice towels applied to the legs and torso) cooling and supplementing the pre-cool with internal cooling via ice ingestion during exercise. The power output in the combined internal and external trial was "likely higher" (+2.5 ± 1.9%) and the thermal comfort was "very likely lower" than control but rectal temperature was unaffected and there was no additional benefit of combining the interventions compared to internal per-cooling alone. Ross et al. [75] observed that internal (ice-slurry ingestion) and external (ice-towel application) cooling increased power output (by ~3%; 8 W) and cycling performance (~1:06 min) compared to control and although rectal temperature and heart rate were lower following pre-cooling the differences had disappeared by the time the performance test commenced. In contrast, Brade et al. [76] reported that total work during a 90 min intermittent exercise bout was reduced following internal pre-cooling but that there was no difference between external (ice vest) or combined (ice vest and ice-slurry ingestion) cooling compared to control despite the combined approach having the greatest effect on reducing physiological strain.

7.3 Timing

7.3.1 Pre-cooling

Cooling prior to exercise in the heat can increase the heat storage capacity and reduce the thermal strain that the participant experiences before and/or during the exercise bout. External pre-cooling lowers the temperature of the peripheral blood, which cools the core body temperature when it circulates. In some cases the reduction in core body temperature is delayed [23, 55, 77–79] due to the *after-drop* phenomenon which can occur if the cooling is too severe and severe vasoconstriction of the peripheral blood vessels occurs almost immediately. The vasoconstriction quickly shifts the warmer peripheral blood to the core and delays the transfer of the cooling of the skin to the core. Pre-cooling can reduce core body temperature prior to, or during the early stages of, exercise but the benefit does not always last for the duration of the exercise bout [21]. Some pre-cooling studies report that the reduction in core body temperature can last the duration of the trial (≤ 90 min) [79–82];

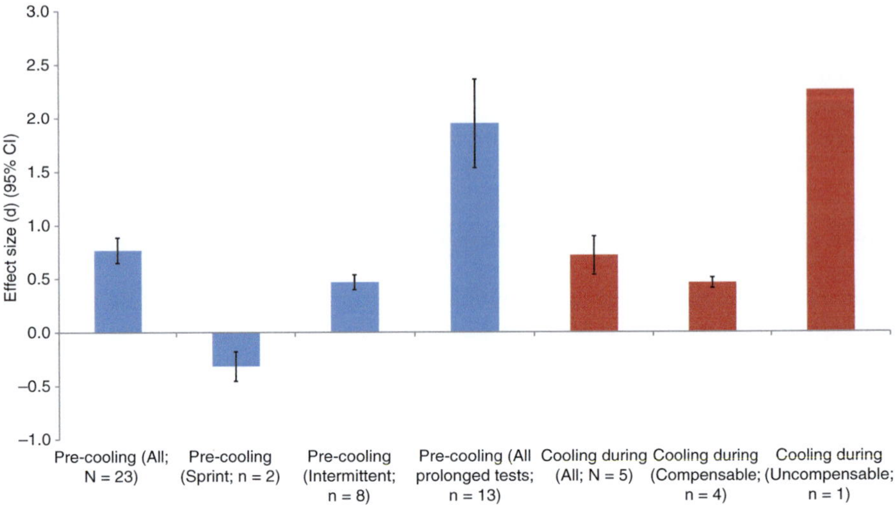

Fig. 7.4 The effect of pre- and per-cooling on exercise performance and capacity in the heat (Adapted from Tyler et al. [21])

however, this is not always the case due to either insufficient cooling or the adoption of a faster pacing strategy following the cooling. In addition to lowering body temperatures, pre-cooling can lower heart rate due to the redistribution of blood from the periphery to the core. As with physiological change, the effectiveness of pre-cooling on improving exercise performance in the heat is dependent upon the heat strain experienced [20] and the magnitude of cooling provided. Pre-cooling can enhance subsequent exercise performance but it does not always improve all forms of exercise performance (Fig. 7.4).

Single sprint performance appears to be impaired following pre-cooling—especially when the temperature of the area superior to the working muscles is lowered. Pre-cooling the torso can improve sprint cycle performance in the heat (+4%) [13], but does not always do so (−2%) [6], whereas pre-cooling the torso and thighs in combination or legs alone impairs peak power output by 7% [6] and 30%, respectively [83]. It is likely that muscle temperature was lower in the studies that directly cooled the legs [6, 83] but this was not measured. Reductions in muscle temperature can extend the time-to-peak tension, decrease voluntary power output [84], and slow the rate of anaerobic metabolism [6, 85] and so it is unsurprising that short duration, explosive sprint performance is impaired following pre-cooling.

Although pre-cooling impairs subsequent sprint performance, it appears to be beneficial to subsequent intermittent sprint exercise performance in the heat overall (Cohen's $d = 0.47$; [21]). The intermittent sprint exercises investigated typically last at least 30 min and raise core body temperatures more than the shorter single-sprints [21] and the greater physiological strain experienced during intermittent sprint exercise explains why pre-cooling is often beneficial. Although pre-cooling prior to intermittent sprint exercise is, on average, beneficial [21], individual study data regarding the effectiveness of cooling prior to intermittent sprint activity are equivocal, even within laboratories. For example, Minett et al. [86, 87] observed a dose–response relationship between the volume and duration of pre-cooling and subsequent improvements in physiological, perceptual, and performance outcomes ($d = 0.00$–2.14) but subsequently reported no improvements in performance, despite physiological and perceptual improvements, using the same pre-cooling method [88]. The distance covered in a simulated cricket bowling test was unaltered by pre-cooling [88] but the distance covered in a longer-duration, intermittent protocol in which higher core temperatures were observed was increased [87].

These data suggest that pre-cooling is more beneficial when thermal strain is high and so it is unsurprising that the greatest pre-cooling benefits are observed when used prior to prolonged exercise in the heat (Fig. 7.4) [21]. The majority of the literature report improvements in prolonged exercise performance and capacity following pre-cooling; however, Mitchell et al. [81] reported that exercise capacity at 100% maximal oxygen uptake was impaired (−7.5%; $d = -0.59$) following pre-cooling using fans and mist spray. The pre-cooling intervention reduced core temperature (by ~1.5 °C) but caused discomfort to the participants who reported lower levels of thermal comfort and a "lack of spring" and a feeling of "heaviness" in the cooling trials when interviewed (p. 123).

7.3.2 Per-cooling

The physiological and perceptual changes caused by pre-cooling are often lost or diminished during exercise resulting in participants finishing the bout under similar levels of thermal strain to the control trials [4, 16, 55, 77–79]. In order to try to prolong the duration that the participants experience a lower thermal strain, cooling during exercise (per-cooling) has been investigated. Per-cooling is an attractive idea if sufficient cooling can be provided; however, there are a number of practical considerations that can restrict its use, for example, the rules and regulations of the sport, excess weight, and skin irritation [16].

Per-cooling can improve exercise performance and capacity in the heat (Hedges' $g = 0.40$; Cohen's $d = 0.76$ [20, 21]), with and without physiological alterations, and offers the greatest physiological benefit when the thermal challenge is very high (Fig. 7.4). An ice vest worn under nuclear, biological, and chemical protective clothing reduced core and skin temperatures and improved exercise capacity by ~12% [25]. In less thermally challenging settings, per-cooling the torso can also improve cycling capacity (+17–21%) in the heat—this time without altering rectal temperature, skin temperature, or heart rate but with improvements in thermal sensations [14, 15]. Per-cooling the head and neck regions also has a limited effect on physiological strain [14, 26, 30–36]; however, it can improve running performance (~6%) and capacity (~12%) in hot environments [34–36]. Statistically significant improvements are not always observed with neck per-cooling but the effect sizes indicate a likely benefit ($p = 0.09$, $d = 0.62$ [32]; $p = 0.12$, $d = 0.31$ [14]). Athletes are often seen cooling the neck region at events such as Tennis Grand Slams and although cooling this region appears to be an effective practical per-cooling intervention for improving exercise performance it should not be used if a reduction in thermal strain is desired or required.

To date, internal per-cooling has received limited attention but the data are positive. Ingesting cold (4 °C) fluids can improve cycling capacity by 13% (7 min) [57] and ice-slurry ingestion can improve performance by ~ 2.5% [74]. Internal per-cooling may [57] or may not [74] reduce physiological strain while improving thermal comfort [74] during exercise; however, it is likely that at least some of cooling benefit is counteracted by the increased heat production occurring as a result of the improved work. Internal per-cooling may enable athletes to adopt a faster self-selected exercise intensity in the heat, which would be advantageous if done within safe limits. Internal per-cooling does not consistently reduce core body temperature [57, 63–65, 74] and recent data have recommended that ice slurries should not be ingested during exercise in the heat—especially by novice athletes who are more susceptible to heat illness [65].

7.3.3 Pre- and Per-cooling

Combining pre- and per-cooling strategies is advantageous to exercise performance; however, the benefit is similar to pre- or per-cooling alone [20, 89]. Lee et al. [56] reported that cold fluid ingested before and during exercise reduced physiological and perceptual strain and improved cycling capacity in the heat by ~23% but Riera et al. [71] and Tran Trong et al. [72] reported no change in physiological or perceptual strain and only observed a benefit with colder fluid when internally pre- and per-cooling. Combining internal and external pre- and per-cooling can lower physiological and perceptual strain and improve performance compared to no cooling but the combination may not be better than per-cooling alone [74]. Schulze et al. [74] pre-cooled individuals using a combination of ice slurry and ice towels for 30 min prior to cycling in the heat during which time ad libitum ice-slurry per-cooling was allowed. Participants did better with the cooling strategies but the combined approach was no better than the per-cooling intervention alone.

7.3.4 Post-exercise Cooling

Exercise-induced hyperthermia can cause irreversible, and potentially fatal, heat illnesses if treatment is not rapidly administered [90, 91]. Heat illnesses are relatively common [1] and range in severity from muscle cramps to exertional heat stroke. Exertional heat stroke is one of the most common causes of sudden death in athletes [92] and so appropriate prevention methods (e.g. appropriate pre-exercise screening and/or following a successful period of heat acclimation/acclimatisation) should be undertaken [93]. Supine rest in the shade can usually treat minor heat illnesses (additional skin cooling and/or rehydration can be administered if required) but more severe heat illnesses require the rapid reduction of body temperature to

<38.9 °C [93]. Ideally core body temperature should be reduced within 30 min [94]—a period that has been termed the "golden half hour" [95].

Post-exercise cooling interventions such as ice sheets, ice towels, cold showers, forearm immersion, and cooling garments are often used in sporting and military settings due to their ease of application; however, their effectiveness is limited. McDermott et al. [96] classified cooling effectiveness based upon cooling rates and suggested that an *ideal* cooling intervention would cool at a rate of at least 0.155 $°C{\cdot}min^{-1}$ and an *acceptable* one would cool at a rate of 0.078–0.155 $°C{\cdot}min^{-1}$. The aforementioned practical interventions often fall below this acceptable cooling rate. Ice sheet cooling and passive rest offer similar cooling rates (~0.05 $°C{\cdot}min^{-1}$), despite greater reductions in the first 6 min of the 15 min cooling period with the sheet [97]. This rate of cooling is similar to that reported using other practical inventions such as ice-towel application (0.06–0.11 $°C{\cdot}min^{-1}$) [98, 99], cold showers (0.07 $°C{\cdot}min^{-1}$) [97], chemical cold packs (0.02–0.06 $°C{\cdot}min^{-1}$) [100], and forearm or hand-cold-water immersion (0.01–0.04 $°C{\cdot}min^{-1}$) [101–105]. The most effective post-exercise interventions involve whole-body immersion in water. Armstrong et al. [99] directly compared the ice towels with cold-water immersion and reported that both interventions were at least *acceptable* but that the cooling rate was almost twice as fast when using cold-water immersion (0.20 $°C{\cdot}min^{-1}$) compared to the ice towels (0.11 $°C{\cdot}min^{-1}$). Zhang et al. [10] reported that the mean cooling rate of water immersion was 0.08 ± 0.03 $°C{\cdot}min^{-1}$—double that of passive recovery alone (0.04 $°C{\cdot}min^{-1}$) and *acceptable* but not *ideal* [96]. Using the mean cooling rates the calculated time to cool the core body temperature by 1 °C ranges from 4 to 24 min in otherwise healthy, moderately hyperthermic individuals depending on the core body temperature, immersion temperature, and immersion depth (Fig. 7.5 [10]). For extremely

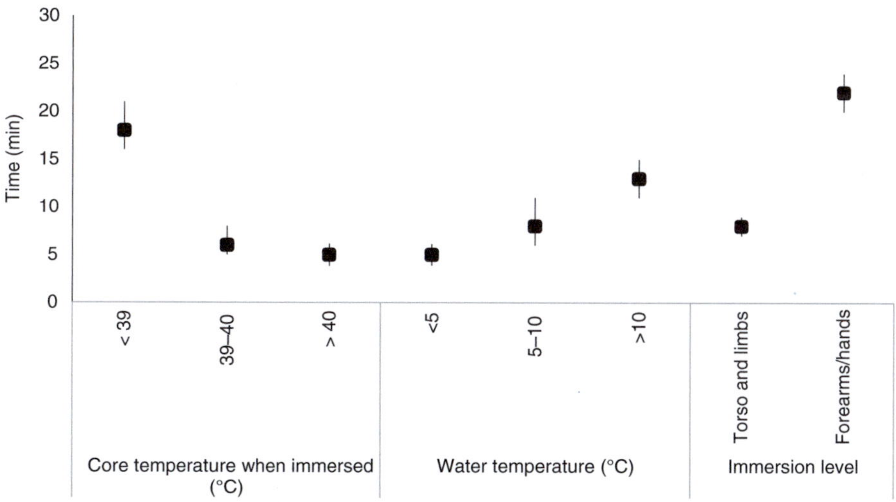

Fig. 7.5 Time taken to cool the core body temperature by 1 °C. Mean, minimum, and maximum durations. Redrawn from data contained in Zhang et al. [10])

hyperthermic individuals the cooling rates observed can be ~20% of the expected rate [91] due to complications such as circulatory failure and/or endogenous heat production and so the rates observed in the field may be lower than those reported in the experimental literature.

It is often suggested that the optimal water temperature for water immersion is "the colder the better" (p. 2469) [10]. Very fast cooling rates have been observed using water of ~2 °C [106–108]; however, warmer water may be just as effective [109]. Cooling rates observed using tarp-cooling (the athlete is placed on a waterproof sheet (i.e. a tarpaulin or tarp) and three of more people hold up the edges to create a bowl-shape before cold water and/or ice is poured into the tarp and over the athlete) are faster with water at 9.2 °C (0.17 °C·min^{-1} [110]) than 2.0 °C (0.14 °C·min^{-1} [111]) and Taylor et al. [109] reported that oesophageal temperature could be rapidly reduced using even warmer water (26 °C). Hyperthermic participants (core body temperature of 39.5 °C) were cooled either using cool air (~23 °C), cool water (14 °C), or temperate water (26 °C) [109]. The resultant cooling rates were 0.10 °C·min^{-1}, 0.88 °C·min^{-1}, and 0.71 °C·min^{-1} for cool air, 14 °C water, and 26 °C, respectively. When converted into estimated time to reach a core body temperature of 37.5 °C the 14 °C (2.2 ± 0.7 min) and 26 °C (2.9 ± 1.2 min) water immersion trials were both about 20 min quicker than the passive control (22.8 ± 16.4 min) [109]. The small differences in cooling times between 14 °C and 26 °C water immersion were subsequently repeated by the same laboratory group (~45 s versus ~54 s) [2] but are much shorter than reported elsewhere with similar water temperatures (20 °C; 17.3 min [107]). The differences appear to be due to the different sites of measurement (Rectal [107] v oesophageal [2, 109]). Rectal cooling rates seem dependent upon water temperature (2 °C = 0.35 °C·min^{-1}; 8 °C = 0.19 °C·min^{-1}; 14 °C = 0.15 °C·min^{-1}; 20 °C = 0.19 °C·min^{-1}); however, cooling rates are similar regardless of water temperature when measured at the oesophagus (0.6–1.0 °C·min^{-1}) [112]. Rectal temperature is sometimes considered impractical in a field setting [113] but it has been successfully used [114] and far more likely to be recorded in the field than oesophageal temperature. When using rectal temperature, it would be prudent to remember that the reduction in oesophageal temperature is likely to exceed the drop recorded rectally, which may lead to an actual state of hypothermia. Proulx et al. [112] reported that when rectal temperature was reduced to 37.7 °C, oesophageal temperature was 33.9 °C and recommended that cooling to rectal temperatures of 38.6 °C and 37.8 °C, respectively, when using water at either 2–8 °C or 14–20 °C would effectively reduce body temperature without causing hypothermia.

Post-exercise water-immersion can be logistically difficult or even impossible and so researchers have investigated approaches with improved practicality. One such intervention is tarp-cooling. Luhring et al. [111] and Hosokawa et al. [110] reported that tarp-assisted cooling with agitated cold (2 °C and 9.2 °C, respectively) water (the agitation was to avoid the formation of a barrier of warm water around the body) was effective at reducing core body temperature (cooling rates = 0.14 °C·min^{-1} and 0.17 °C·min^{-1}, respectively). While more practical than full cold-water immersion, tarp-cooling still requires water immersion, and this may be problematic.

Webster et al. [19] reported that ice vests can reduce core temperature following exercise; however, there appears to be little difference between control (~0.03 °C·min^{-1}) and vest (~0.03–0.04 °C·min^{-1}) and the cooling rates observed were similar to those reported by Maroni et al. [115] following 30 min of passive rest with (0.05 °C·min^{-1}) and without (0.04 °C·min^{-1}) an ice vest. The cooling rate provided by ice vests do not reach the *acceptable* rate classified by McDermott et al. [96] and this is fairly typical of non-immersion post-cooling. As highlighted above, ice-towel application may or may not offer *acceptable* rates [98, 99] but cold showers [97], chemical cold pack application [100], and forearm or hand water immersion [101–105] do not. Most post-exercise cooling interventions can elicit beneficial physiological and/or perceptual changes and are more effective than no cooling; however, water immersion is the most effective method of post-exercise cooling for the reduction in core body temperature in hyperthermic individuals [10, 96, 99, 116].

7.4 Summary

Cooling prior to exercise in the heat using cold-water immersion, ice packs, and ice vests can increase the heat storage capacity and reduce the physiological and perceptual strain experienced. The effectiveness of pre-cooling is dependent upon the heat strain experienced and the magnitude of cooling provided. Pre-cooling the torso can improve sprint cycle performance but when the working muscles are pre-cooled peak power output is impaired. In longer, more thermally challenging events, pre-cooling is more often beneficial and the greatest pre-cooling benefits are observed when used prior to prolonged exercise in the heat.

The benefits of pre-cooling are often lost or diminished during exercise and so cooling during exercise (per-cooling) is often adopted. Although there are a number of practical considerations that can restrict its use (e.g. rules and regulations of the sport, excess weight, and skin irritation), per-cooling can improve exercise performance and capacity in the heat with and without physiological alterations. Per-cooling can be practically applied using ice vests, neck cooling collars, and cold fluid ingestion. Combining pre- and per-cooling does not appear to be more effective than pre- or per-cooling alone.

Exercise-induced hyperthermia can cause irreversible, and potentially fatal, heat illnesses if treatment is not rapidly administered and so post-exercise cooling should be rapidly administered if required. Severe heat illnesses require the rapid reduction of body temperature to <38.9 °C within 30 min. Interventions such as ice sheets, ice towels, cold showers, forearm immersion, and cooling garments are often used but the most effective post-exercise interventions involve whole-body immersion in water. The water does not need to be very cold—26 °C water can effectively lower core body temperature post-exercise.

7.5 Recommendations

- Pre-cooling is advantageous to subsequent exercise in the heat but direct cooling of locomotor muscles should be avoided.
- Pre-cooling with garments that do not cool the locomotor muscles (e.g. ice vests) during warm-up bouts may be beneficial.
- Practical pre-cooling strategies (e.g. ice vests) can be beneficial, but water immersion causes larger reductions in actual and perceived thermal strain. Cold-water immersion is disadvantageous for explosive activities if muscle temperature is reduced.
- Per-cooling may disrupt pacing strategies and/or impair the ability to self-detect thermal strain, so close physiological monitoring of the athlete should take place if cooling during exercise in the heat. Ideally, this should involve core body temperature assessment but heart rate monitoring is more practical and also offers useful insight.
- Athletes should be familiarised with pre- and per-cooling strategies prior to use.
- Hyperthermia requires rapid treatment. Whole-body water immersion is the most effective mode of cooling and the temperature of the water need not be very cold, but should be cold enough (< 20 °C) to reduce body temperature at a rate of at least 0.078 $°C \cdot min^{-1}$.

References

1. Periard JD, Racinais S, Timpka T, Dahlstrom O, Spreco A, Jacobsson J, et al. Strategies and factors associated with preparing for competing in the heat: a cohort study at the 2015 IAAF world athletics championships. Br J Sports Med. 2017;51(4):264–70.
2. Caldwell JN, van den Heuvel AMJ, Kerry P, Clark MJ, Peoples GE, Taylor NAS. A vascular mechanism to explain thermally mediated variations in deep-body cooling rates during the immersion of profoundly hyperthermic individuals. Exp Physiol. 2018;103(4):512–22.
3. Gonzalez-Alonso J, Teller C, Andersen SL, Jensen FB, Hyldig T, Nielsen B. Influence of body temperature on the development of fatigue during prolonged exercise in the heat. J Appl Physiol. 1999;86(3):1032–9.
4. Booth J, Marino FE, Ward JJ. Improved running performance in hot humid conditions following whole body precooling. Med Sci Sports Exerc. 1997;29(7):943–9.
5. Siegel R, Mate J, Watson G, Nosaka K, Laursen PB. Pre-cooling with ice slurry ingestion leads to similar run times to exhaustion in the heat as cold water immersion. J Sports Sci. 2012;30(2):155–65.
6. Sleivert GG, Cotter JD, Roberts WS, Febbraio MA. The influence of whole-body vs. torso pre-cooling on physiological strain and performance of high-intensity exercise in the heat. Comp Biochem Physiol A Mol Integr Physiol. 2001;128(4):657–66.
7. Vaile J, Halson S, Gill N, Dawson B. Effect of cold water immersion on repeat cycling performance and thermoregulation in the heat. J Sports Sci. 2008;26(5):431–40.
8. Tyler CJ, Reeve T, Cheung SS. Cold-induced vasodilation during single digit immersion in 0 degrees C and 8 degrees C water in men and women. PLoS One. 2015;10(4):e0122592.
9. Sendowski I, Savourey G, Besnard Y, Bittel J. Cold induced vasodilatation and cardiovascular responses in humans during cold water immersion of various upper limb areas. Eur J Appl Physiol Occup Physiol. 1997;75(6):471–7.

10. Zhang Y, Davis JK, Casa DJ, Bishop PA. Optimizing cold water immersion for exercise-induced hyperthermia: a meta-analysis. Med Sci Sports Exerc. 2015;47(11):2464–72.
11. Racinais S, Blonc S, Oksa J, Hue O. Does the diurnal increase in central temperature interact with pre-cooling or passive warm-up of the leg? J Sci Med Sport. 2009;12(1):97–100.
12. Goosey-Tolfrey V, Swainson M, Boyd C, Atkinson G, Tolfrey K. The effectiveness of hand cooling at reducing exercise-induced hyperthermia and improving distance-race performance in wheelchair and able-bodied athletes. J Appl Physiol (1985). 2008;105(1):37–43.
13. Marsh D, Sleivert G. Effect of precooling on high intensity cycling performance. Br J Sports Med. 1999;33(6):393–7.
14. Cuttell SA, Kiri V, Tyler C. A comparison of 2 practical cooling methods on cycling capacity in the heat. J Athl Train. 2016;51(7):525–32.
15. Luomala MJ, Oksa J, Salmi JA, Linnamo V, Holmer I, Smolander J, et al. Adding a cooling vest during cycling improves performance in warm and humid conditions. J Therm Biol. 2012;37:47–55.
16. Arngrimsson SA, Petitt DS, Stueck MG, Jorgensen DK, Cureton KJ. Cooling vest worn during active warm-up improves 5-km run performance in the heat. J Appl Physiol. 2004;96(5):1867–74.
17. Clarke ND, Maclaren DP, Reilly T, Drust B. Carbohydrate ingestion and pre-cooling improves exercise capacity following soccer-specific intermittent exercise performed in the heat. Eur J Appl Physiol. 2011;111(7):1447–55.
18. Price MJ, Boyd C, Goosey-Tolfrey VL. The physiological effects of pre-event and midevent cooling during intermittent running in the heat in elite female soccer players. Appl Physiol Nutr Metab. 2009;34(5):942–9.
19. Webster J, Holland EJ, Sleiverts G, Laing RM, Niven BE. A light-weight cooling vest enhances performance of athletes in the heat. Ergonomics. 2005;48(7):821–37.
20. Bongers CC, Thijssen DH, Veltmeijer MT, Hopman MT, Eijsvogels TM. Precooling and percooling (cooling during exercise) both improve performance in the heat: a meta-analytical review. Br J Sports Med. 2015;49(6):377–84.
21. Tyler CJ, Sunderland C, Cheung SS. The effect of cooling prior to and during exercise on exercise performance and capacity in the heat: a meta-analysis. Br J Sports Med. 2015;49(1):7–13.
22. Eijsvogels TM, Bongers CC, Veltmeijer MT, Moen MH, Hopman M. Cooling during exercise in temperate conditions: impact on performance and thermoregulation. Int J Sports Med. 2014;35(10):840–6.
23. Bogerd N, Perret C, Bogerd CP, Rossi RM, Daanen HA. The effect of pre-cooling intensity on cooling efficiency and exercise performance. J Sports Sci. 2010;28(7):771–9.
24. Cureton KJ, Sparling PB, Evans BW, Johnson SM, Kong UD, Purvis JW. Effect of experimental alterations in excess weight on aerobic capacity and distance running performance. Med Sci Sports. 1978;10(3):194–9.
25. Kenny GP, Schissler AR, Stapleton J, Piamonte M, Binder K, Lynn A, et al. Ice cooling vest on tolerance for exercise under uncompensable heat stress. J Occup Environ Hyg. 2011;8(8):484–91.
26. Gordon NF, Bogdanffy GM, Wilkinson J. Effect of a practical neck cooling device on core temperature during exercise. Med Sci Sports Exerc. 1990;22(2):245–9.
27. Shvartz E. Effect of neck versus chest cooling on responses to work in heat. J Appl Physiol. 1976;40(5):668–72.
28. Tikuisis P, Meunier P, Jubenville CE. Human body surface area: measurement and prediction using three dimensional body scans. Eur J Appl Physiol. 2001;85(3–4):264–71.
29. Cotter JD, Taylor NA. The distribution of cutaneous sudomotor and alliesthesial thermosensitivity in mildly heat-stressed humans: an open-loop approach. J Physiol. 2005;565(Pt 1):335–45.
30. Ando S, Komiyama T, Sudo M, Kiyonaga A, Tanaka H, Higaki Y. The effects of temporal neck cooling on cognitive function during strenuous exercise in a hot environment: a pilot study. BMC Res Notes. 2015;8:202.

31. Bulbulian R, Shapiro R, Murphy M, Levenhagen D. Effectiveness of a commercial head-neck cooling device. J Strength Cond Res. 1999;13(3):198–205.
32. Lee JK, Koh AC, Koh SX, Liu GJ, Nio AQ, Fan PW. Neck cooling and cognitive performance following exercise-induced hyperthermia. Eur J Appl Physiol. 2014;114(2):375–84.
33. Sunderland C, Stevens R, Everson B, Tyler CJ. Neck-cooling improves repeated sprint performance in the heat. Front Physiol. 2015;6:314.
34. Tyler CJ, Wild P, Sunderland C. Practical neck cooling and time-trial running performance in a hot environment. Eur J Appl Physiol. 2010;110(5):1063–74.
35. Tyler CJ, Sunderland C. Cooling the neck region during exercise in the heat. J Athl Train. 2011;46(1):61–8.
36. Tyler CJ, Sunderland C. Neck cooling and running performance in the heat: single versus repeated application. Med Sci Sports Exerc. 2011;43(12):2388–95.
37. Nunneley SA, Troutman SJ Jr, Webb P. Head cooling in work and heat stress. Aerosp Med. 1971;42(1):64–8.
38. Junge N, Jorgensen R, Flouris AD, Nybo L. Prolonged self-paced exercise in the heat - environmental factors affecting performance. Temperature (Austin). 2016;3(4):539–48.
39. Morrison SA, Cheung S, Cotter JD. Importance of airflow for physiologic and ergogenic effects of precooling. J Athl Train. 2014;49(5):632–9.
40. Saunders AG, Dugas JP, Tucker R, Lambert MI, Noakes TD. The effects of different air velocities on heat storage and body temperature in humans cycling in a hot, humid environment. Acta Physiol Scand. 2005;183(3):241–55.
41. Adams WC, Mack GW, Langhans GW, Nadel ER. Effects of varied air velocity on sweating and evaporative rates during exercise. J Appl Physiol (1985). 1992;73(6):2668–74.
42. Mora-Rodriguez R, del CJ, Aguado-Jimenez R, Estevez E. Separate and combined effects of airflow and rehydration during exercise in the heat. Med Sci Sports Exerc. 2007;39(10):1720–6.
43. Otani H, Kaya M, Tamaki A, Watson P, Maughan RJ. Air velocity influences thermoregulation and endurance exercise capacity in the heat. Appl Physiol Nutr Metab. 2018;43(2):131–8.
44. Schranner D, Scherer L, Lynch GP, Korder S, Brotherhood JR, Pluim BM, et al. In-play cooling interventions for simulated match-play tennis in hot/humid conditions. Med Sci Sports Exerc. 2017;49(5):991–8.
45. Barwood MJ, Corbett J, White D, James J. Early change in thermal perception is not a driver of anticipatory exercise pacing in the heat. Br J Sports Med. 2012;46(13):936–42.
46. Barwood MJ, Corbett J, White DK. Spraying with 0.20% L-menthol does not enhance 5 km running performance in the heat in untrained runners. J Sports Med Phys Fitness. 2014;54(5):595–604.
47. Barwood MJ, Corbett J, Thomas K, Twentyman P. Relieving thermal discomfort: effects of sprayed L-menthol on perception, performance, and time trial cycling in the heat. Scand J Med Sci Sports. 2015;25(Suppl 1):211–8.
48. Gillis DJ, House JR, Tipton MJ. The influence of menthol on thermoregulation and perception during exercise in warm, humid conditions. Eur J Appl Physiol. 2010;110(3):609–18.
49. Gillis DJ, Barwood MJ, Newton PS, House JR, Tipton MJ. The influence of a menthol and ethanol soaked garment on human temperature regulation and perception during exercise and rest in warm, humid conditions. J Therm Biol. 2016;58:99–105.
50. Schlader ZJ, Simmons SE, Stannard SR, Mundel T. The independent roles of temperature and thermal perception in the control of human thermoregulatory behavior. Physiol Behav. 2011;103(2):217–24.
51. Kounalakis SN, Botonis PG, Koskolou MD, Geladas ND. The effect of menthol application to the skin on sweating rate response during exercise in swimmers and controls. Eur J Appl Physiol. 2010;109(2):183–9.
52. Johnson CD, Melanaphy D, Purse A, Stokesberry SA, Dickson P, Zholos AV. Transient receptor potential melastatin 8 channel involvement in the regulation of vascular tone. Am J Physiol Heart Circ Physiol. 2009;296(6):H1868–77.
53. Yosipovitch G, Szolar C, Hui XY, Maibach H. Effect of topically applied menthol on thermal, pain and itch sensations and biophysical properties of the skin. Arch Dermatol Res. 1996;288(5–6):245–8.

54. Tajino K, Matsumura K, Kosada K, Shibakusa T, Inoue K, Fushiki T, et al. Application of menthol to the skin of whole trunk in mice induces autonomic and behavioral heat-gain responses. Am J Physiol Regul Integr Comp Physiol. 2007;293(5):R2128–35.
55. Hasegawa H, Takatori T, Komura T, Yamasaki M. Combined effects of pre-cooling and water ingestion on thermoregulation and physical capacity during exercise in a hot environment. J Sports Sci. 2006;24(1):3–9.
56. Lee JK, Shirreffs SM, Maughan RJ. Cold drink ingestion improves exercise endurance capacity in the heat. Med Sci Sports Exerc. 2008;40(9):1637–44.
57. Mundel T, King J, Collacott E, Jones DA. Drink temperature influences fluid intake and endurance capacity in men during exercise in a hot, dry environment. Exp Physiol. 2006;91(5):925–33.
58. Ihsan M, Landers G, Brearley M, Peeling P. Beneficial effects of ice ingestion as a precooling strategy on 40-km cycling time-trial performance. Int J Sports Physiol Perform. 2010;5(2):140–51.
59. Stanley J, Leveritt M, Peake JM. Thermoregulatory responses to ice-slush beverage ingestion and exercise in the heat. Eur J Appl Physiol. 2010;110(6):1163–73.
60. Zimmermann M, Landers G, Wallman KE, Saldaris J. The effects of crushed ice ingestion prior to steady state exercise in the heat. Int J Sport Nutr Exerc Metab. 2017;27(3):220–7.
61. Snipe RMJ, Costa RJS. Does the temperature of water ingested during exertional-heat stress influence gastrointestinal injury, symptoms, and systemic inflammatory profile? J Sci Med Sport. 2018;21(8):771–6.
62. Burdon C, O'Connor H, Gifford J, Shirreffs S, Chapman P, Johnson N. Effect of drink temperature on core temperature and endurance cycling performance in warm, humid conditions. J Sports Sci. 2010;28(11):1147–56.
63. Burdon CA, Hoon MW, Johnson NA, Chapman PG, O'Connor HT. The effect of ice slushy ingestion and mouthwash on thermoregulation and endurance performance in the heat. Int J Sport Nutr Exerc Metab. 2013;23(5):458–69.
64. Lee JK, Maughan RJ, Shirreffs SM. The influence of serial feeding of drinks at different temperatures on thermoregulatory responses during cycling. J Sports Sci. 2008;26(6):583–90.
65. Morris NB, Coombs G, Jay O. Ice slurry ingestion leads to a lower net heat loss during exercise in the heat. Med Sci Sports Exerc. 2016;48(1):114–22.
66. Jay O, Morris NB. Does cold water or ice slurry ingestion during exercise elicit a net body cooling effect in the heat? Sports Med. 2018;48(Suppl 1):17–29.
67. Stevens CJ, Mauger AR, Hassmen P, Taylor L. Endurance performance is influenced by perceptions of pain and temperature: theory, applications and safety considerations. Sports Med. 2018;48(3):525–37.
68. Mundel T, Jones DA. The effects of swilling an L(−)-menthol solution during exercise in the heat. Eur J Appl Physiol. 2010;109(1):59–65.
69. Stevens CJ, Thoseby B, Sculley DV, Callister R, Taylor L, Dascombe BJ. Running performance and thermal sensation in the heat are improved with menthol mouth rinse but not ice slurry ingestion. Scand J Med Sci Sports. 2016;26(10):1209–16.
70. Stevens CJ, Bennett KJ, Sculley DV, Callister R, Taylor L, Dascombe BJ. A comparison of mixed-method cooling interventions on preloaded running performance in the heat. J Strength Cond Res. 2017;31(3):620–9.
71. Riera F, Trong TT, Sinnapah S, Hue O. Physical and perceptual cooling with beverages to increase cycle performance in a tropical climate. PLoS One. 2014;9(8):e103718.
72. Tran TT, Riera F, Rinaldi K, Briki W, Hue O. Ingestion of a cold temperature/menthol beverage increases outdoor exercise performance in a hot, humid environment. PLoS One. 2015;10(4):e0123815.
73. Pires W, Veneroso CE, Wanner SP, Pacheco DAS, Vaz GC, Amorim FT, et al. Association between exercise-induced hyperthermia and intestinal permeability: a systematic review. Sports Med. 2017;47(7):1389–403.
74. Schulze E, Daanen HA, Levels K, Casadio JR, Plews DJ, Kilding AE, et al. Effect of thermal state and thermal comfort on cycling performance in the heat. Int J Sports Physiol Perform. 2015;10(5):655–63.

75. Ross ML, Garvican LA, Jeacocke NA, Laursen PB, Abbiss CR, Martin DT, et al. Novel precooling strategy enhances time trial cycling in the heat. Med Sci Sports Exerc. 2011;43(1):123–33.
76. Brade C, Dawson B, Wallman K. Effects of different precooling techniques on repeat sprint ability in team sport athletes. Eur J Sport Sci. 2014;14(Suppl 1):S84–91.
77. Duffield R, Green R, Castle P, Maxwell N. Precooling can prevent the reduction of self-paced exercise intensity in the heat. Med Sci Sports Exerc. 2010;42(3):577–84.
78. Kay D, Taaffe DR, Marino FE. Whole-body pre-cooling and heat storage during self-paced cycling performance in warm humid conditions. J Sports Sci. 1999;17:937–44.
79. Quod MJ, Martin DT, Laursen PB, Gardner AS, Halson SL, Marino FE, et al. Practical precooling: effect on cycling time trial performance in warm conditions. J Sports Sci. 2008;26(14):1477–87.
80. Cotter JD, Sleivert GG, Roberts WS, Febbraio MA. Effect of pre-cooling, with and without thigh cooling, on strain and endurance exercise performance in the heat. Comp Biochem Physiol A Mol Integr Physiol. 2001;128(4):667–77.
81. Mitchell JB, McFarlin BK, Dugas JP. The effect of pre-exercise cooling on high intensity running performance in the heat. Int J Sports Med. 2003;24(2):118–24.
82. Yeargin SW, Casa DJ, McClung JM, Knight JC, Healey JC, Goss PJ, et al. Body cooling between two bouts of exercise in the heat enhances subsequent performance. J Strength Cond Res. 2006;20(2):383–9.
83. Crowley GC, Garg A, Lohn MS, Van SN, Wade AJ. Effects of cooling the legs on performance in a standard Wingate anaerobic power test. Br J Sports Med. 1991;25(4):200–3.
84. Bigland-Ritchie B, Thomas CK, Rice CL, Howarth JV, Woods JJ. Muscle temperature, contractile speed, and motoneuron firing rates during human voluntary contractions. J Appl Physiol. 1992;73(6):2457–61.
85. Febbraio MA, Snow RJ, Stathis CG, Hargreaves M, Carey MF. Effect of heat stress on muscle energy metabolism during exercise. J Appl Physiol. 1994;77(6):2827–31.
86. Minett GM, Duffield R, Marino FE, Portus M. Volume-dependent response of precooling for intermittent-sprint exercise in the heat. Med Sci Sports Exerc. 2011;43(9):1760–9.
87. Minett GM, Duffield R, Marino FE, Portus M. Duration-dependant response of mixed-method pre-cooling for intermittent-sprint exercise in the heat. Eur J Appl Physiol. 2012;112(10):3655–66.
88. Minett GM, Duffield R, Kellett A, Portus M. Mixed-method pre-cooling reduces physiological demand without improving performance of medium-fast bowling in the heat. J Sports Sci. 2012;30(9):907–15.
89. Bongers CC, Hopman MT, Eijsvogels TM. Cooling interventions for athletes: an overview of effectiveness, physiological mechanisms, and practical considerations. Temperature (Austin). 2017;4(1):60–78.
90. Epstein Y, Roberts WO, Golan R, Heled Y, Sorkine P, Halpern P. Sepsis, septic shock, and fatal exertional heat stroke. Curr Sports Med Rep. 2015;14(1):64–9.
91. Rae DE, Knobel GJ, Mann T, Swart J, Tucker R, Noakes TD. Heatstroke during endurance exercise: is there evidence for excessive endothermy? Med Sci Sports Exerc. 2008;40(7):1193–204.
92. Kerr ZY, Casa DJ, Marshall SW, Comstock RD. Epidemiology of exertional heat illness among U.S. high school athletes. Am J Prev Med. 2013;44(1):8–14.
93. Casa DJ, DeMartini JK, Bergeron MF, Csillan D, Eichner ER, Lopez RM, et al. National Athletic Trainers' association position statement: exertional heat illnesses. J Athl Train. 2015;50(9):986–1000.
94. Adams WM, Hosokawa Y, Casa DJ. The timing of exertional heat stroke survival starts prior to collapse. Curr Sports Med Rep. 2015;14(4):273–4.
95. Casa DJ, Armstrong LE, Kenny GP, O'Connor FG, Huggins RA. Exertional heat stroke: new concepts regarding cause and care. Curr Sports Med Rep. 2012;11(3):115–23.
96. McDermott BP, Casa DJ, Ganio MS, Lopez RM, Yeargin SW, Armstrong LE, et al. Acute whole-body cooling for exercise-induced hyperthermia: a systematic review. J Athl Train. 2009;44(1):84–93.

97. Butts CL, Spisla DL, Adams JD, Smith CR, Paulsen KM, Caldwell AR, et al. Effectiveness of ice-sheet cooling following exertional hyperthermia. Mil Med. 2017;182(9):e1951–7.
98. DeMartini JK, Ranalli GF, Casa DJ, Lopez RM, Ganio MS, Stearns RL, et al. Comparison of body cooling methods on physiological and perceptual measures of mildly hyperthermic athletes. J Strength Cond Res. 2011;25(8):2065–74.
99. Armstrong LE, Crago AE, Adams R, Roberts WO, Maresh CM. Whole-body cooling of hyperthermic runners: comparison of two field therapies. Am J Emerg Med. 1996;14(4):355–8.
100. Lissoway JB, Lipman GS, Grahn DA, Cao VH, Shaheen M, Phan S, et al. Novel application of chemical cold packs for treatment of exercise-induced hyperthermia: a randomized controlled trial. Wilderness Environ Med. 2015;26(2):173–9.
101. Adams EL, Vandermark LW, Pryor JL, Pryor RR, VanScoy RM, Denegar CR, et al. Effects of heat acclimation on hand cooling efficacy following exercise in the heat. J Sports Sci. 2017;35(9):828–34.
102. Adams WM, Hosokawa Y, Adams EL, Belval LN, Huggins RA, Casa DJ. Reduction in body temperature using hand cooling versus passive rest after exercise in the heat. J Sci Med Sport. 2016;19(11):936–40.
103. Grahn DA, Dillon JL, Heller HC. Heat loss through the glabrous skin surfaces of heavily insulated, heat-stressed individuals. J Biomech Eng. 2009;131(7):071005.
104. Hostler D, Reis SE, Bednez JC, Kerin S, Suyama J. Comparison of active cooling devices with passive cooling for rehabilitation of firefighters performing exercise in thermal protective clothing: a report from the Fireground rehab evaluation (FIRE) trial. Prehosp Emerg Care. 2010;14(3):300–9.
105. Kuennen MR, Gillum TL, Amorim FT, Kwon YS, Schneider SM. Palm cooling to reduce heat strain in subjects during simulated armoured vehicle transport. Eur J Appl Physiol. 2010;108(6):1217–23.
106. Friesen BJ, Carter MR, Poirier MP, Kenny GP. Water immersion in the treatment of exertional hyperthermia: physical determinants. Med Sci Sports Exerc. 2014;46(9):1727–35.
107. Proulx CI, Ducharme MB, Kenny GP. Effect of water temperature on cooling efficiency during hyperthermia in humans. J Appl Physiol (1985). 2003;94(4):1317–23.
108. Tan PM, Teo EY, Ali NB, Ang BC, Iskandar I, Law LY, et al. Evaluation of various cooling systems after exercise-induced hyperthermia. J Athl Train. 2017;52(2):108–16.
109. Taylor NA, Caldwell JN, Van den Heuvel AM, Patterson MJ. To cool, but not too cool: that is the question--immersion cooling for hyperthermia. Med Sci Sports Exerc. 2008;40(11): 1962–9.
110. Hosokawa Y, Adams WM, Belval LN, Vandermark LW, Casa DJ. Tarp-assisted cooling as a method of whole-body cooling in hyperthermic individuals. Ann Emerg Med. 2017;69(3):347–52.
111. Luhring KE, Butts CL, Smith CR, Bonacci JA, Ylanan RC, Ganio MS, et al. Cooling effectiveness of a modified cold-water immersion method after exercise-induced hyperthermia. J Athl Train. 2016;51(11):946–51.
112. Proulx CI, Ducharme MB, Kenny GP. Safe cooling limits from exercise-induced hyperthermia. Eur J Appl Physiol. 2006;96(4):434–45.
113. Mazerolle SM, Scruggs IC, Casa DJ, Burton LJ, McDermott BP, Armstrong LE, et al. Current knowledge, attitudes, and practices of certified athletic trainers regarding recognition and treatment of exertional heat stroke. J Athl Train. 2010;45(2):170–80.
114. DeMartini JK, Casa DJ, Stearns R, Belval L, Crago A, Davis R, et al. Effectiveness of cold water immersion in the treatment of exertional heat stroke at the Falmouth road race. Med Sci Sports Exerc. 2015;47(2):240–5.
115. Maroni T, Dawson B, Barnett K, Guelfi K, Brade C, Naylor L, et al. Effectiveness of hand cooling and a cooling jacket on post-exercise cooling rates in hyperthermic athletes. Eur J Sport Sci. 2018;18(4):1–9.
116. Casa DJ, McDermott BP, Lee EC, Yeargin SW, Armstrong LE, Maresh CM. Cold water immersion: the gold standard for exertional heatstroke treatment. Exerc Sport Sci Rev. 2007;35(3):141–9.

Heat Acclimation

8

Sébastien Racinais, Michael Sawka, Hein Daanen, and Julien D. Périard

8.1 Introduction

Acute heat exposure impairs aerobic exercise capacity via thermoregulatory mediated cardiovascular adjustments, hyperthermia-induced skeletal muscle metabolism alterations, and central nervous system perturbations [1]. However, repeated exposures to heat allow for a physiological conditioning known as heat acclimation when exposed to hot rooms, saunas, or baths, and acclimatisation when exposed to naturally hot areas/environments [2]. These heat exposures induce numerous integrated physiological adaptations that improve performance in the heat and reduce the risk of serious heat illness [3]. This chapter covers the adaptations related to heat acclimation, including the increase in sweat rate and plasma volume, which contribute to decrease in heart rate and core temperature during exercise at a given intensity. It also covers the resulting benefit for exercise capacity in the heat, along with the potential implication for exercising in cooler environments. Lastly, the chapter covers the methods and kinetics of induction, as well as the kinetics of decay. The term heat acclimation will be used as a generic term for both acclimation and acclimatisation.

S. Racinais (✉)
Athlete Health and Performance Research Centre, Aspetar, Doha, Qatar
e-mail: sebastien.racinais@aspetar.com

M. Sawka
School of Applied Physiology, Georgia Institute of Technology, Atlanta, GA, USA

H. Daanen
Department of Human Movement Sciences, Faculty of Behavioural and Movement Sciences, Vrije Universiteit, Amsterdam, The Netherlands

J. D. Périard
University of Canberra Research Institute for Sport and Exercise (UCRISE), Bruce, ACT, Australia

J. D. Périard, S. Racinais (eds.), *Heat Stress in Sport and Exercise*,
https://doi.org/10.1007/978-3-319-93515-7_8

8.1.1 Historical Perspectives

In one of the first reports on human heat adaptations in 1768, Lind indicated that when relocating to East and West Indian climates, Europeans were at first affected by hot environments but, by length of time, habituated and lived comfortably [4]. In his book 'An Essay on Diseases Incidental to Europeans in Hot Climates', Lind did not yet employ the terminology 'acclimation', but recommended 'a plan for all newly arrived Europeans'. This included behavioural strategies such a reducing workload, rescheduling work to avoid the heat of the day, and using shelters. Lind also mentioned that there were probably blood adaptations allowing Europeans to 'enjoy a pretty good state of health' once adapted.

Despite the early report from Lind on adaptations to hot environments [4], research pertaining to heat acclimation only gained importance in the early twentieth century, likely in response to requirements related to the industrial revolution. Within this context, the mining industry published some of the early discoveries relating to heat acclimation. For example, it was described in 1922 that trained miners sweated more than new workers in a hot mine (4.5–8.5 L versus 2.5 L per 5 h work shift) [5]. Some mines therefore initiated acclimation programs for new workers as early as 1926 [6]. In the quest to optimise work efficiency and safety in the mining industry, Dreosti [7] performed one of the first scientific acclimation studies. He demonstrated that within 14 days of heat acclimation, heart rate was lowered and sweat rate increased, enhancing work performance during a rock-shovelling test, with the progressive beneficial effects being maintained for ~1 month [8]. Dreosti [7] also reported that resting oral temperature was lower in acclimatised than un-acclimatised workers and that even a single previous exposure was sufficient to attenuate the rise in oral temperature in hot environments. Subsequent observations confirmed that acclimatised mine workers had a lower temperature and heart rate (after work) and an increased sweat loss compared with their un-acclimatised counterparts; a finding that, in association with an improved ability to work, was considered characteristic of heat acclimatisation [9]. It was also noted that the rate of heat acclimatisation seemed relatively rapid, as clear improvements in working capacity occurred in only 3 days [9]. Interestingly, miners originating from hot regions did not appear to have a usually high heat tolerance [10].

8.2 Physiological Adaptations

8.2.1 Sweat Rate

An improved sweat rate response is considered the hallmark indicator of heat acclimation. The increase in sweat rate allows for increased evaporative cooling in environmental conditions with low vapour pressure gradients [11]. Most of the adaptive increase in sweat rate occurs in 3–4 days [11, 12] (Fig. 8.1). In essence, sweating

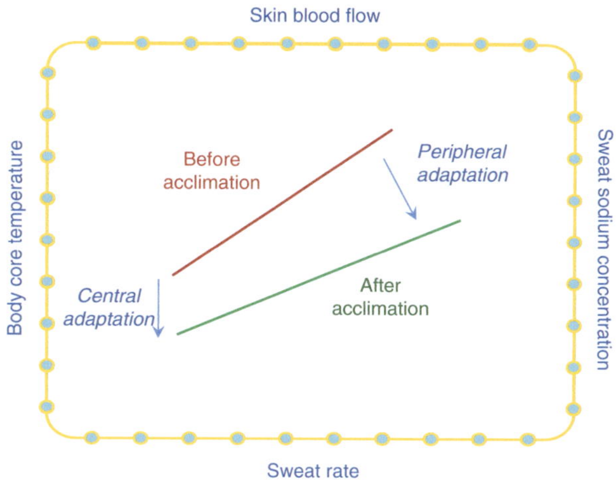

Fig. 8.1 Heat acclimation lowers the core temperate threshold for sweating and increases the rate and sensitivity of response. Heat acclimation decreases also sweat sodium concentration. Based on Periard et al. [3]

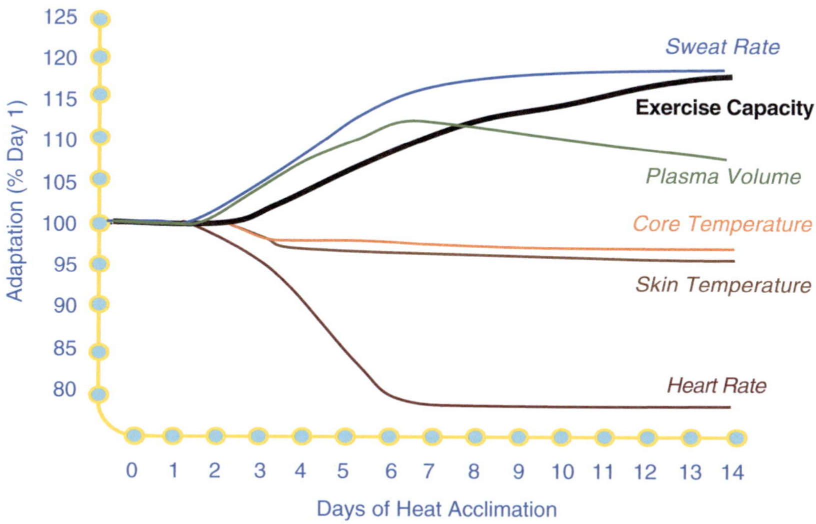

Fig. 8.2 Time course of induction in human adaptations to heat stress. Based on Periard et al. [3]

begins at a lower body temperature and increases to a greater amount for a given body temperature elevation after heat acclimation [13] (Fig. 8.2). In tropical climates, sweat rate is lower and as such less wasteful, allowing for secreted sweat to evaporate and cool the skin [14]. Sweat sodium concentration also decreases with heat acclimation for a given sweating rate [15]. The mechanisms behind the increase in sweat rate and the decrease in sweat sodium concentration with acclimation are reasonably well understood and include neurological and endocrine adaptations [16]. The decrease in sweat sodium concentration can start occurring in 2 days only and is thereafter linear during the first week of heat acclimation [17].

These sweating adaptations are supported by changes in cutaneous vasodilation with an earlier and greater increase in skin blood flow for a given body temperature after heat acclimation [12] (Fig. 8.2).

8.2.2 Blood Volume

Another commonly reported adaptation with heat acclimation is an increase in blood volume, and more specifically plasma volume. Large alterations in blood volume in response to climatic changes have been reported as early as 1922 [18]. In 1940, Bazett et al. [19] published a comprehensive description of the haematological adaptations related to acclimation based on 4 studies. Using various tracers to quantify blood volume, their work showed that heat acclimation induces an increase in both plasma volume and total circulating haemoglobin. However, the plasma volume expansion occurred more rapidly than the increase in haemoglobin, triggering a temporary decrease in haemoglobin concentration and haematocrit [19]. The increase in plasma volume became a standard observation reported in most heat acclimation studies from the 1950s [20], and typically occurs after 3–4 days of heat exposure [21, 22] (Fig. 8.1). Plasma volume expansion is particularly variable between individual with values ranging from 3% to 27% [23–27], whereas erythrocyte volume is considered to remain unchanged [28].

The expansion of plasma volume in later studies was reported to be a transient phenomenon [29, 30], peaking around the fifth day. However, recent studies have suggested that plasma volume could remain expanded if the adaptation stimulus was maintained constant by clamping core temperature during acclimation [24, 25]. Plasma volume maintenance may be facilitated by the oncotic effect of an increase in intravascular protein content [21, 30, 31], potentially due to an increase in protein synthesis [32, 33], a reduction in protein loss through the cutaneous capillaries in response to an acclimation-induced decrease in skin blood flow [34], and a reduction in the permeability of cutaneous capillaries to large molecules [35, 36]. Plasma volume increase and maintenance can also be facilitated by the increase in extracellular fluid due to sodium preservation [24, 25, 37]. Notwithstanding, whether plasma volume remains expanded when the stimulus for adaptation is constant requires further investigation.

Functionally, the expansion in plasma volume may improve vascular filling pressure to support cardiovascular stability (i.e. increased stroke volume and arterial blood pressure) [21], as well as increase the specific heat of blood [38], thus improving heat transfer from the core to the skin [16]. However, an acute artificial increase in plasma volume does not appear to improve thermoregulatory control [22, 39] and, therefore, the importance of plasma volume expansion in heat acclimation is still debated. Even if it is probably not the main physiological mechanism improving exercise capacity in the heat, plasma volume changes might still be a marker of short-term acclimation as the changes in haematocrit during a heat-response test appear to correlate to the changes in physical performance in the heat [40, 41] (Fig. 8.3).

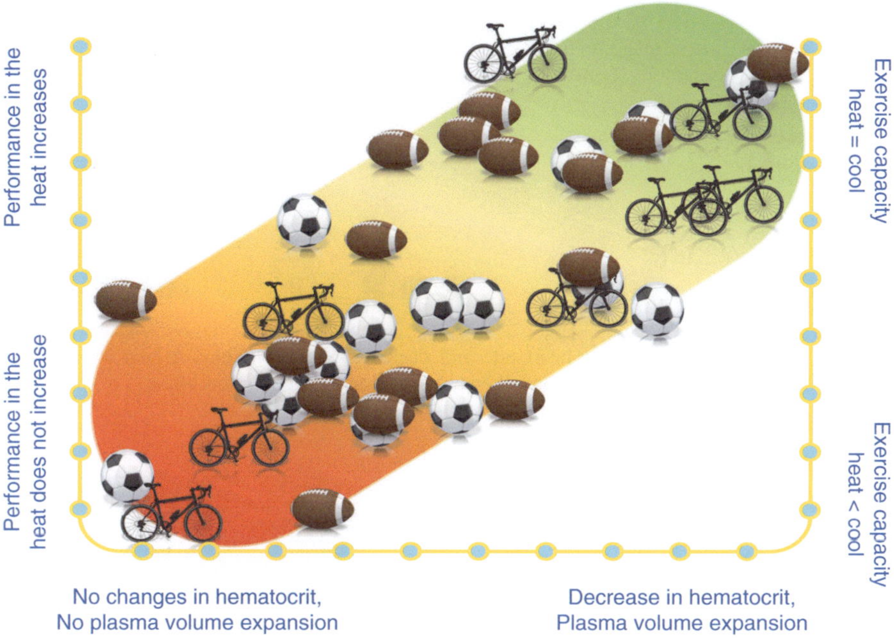

Fig. 8.3 Relationship between haematological changes and changes in exercise capacity in the heat after short-term heat acclimation. Symbols represent individual data from soccer players [40], Australian football players [41], and cyclist [27]

8.2.3 Cardiovascular Adaptations

As detailed in Chap. 3, an increase in core and skin temperatures during exercise in the heat is associated with increased cardiovascular strain [42], with a significant increase in heart rate along with decline in stroke volume, mean arterial pressure, and potentially cardiac output [43]. Some investigators believe that a high heart rate is the primary contributor to reduced stroke volume during prolonged exercise in the heat [44].

The increases in sweat rate and plasma volume support heat dissipation at the level of the skin. As such, it has been well described that heat acclimation improves evaporative cooling and lowers skin temperature more than rectal temperature, therefore increasing the internal thermal gradient and facilitating heat flow to the surface [11, 45]. This ultimately improves heat transfer via blood flow, 'sparing' blood for the rest of the circulation [11] and relieving the circulatory strain [45]. Indeed, an increased plasma volume will reduce heart rate and better sustain stroke volume during exercise in the heat [22], improving the ability to sustain blood pressure and cardiac output requirements (Fig. 8.4). Thus, a decrease in heart rate at a given work rate is a sign of a better sustained stroke volume and perhaps cardiac output during exercise-heat stress. The decrease in heart rate is probably due in-part to the increase in plasma volume supporting greater venous return, cardiac

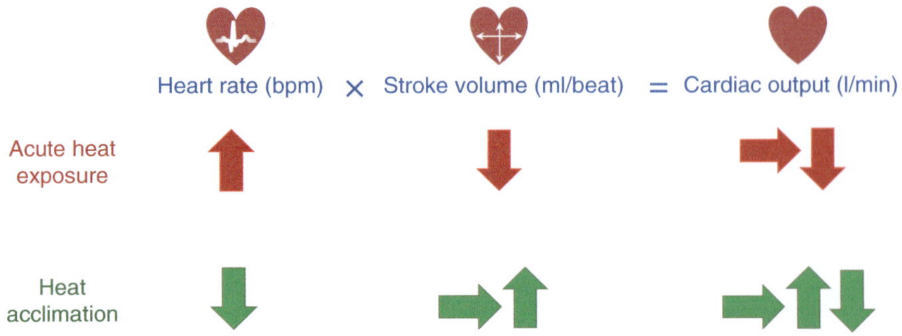

Fig. 8.4 Hyperthermia increases heart rate and decreases stroke volume. A heat acclimation-induced increase in plasma volume may allow to partly compensate for this effect

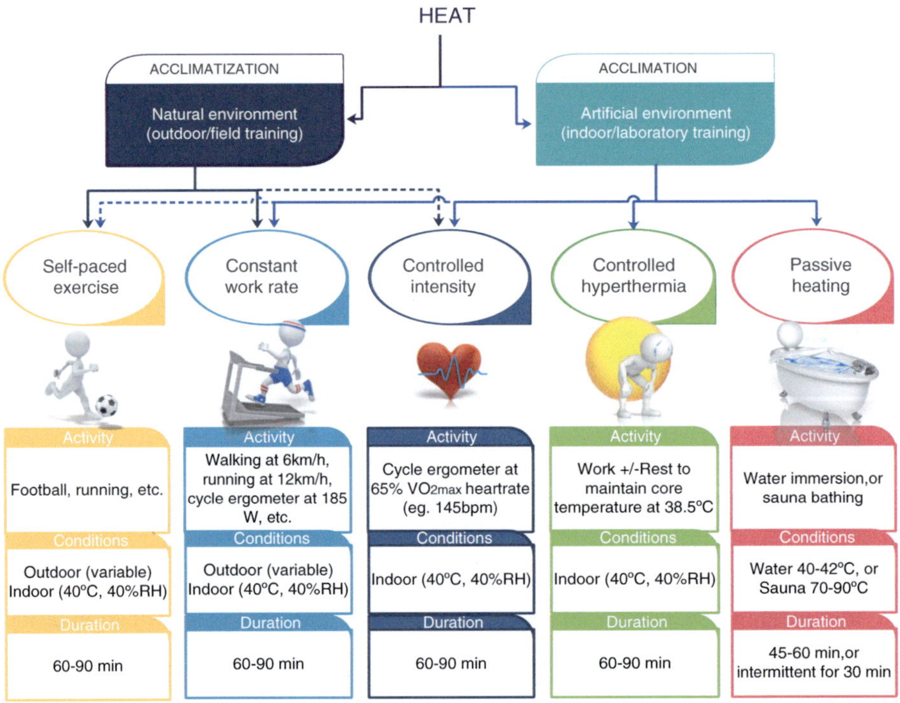

Fig. 8.5 Overview of heat acclimation methods. Based on Daanen et al. [121]

preloading, and improving ventricular filling [23, 46], but alterations in body temperature and sympathetic activity can also contribute. Accordingly, a decrease in heart rate for a given exercise intensity in the heat is considered a hallmark of heat acclimation [26, 47]. It has therefore been proposed to utilise heart rate during heat acclimation to maintain a given level of cardiovascular strain during daily exercise-heat exposures (i.e. controlled heart-rate protocol) [3] (Fig. 8.5).

8.2.4 Thermal Tolerance

In addition to the various sudomotor, vasomotor, haematological, thermoregulatory, and cardiovascular adaptations that contribute to reduce heat strain, repeated heat exposure also improves thermal tolerance. Whilst heat acclimation reduces heat strain, thermal tolerance is a cellular adaptation allowing the cells and organisms to survive a higher level of strain [48–50]. Thermal tolerance largely depends on the heat shock protein (HSP) response [51] with HSP72 being particularly responsive to heat stress and exercise [52]. Briefly, HSP binds to denatured or nascent cellular polypeptides to provide protection and support repair from various stressors including heat stress, fever, hypoxia, ischemia, viral infection, energy depletion, and acidosis [53]. Human studies have highlighted the complexity of the HSP response. For example, heat acclimation has been shown to increase basal levels of intracellular Hsp72 and HSP90, with individuals demonstrating the greatest physiological adaptations exhibiting reduced post-exercise expression (measured ex vivo via water bath incubation) [54]. In contrast, basal levels of extracellular HSP72 decreased (with an increase in the post-exercise response) after a couple of days of heat acclimation [55], but not after ~10 days [56, 57]. A potential increase in basal levels (with a decrease in the post-exercise response) after 15 days of heat acclimation has also been noted [58]. In addition to the HSP response, heat stress also up-regulates or down-regulates numerous gene expressions [59]. Globally, these adaptations allow the heat-acclimated phenotype to be more tolerant to heat as well as to other stressors [60].

Even if heat acclimation is a broader response than thermal tolerance, and it is possible to induce some level of heat acclimation without inducing significant intracellular Hsp70 responses [61], it is noteworthy that HSP responses may participate in both heat acclimation and thermal tolerance in parallel. For example, changes in HSP90 within cutaneous tissues might contribute to improved vasodilatory effects [62, 63].

8.2.5 Muscle Adaptations

Heat acclimation has been purported to act at the muscle level, possibly altering whole-body [64] and skeletal muscle metabolism [65, 66]. For example, the basal metabolic rate of young men in Asia has been reported to decrease during the warmer months of the year [67]. Heat acclimation has also been suggested to decrease oxygen uptake [68] and muscle glycogen utilisation (40–50%) [69, 70] during submaximal exercise in the heat, albeit the glycogen-sparing effect of heat acclimation may be small [65]. Heat acclimation may also reduce blood and muscle lactate accumulation during submaximal exercise [66], and increase power output at lactate threshold [71]. This could be linked to improved lactate removal due to the increase in total body water increasing splanchnic circulation [72], and to delayed lactate accumulation due to an increase in cardiac output and a decrease in metabolic rate [65, 68]. Heat acclimation may also improve muscle aerobic

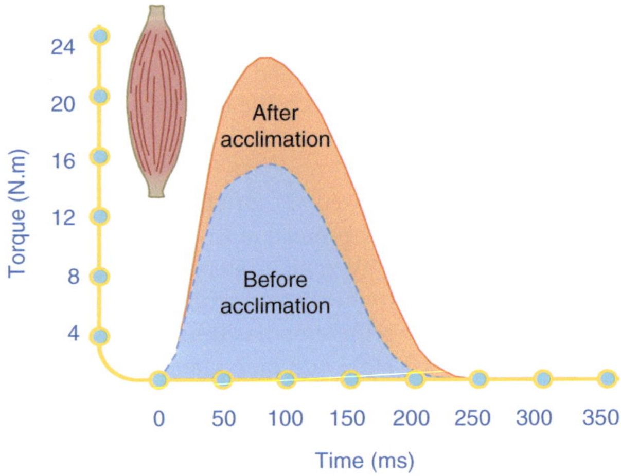

Fig. 8.6 Repeated heat exposure increases peak twitch amplitude in human plantar flexors. Based on data from Racinais et al. [124]

metabolism through mitochondrial adaptations [73] and muscle capillary growth [74]. Lastly, heat acclimation has been shown to facilitate hypertrophy in cultured [75, 76] or animals [77] muscle cells, and thus increase muscle cross-sectional area in humans [78] (Fig. 8.6).

8.2.6 Neural Adaptations

Acute hyperthermia has been shown to decrease neural drive to the muscle [79]. As detailed in Chap. 4, this impairment is due to peripheral failures in neural drive transmission and supraspinal alterations when contractions are prolonged [80]. In vitro studies have suggested that heat acclimation could reverse the peripheral failure in neural drive transmission induced by hyperthermia [81, 82]. However, in humans, the peripheral failures in neural drive transmission (estimated through M-wave and H-reflex evoked potentials) induced by hyperthermia are not reverted by heat acclimation [83]. Importantly, this suggests that the decrease in neural drive transmission in humans might not be linked to synaptic failure, but rather be a side effect of the increase in axonal conduction velocity which shortens depolarisation time [84, 85]. Conversely, heat acclimation partly restores the ability to sustain neural drive during prolonged contractions [83] (refer to the Chap. 4, Fig. 4.4). As this recovery is not accompanied by any spinal or peripheral nervous system adjustments, it suggests a supraspinal adaptation to heat acclimation [83]. The benefit of this adaptation for whole-body exercise capacity remains unclear however, as the limiting factor for prolonged exercise in the heat is more cardiovascular than neural [86, 87].

8.2.7 Perceptual and Cognitive Adaptations

Acute hyperthermia increases thermal discomfort as well as the negative/positive affects ratio [83], potentially affecting exercise capacity [88] and cognitive function [89, 90] in the heat. Heat acclimation may not affect thermal comfort per se at rest [83, 91], but it can improve the ratings of thermal sensation/comfort during training and competition in the heat for both team-sports and endurance athletes [91–93]. It is however unclear to which extend this improvement participates in the physical performance improvement associated with acclimation. Heat acclimation can also protect various cognitive tasks such as psychomotor performance [94], attention tasks [95] and planning task [96] from the deleterious effects of hyperthermia.

8.3 Exercise Capacity and Physical Performance

8.3.1 In Hot Environments

From the initial studies in the mining industry, heat acclimation demonstrated an improved work capacity under heat stress conditions [97]. This improvement has been consistently verified using either natural (acclimatisation) or artificial (acclimation) heat exposures, with a recent meta-analysis calculating an average performance improvement of 15% [91].

The decrease in VO_{2max} associated with the development of hyperthermia is likely one of the main limiting factors for prolonged exercise performance in hot ambient conditions [98]. Heat acclimation can improve VO_{2max} in temperate and hot ambient conditions; however, it cannot fully restore the initial decrease in VO_{2max} imposed by heat stress [71, 99].

The increase in VO_{2max} with acclimation is associated with a proportional improvement in exercise performance in the heat (e.g. cycling time trial) [71]. In some circumstances, heat acclimation may be sufficient to restore a similar level of performance to that observed in cooler environments [100, 101]. For example, power output during an outdoor cycling time trial (43 km) in 37 °C was ~16% lower compared to a time trial undertaken in 8 °C, with half of this decrease restored after 1 week of heat acclimation, and the decrease almost fully compensated for after 2 weeks of heat exposure [100] (Fig. 8.7). The magnitude to which heat acclimation can improve performance in the heat depends on several factors, including the severity of the environmental conditions (e.g. ambient temperature, relative humidity, wind velocity) and the type of performance (e.g. duration, intensity). Notwithstanding, heat acclimation consistently improves submaximal intensity exercise capacity in the heat.

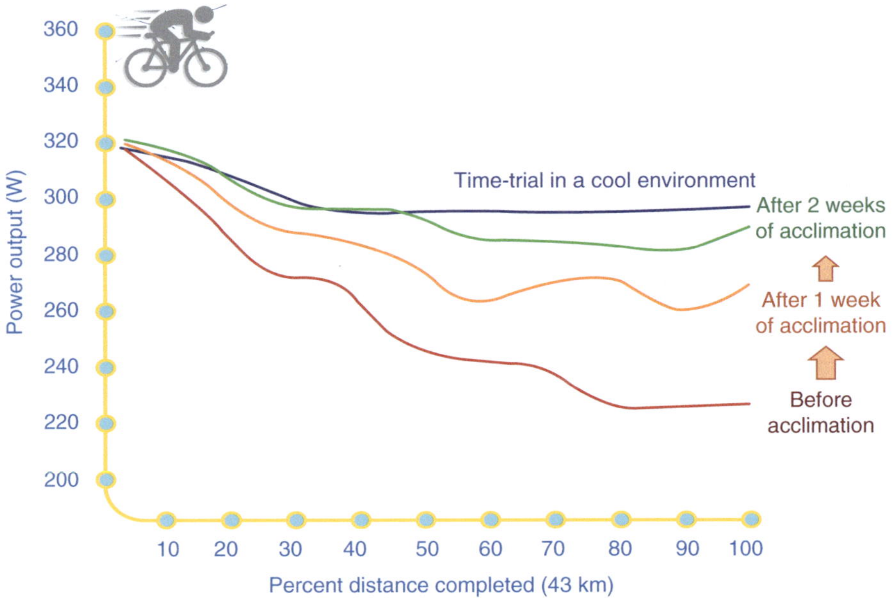

Fig. 8.7 Power output during a 43 km cycling time trial in cool ambient conditions (blue line) and in hot ambient conditions on the first (red line), the 6th (orange line), and the 14th (green line) days of heat exposure. Based on data from Racinais et al. [100]

8.3.2 In Temperate Environments

Whilst it is well accepted that heat acclimation improves both VO_{2max} and exercise capacity in the heat, there is conflicting evidence regarding its effect on exercise capacity in cooler environments. On one hand, several studies did not observe any improvement in VO_{2max} in cool environment following 5 days of isothermic heat acclimation with permissive dehydration [102], 10 days of constant intensity heat acclimation [101], or 14 days of heat acclimatisation [103]. On the other hand, other studies have reported an increase in VO_{2max} in cool conditions following heat acclimation in unfit (23%) or untrained (13%) individuals [104], but also in trained athletes (3–5%) [71, 99]. Whether heat acclimation improves VO_{2max} in cool environments therefore remains a topic of contention.

In addition to potential improvements in VO_{2max}, decreases in glycogen utilisation [65, 69, 70] and blood and muscle lactate accumulation [66] may also improve exercise capacity in cool environments following heat acclimation. The mechanisms for these adaptations are unclear, but could include an increase in total body water facilitating lactate removal through increased splanchnic circulation [72], and/or an increase in cardiac output and decrease in metabolic rate delaying lactate accumulation [65, 105]. This can translate in an increased power output at lactate threshold [71, 102], VO_{2max} [71, 99], and performance [71]. For example, heat acclimation (via sauna bathing) improves run time to exhaustion in competitive runners with the improvement being correlated to the increase in plasma volume and total blood volume [106].

Several studies have also reported that heat acclimation increased fitness level in team-sports athlete following pre- and in-season training camps in the heat [41, 107, 108]. This is particularly interesting as obtaining rapid fitness improvements without performing specific aerobic training sessions may save time for technical training [41, 107, 108]. In summary, even if the benefits of heat acclimation in cooler environment are debated, there is no evidence to indicate that heat acclimation impairs performance in cooler environments. As such, heat acclimation should be implemented before any event with potentially hot, even if uncertain conditions.

8.4 Heat Acclimation and Decay

8.4.1 Kinetics of Adaptation

The kinetics of heat acclimation adaptation and decay were one of the primary interests of the occupational studies stemming from the mining industry and military [109, 110]. Robinson and his collaborators [111] reported in 1943 that daily walking in a hot room for 1–1.5 h decreased heart rate, skin temperature, and rectal temperature, with the decrement occurring rapidly in the first 7 days of acclimation, and then more slowly up to 23 days. As early as 1951, Ladell [12] suggested that heat acclimation occurred in two phases with an initial decrease in heart rate and temperature threshold for sweating (within 2–3 days), followed by an increase in sweat rate at a given temperature and a better resistance to fatigue. Heat acclimation is now considered as a relatively rapid process that begins from the first day of exposure, with 75–80% of the adaptations occurring in the first 4–7 days [3, 112] (Fig. 8.1).

Importantly, the kinetics of adaptation are based on daily heat exposure. Indeed, 10 days of daily heat exposure was reported to induce similar adaptations than 1 exposure every 3 day, but in the third of the time [113]. In addition, even if 1 exposure every 3 days allows for some adaptations, interspacing the exposure by 1 week does not allow for adaptations to develop [114]. Thus, if and when allowing for rest days during a pre-competition taper to avoid fatigue and hypohydration [115], it must be accounted for that intermittent heat exposure allows for lesser adaptations compared to daily exposure [116].

Of note, the constraints of the competition calendar in international sports have led recent research to focus on short-term heat acclimation [115, 117, 118]. However, even if athletes benefit from only few days of acclimation [92, 115, 119], they may require 7–10 days to achieve near complete cardiovascular and sudomotor adaptations [23, 27, 71], and two or more weeks to optimise aerobic performance in hot environments [100].

8.4.2 Heat Acclimation Induction

A classic research from 1963 originally suggested that 100 min of daily exercise in the heat was suitable to induce heat acclimation [120]. The panel of induction methods has diversified over the years and includes active, passive, or a mixture of active and passive hyperthermia [121] (Fig. 8.5). The rule of thumb to induce heat acclimation is to increase whole-body temperature, induce profuse sweating, and elevate skin blood flow [122]. This can be achieved through: (1) passive heat exposure, (2) self-paced exercise, (3) constant work rate exercise, (4) controlled hyperthermia, or (5) controlled heart-rate approaches (Fig. 8.5).

Passive exposure has been used in scientific studies to isolate the effect of repeated heat exposure without training [83, 123, 124]. However, heat acclimation before competing in the heat should be as specific as possible and as such, passive exposure is likely not as efficient as training in the heat. Notwithstanding, repeated passive heat exposures allow to partially develop a heat-acclimated phenotype [125, 126] and may be practical for athletes living in cold countries. Passive exposure may also include maintaining an elevated temperature post-training [106, 127], or even increasing temperature before active acclimation [128].

Self-paced exercise in the heat has been used for decades [129, 130], and includes working or training in natural heat. Self-paced heat acclimation has been largely replaced by more controlled methods (see below) in the recent years, but is still the primary method for team-sports athletes acclimatising through their natural outdoor training [40, 41, 107]. For endurance sports, such as cycling, choosing to naturally acclimatise through self-paced exercise, intensity blocks should be prioritised towards the beginning of the session to maintain absolute exercise intensity.

Acclimation via constant work rate exercise has been the primary model from which much of our knowledge on acclimation kinetics (see above) is based [13, 23, 111, 131]. In this model, there are fixed endogenous and exogenous thermal loads (e.g. cycling at 160 W in 35 °C, 60% RH), representing a constant forcing function. However, the relative stress imposed by of those fixed load will progressively diminish as acclimation develops [11, 132–134]. This relative decrease has been suggested as one of the reasons for the decline in plasma volume following its initial increase [24, 25]. It is important to note however, that whilst the training stimulus or forcing function may decrease as adaptations develop, a recent meta-analysis concluded that the constant work rate acclimation leads to similar adaptations than clamping core temperature to maintain the forcing function (described below) [91].

Acclimation by isothermic controlled hyperthermia (i.e. clamping core temperature at 38.5 °C) has been proposed to provide a constant forcing function and thus maintain the acclimation stimulus as adaptations develop [24, 117, 135, 136]. This concept has recently seen a regain in interest, but it is not new and was proposed half a century ago [132, 135]. In general, isothermic protocols may not allow appropriate training for athletes as the exercise intensity dramatically decreases when the target core temperature is reached, with exercise completely ceasing in some circumstances [137], thus reducing one of the main drivers for adaptation (i.e. sweating).

Training in the heat at a given heart rate has therefore been suggested as a more suitable method for athletes [3]. Heart rate represents a global index of the cardiovascular stress imposed by the thermal stress and exercise (see above), and decreases rapidly with acclimation (Fig. 8.1). Thus training at a clamped heart rate accounts for the acute effect of heat stress via power output adjustments, but then allows for maintaining a given level of cardiovascular strain as acclimation develops and power output increases at the pre-determined heart rate [3].

8.4.3 Individual Responses

There are large inter-individual differences in heat acclimation responses [40, 138], with some athletes acclimating in 1 week and others requiring 2 weeks or longer [121]. Although heat acclimation improves exercise capacity in the heat in a variety of populations including mine workers, military personnel, and athletes, an individualised strategy may be required in elite athletes to truly optimise performance. For example, when analysing a soccer game in dry heat after 1 week of heat acclimation, some football players can have the same physical activity than they had 1 week before in temperate environment, whereas other players from the same team have to dramatically reduce their physical activity [40]. Anecdotally, although the changes in performance following short-term acclimation may correlate to the changes in haematocrit during a heat-response test (Fig. 8.3), there is currently no test in temperate environment or resting blood measures allowing to quantify heat acclimation [27, 40, 41].

8.4.4 Decay and Re-acclimation

Early studies on miner usually working in hot ambient conditions [110, 139] or on soldier undergoing 12 days of acclimation [140] suggest that most adaptations are lost within 6–7 days without heat exposure. However, Robinson et al. [111] reported that 3 of 5 participants could still benefit from the adaptations induced by 23 days of heat acclimation 2–3 weeks after the end of the procedure, whereas two other participants lost most of their adaptations. This support the early study from Dreosti [7] suggesting that some heat acclimation benefits could be maintained for up to 1 month after 14 days of heat acclimation. Although those early reports are informative, there is limited research available on the kinetics of heat acclimation decay. A recent systematic review and meta-analyses reported that despite athletes maintaining some adaptations for several weeks after acclimation, the magnitude of those adaptations decreased daily by ~2.5% for heart rate and body core temperature during exercise [121]. Sweat rate showed a fast decay, but too little data was available to draw firm conclusions. Noteworthy is that the rate of decay may likely be slowed by training and regular heat exposure post-acclimation. Importantly, re-acclimation during this period appears to be faster than the initial rate of acclimation

[121]. Thus, 4–5 days of re-acclimation could be enough to regain complete heat acclimation if undertaken within 1 month [141].

8.5 Conclusion

Repeated exercise-heat exposures induce heat acclimation, characterised by a series of physiological adaptations improving thermoregulation and reducing physiological strain. These adaptations include an increase in sweat rate, a decrease in heart rate, and a decrease in core temperatures at a given exercise intensity. To obtain these adaptions, it is necessary to increase both whole-body temperature (i.e. core, skin, muscle, and tissue temperature), as well as stimulate profuse sweating. The adaptations improve exercise capacity in warm-hot environments and their transfer to temperate environments remains debated. Most adaptations can be obtained within a few days of daily heat exposure, but exercise capacity in the heat may improve optimally after ~2 weeks.

References

1. Nybo L, Rasmussen P, Sawka MN. Performance in the heat-physiological factors of importance for hyperthermia-induced fatigue. Compr Physiol. 2014;4(2):657–89.
2. Armstrong L, Maresh C. The induction and decay of heat acclimatisation in trained athletes. Sports Med. 1991;12(5):302.
3. Périard JD, Racinais S, Sawka MN. Adaptations and mechanisms of human heat acclimation: applications for competitive athletes and sports. Scand J Med Sci Sports. 2015;25:20–38.
4. Lind J. An essay on diseases incidental to Europeans in hot climates, with the method of preventing their fatal consequences. Printed by William Duane; 1768.
5. Kuno Y. Human perspiration. Springfield: Thomas; 1956.
6. Pogue TE. The evolution of research collaboration in South African gold mining: 1886–1933. Doctoral thesis. Universiteit Maastricht. 2006 June 29; p. 1–219.
7. Dreosti AO. The results of some investigations into the medical aspects of deep mining on the Witwatersrand. J Chem Metal Mining Soc S Afr. 1935;6:102–29.
8. Buskirk ER. The temperature regulatory system. Exercise physiology - people and ideas. Oxford: Oxford University Press; 2003. p. 423–51.
9. Weiner JS. Observations on the working ability of Bantu mineworkers with reference to acclimatization to hot humid conditions. Br J Ind Med. 1950;7(1):17–26.
10. Ladell WS. Physiological observations on men working in supposedly limiting environments in a West African gold-mine. Br J Ind Med. 1955;12(2):111–25.
11. Eichna LW, Park CR, Nelson N, Horvath SM, Palmes ED. Thermal regulation during acclimatization in a hot, dry (desert type) environment. Am J Physiol. 1950;163(3):585–97.
12. Ladell WS. Assessment of group acclimatization to heat and humidity. J Physiol. 1951;115(3):296–312.
13. Nadel ER, Pandolf KB, Roberts MF, Stolwijk JA. Mechanisms of thermal acclimation to exercise and heat. J Appl Physiol. 1974;37(4):515–20.
14. Gonzalez RR, Pandolf KB, Gagge AP. Heat acclimation and decline in sweating during humidity transients. J Appl Physiol. 1974;36(4):419–25.
15. Allan JR, Wilson CG. Influence of acclimatization on sweat sodium concentration. J Appl Physiol. 1971;30(5):708–12.

16. Sawka MN, Leon LR, Montain SJ, Sonna LA. Integrated physiological mechanisms of exercise performance, adaptation, and maladaptation to heat stress. Compr Physiol. 2011;1(4):1883–928.
17. Buono MJ, Kolding M, Leslie E, Moreno D, Norwood S, Ordille A, et al. Heat acclimation causes a linear decrease in sweat sodium ion concentration. J Therm Biol. 2018;71:237–40.
18. Barcroft J, Meakins JC, Davies HW, Duncan Scott JM, Fetter WJ. Observations upon the effect of high altitude on the physiological processes of the human body. On the relation of external temperature to blood volume. Phil Tr Roy Soc. 1922;211:455–64.
19. Bazett HC, Sunderman FW, Doupe J, Scott JC. Climatic effects on the volume and composition of blood in man. Am J Physiol-Leg Content. 1940;129(1):69–83.
20. Glaser EM. Acclimatization to heat and cold. J Physiol. 1950;110(3–4):330–7.
21. Senay LC, Mitchell D, Wyndham CH. Acclimatization in a hot, humid environment: body fluid adjustments. J Appl Physiol. 1976;40(5):786–96.
22. Sawka MN, Coyle EF. Influence of body water and blood volume on thermoregulation and exercise performance in the heat. Exerc Sport Sci Rev. 1999;27:167–218.
23. Nielsen B, Hales JR, Strange S, Christensen NJ, Warberg J, Saltin B. Human circulatory and thermoregulatory adaptations with heat acclimation and exercise in a hot, dry environment. J Physiol. 1993;460:467–85.
24. Patterson MJ, Stocks JM, Taylor NAS. Sustained and generalized extracellular fluid expansion following heat acclimation. J Physiol. 2004;559(Pt 1):327–34.
25. Patterson MJ, Stocks JM, Taylor NAS. Whole-body fluid distribution in humans during dehydration and recovery, before and after humid-heat acclimation induced using controlled hyperthermia. Acta Physiol (Oxford). 2014;210(4):899–912.
26. Périard JD, Travers GJS, Racinais S, Sawka MN. Cardiovascular adaptations supporting human exercise-heat acclimation. Auton Neurosci. 2016;196:52–62.
27. Karlsen A, Nybo L, Norgaard SJ, Jensen MV, Bonne T, Racinais S. Time course of natural heat acclimatization in well-trained cyclists during a 2-week training camp in the heat. Scand J Med Sci Sports. 2015;25:240–9.
28. Sawka MN, Convertino VA, Eichner ER, Schnieder SM, Young AJ. Blood volume: importance and adaptations to exercise training, environmental stresses, and trauma/sickness. Med Sci Sports Exerc. 2000;32(2):332–48.
29. Wyndham CH, Williams CG, Morrison JF, Heyns AJ, Siebert J. Tolerance of very hot humid environments by highly acclimatized Bantu at rest. Br J Ind Med. 1968;25(1):22–39.
30. Senay LC. Effects of exercise in the heat on body fluid distribution. Med Sci Sports. 1979;11(1):42–8.
31. Harrison MH, Edwards RJ, Graveney MJ, Cochrane LA, Davies JA. Blood volume and plasma protein responses to heat acclimatization in humans. J Appl Physiol. 1981;50(3):597–604.
32. Horowitz M, Adler JH. Plasma volume regulation during heat stress: albumin synthesis vs capillary permeability. A comparison between desert and non-desert species. Comp Biochem Physiol A Comp Physiol. 1983;75(1):105–10.
33. Yang RC, Mack GW, Wolfe RR, Nadel ER. Albumin synthesis after intense intermittent exercise in human subjects. J Appl Physiol. 1998;84(2):584–92.
34. Harrison MH. Effects on thermal stress and exercise on blood volume in humans. Physiol Rev. 1985;65(1):149–209.
35. Senay LC. Movement of water, protein and crystalloids between vascular and extra-vascular compartments in heat-exposed men during dehydration and following limited relief of dehydration. J Physiol. 1970;210(3):617–35.
36. Senay LC. Changes in plasma volume and protein content during exposures of working men to various temperatures before and after acclimatization to heat: separation of the roles of cutaneous and skeletal muscle circulation. J Physiol. 1972;224(1):61–81.
37. Mack GW, Nadel ER. Body fluid balance during heat stress in humans. In: Fregly MJ, Blatteis CM, editors. Handbook of physiology environmental physiology. New York: Oxford University Press; 1996. p. 187–214.

38. Blake AS, Petley GW, Deakin CD. Effects of changes in packed cell volume on the specific heat capacity of blood: implications for studies measuring heat exchange in extracorporeal circuits. Br J Anaesth. 2000;84(1):28–32.
39. Watt MJ, Garnham AP, Febbraio MA, Hargreaves M. Effect of acute plasma volume expansion on thermoregulation and exercise performance in the heat. Med Sci Sports Exerc. 2000;32(5):958–62.
40. Racinais S, Mohr M, Buchheit M, Voss SC, Gaoua N, Grantham J, et al. Individual responses to short-term heat acclimatisation as predictors of football performance in a hot, dry environment. Br J Sports Med. 2012;46(11):810–5.
41. Racinais S, Buchheit M, Bilsborough J, Bourdon PC, Cordy J, Coutts AJ. Physiological and performance responses to a training camp in the heat in professional Australian football players. Int J Sports Physiol Perform. 2014;9(4):598–603.
42. González-Alonso J, Crandall CG, Johnson JM. The cardiovascular challenge of exercising in the heat. J Physiol. 2008;586(1):45–53.
43. Périard JD, Cramer MN, Chapman PG, Caillaud C, Thompson MW. Cardiovascular strain impairs prolonged self-paced exercise in the heat. Exp Physiol. 2011;96(2):134–44. Available from: http://ep.physoc.org/content/96/2/134.long.
44. Coyle EF, González-Alonso J. Cardiovascular drift during prolonged exercise: new perspectives. Exerc Sport Sci Rev. 2001;29(2):88–92.
45. MacDonald DKC, Wyndham CH. Heat transfer in man. J Appl Physiol. 1950;3(6):342–64.
46. Senay LC. An inquiry into the role of cardiac filling pressure in acclimatization to heat. Yale J Biol Med. 1986;59(3):247–56.
47. Sundstroem ES. The physiological effects of tropical climate. Physiol Rev. 1927;7:320–62.
48. Moseley PL. Heat shock proteins and heat adaptation of the whole organism. J Appl Physiol. 1997;83(5):1413–7.
49. Horowitz M. Do cellular heat acclimation responses modulate central thermoregulatory activity? News Physiol Sci. 1998;13:218–25.
50. Maloyan A, Palmon A, Horowitz M. Heat acclimation increases the basal HSP72 level and alters its production dynamics during heat stress. Am J Phys Regul Integr Comp Phys. 1999;276(5 Pt 2):R1506–15.
51. Kuennen MR, Gillum TL, Dokladny K, Bedrick EJ, Schneider SM, Moseley PL. Thermotolerance and heat acclimation may share a common mechanism in humans. Am J Phys Regul Integr Comp Phys. 2011;301(2):R524–33.
52. Locke M. The cellular stress response to exercise: role of stress proteins. Exerc Sport Sci Rev. 1997;25:105–36.
53. Kregel KC. Heat shock proteins: modifying factors in physiological stress responses and acquired thermotolerance. J Appl Physiol. 2002;92(5):2177–86.
54. Mcclung JP, Hasday JD, He J-R, Montain SJ, Cheuvront SN, Sawka MN, et al. Exercise-heat acclimation in humans alters baseline levels and ex vivo heat inducibility of HSP72 and HSP90 in peripheral blood mononuclear cells. Am J Phys Regul Integr Comp Phys. 2008;294(1):R185–91.
55. Marshall HC, Ferguson RA, Nimmo MA. Human resting extracellular heat shock protein 72 concentration decreases during the initial adaptation to exercise in a hot, humid environment. Cell Stress Chaperones. 2006;11(2):129–34.
56. Yamada PM, Amorim FT, Moseley P, Robergs R, Schneider SM. Effect of heat acclimation on heat shock protein 72 and interleukin-10 in humans. J Appl Physiol. 2007;103(4):1196–204.
57. de Castro Magalhães F, Amorim FT, Passos RLF, Fonseca MA, Oliveira KPM, Lima MRM, et al. Heat and exercise acclimation increases intracellular levels of Hsp72 and inhibits exercise-induced increase in intracellular and plasma Hsp72 in humans. Cell Stress Chaperones. 2010;15(6):885–95.
58. Sandström ME, Siegler JC, Lovell RJ, Madden LA, Mcnaughton L. The effect of 15 consecutive days of heat-exercise acclimation on heat shock protein 70. Cell Stress Chaperones. 2008;13(2):169–75.

59. Sonna LA, Sawka MN, Lilly CM. Exertional heat illness and human gene expression. Prog Brain Res. 2007;162:321–46.
60. Horowitz M. Genomics and proteomics of heat acclimation. Front Biosci (Schol Ed). 2010;2:1068–80.
61. Hom LL, Lee EC-H, Apicella JM, Wallace SD, Emmanuel H, Klau JF, et al. Eleven days of moderate exercise and heat exposure induces acclimation without significant HSP70 and apoptosis responses of lymphocytes in college-aged males. Cell Stress Chaperones. 2012;17(1):29–39.
62. Fujii N, Zhang SY, McNeely BD, Nishiyasu T, Kenny GP. Heat shock protein 90 contributes to cutaneous vasodilation through activating nitric oxide synthase in young male adults exercising in the heat. J Appl Physiol. 2017;123:844–50.
63. Carter HH, Green DJ. HSP90: an unappreciated mediator of cutaneous vascular adaptation? J Appl Physiol. 2018;124:521.
64. Sawka MN, Wenger CB, Pandolf KB. Thermoregulatory responses to acute exercise-heat stress and heat acclimation. In: Fregly MJ, Blatteis CM, editors. Handbook of Physiology Environmental Physiology. New York: Oxford University Press; 1996. p. 157–85.
65. Young AJ, Sawka MN, Levine L, Cadarette BS, Pandolf KB. Skeletal muscle metabolism during exercise is influenced by heat acclimation. J Appl Physiol. 1985;59(6):1929–35.
66. Febbraio MA, Snow RJ, Hargreaves M, Stathis CG, Martin IK, Carey MF. Muscle metabolism during exercise and heat stress in trained men: effect of acclimation. J Appl Physiol. 1994;76(2):589–97.
67. Hori S. Adaptation to heat. Jpn J Physiol. 1995;45(6):921–46.
68. Sawka MN, Pandolf KB, Avellini BA, Shapiro Y. Does heat acclimation lower the rate of metabolism elicited by muscular exercise? Aviat Space Environ Med. 1983;54(1):27–31.
69. King DS, Costill DL, Fink WJ, Hargreaves M, Fielding RA. Muscle metabolism during exercise in the heat in unacclimatized and acclimatized humans. J Appl Physiol. 1985;59(5):1350–4.
70. Kirwan JP, Costill DL, Kuipers H, Burrell MJ, Fink WJ, Kovaleski JE, et al. Substrate utilization in leg muscle of men after heat acclimation. J Appl Physiol. 1987;63(1):31–5.
71. Lorenzo S, Halliwill JR, Sawka MN, Minson CT. Heat acclimation improves exercise performance. J Appl Physiol. 2010;109(4):1140–7.
72. Rowell LB, Brengelmann GL, Blackmon JR, Twiss RD, Kusumi F. Splanchnic blood flow and metabolism in heat-stressed man. J Appl Physiol. 1968;24(4):475–84.
73. Tamura Y, Matsunaga Y, Masuda H, Takahashi Y, Takahashi Y, Terada S, et al. Postexercise whole body heat stress additively enhances endurance training-induced mitochondrial adaptations in mouse skeletal muscle. Am J Physiol Regul Integr Comp Physiol. 2014;307(7):R931–43. Available from: http://eutils.ncbi.nlm.nih.gov/entrez/eutils/elink.fcgi?dbfrom=pubmed&id=25080501&retmode=ref&cmd=prlinks.
74. Kuhlenhoelter AM, Kim K, Neff D, Nie Y, Blaize AN, Wong BJ, et al. Heat therapy promotes the expression of angiogenic regulators in human skeletal muscle. Am J Physiol Regul Integr Comp Physiol. 2016;311(2):R377–91. Available from: http://eutils.ncbi.nlm.nih.gov/entrez/eutils/elink.fcgi?dbfrom=pubmed&id=27357800&retmode=ref&cmd=prlinks.
75. Yamashita-Goto K, Ohira Y, Okuyama R, Sugiyama H, Honda M, Sugiura T, et al. Heat stress facilitates stretch-induced hypertrophy of cultured muscle cells. J Gravit Physiol [Internet]. 2002;9(1):P145–6. Available from: http://eutils.ncbi.nlm.nih.gov/entrez/eutils/elink.fcgi?dbfrom=pubmed&id=15002522&retmode=ref&cmd=prlinks.
76. Goto K, Okuyama R, Sugiyama H, Honda M, Kobayashi T, Uehara K, et al. Effects of heat stress and mechanical stretch on protein expression in cultured skeletal muscle cells. Pflugers Arch. 2003;447:247–53.
77. Uehara K, Goto K, Kobayashi T, Kojima A, Akema T, Sugiura T, et al. Heat-stress enhances proliferative potential in rat soleus muscle. Jpn J Physiol [Internet]. 2004;54(3):263–71. Available from: http://eutils.ncbi.nlm.nih.gov/entrez/eutils/elink.fcgi?dbfrom=pubmed&id=15541204&retmode=ref&cmd=prlinks.

78. Goto K, Oda H, Kondo H, Igaki M, Suzuki A, Tsuchiya S, et al. Responses of muscle mass, strength and gene transcripts to long-term heat stress in healthy human subjects. Eur J Appl Physiol. 2011;111:17–27.
79. Nybo L, Nielsen B. Hyperthermia and central fatigue during prolonged exercise in humans. J Appl Physiol. 2001;91(3):1055–60.
80. Racinais S, Gaoua N, Grantham J. Hyperthermia impairs short-term memory and peripheral motor drive transmission. J Physiol. 2008;586(Pt 19):4751–62.
81. Kelty JD, Noseworthy PA, Feder ME, Robertson RM, Ramirez J-M. Thermal preconditioning and heat-shock protein 72 preserve synaptic transmission during thermal stress. J Neurosci. 2002;22(1):RC193.
82. Karunanithi S, Barclay JW, Brown IR, Robertson RM, Atwood HL. Enhancement of presynaptic performance in transgenic Drosophila overexpressing heat shock protein Hsp70. Synapse. 2002;44(1):8–14.
83. Racinais S, Wilson MG, Gaoua N, Périard JD. Heat acclimation has a protective effect on the central but not peripheral nervous system. J Appl Physiol. 2017;123(4):816–24.
84. Bolton CF, Sawa GM, Carter K. The effects of temperature on human compound action potentials. J Neurol Neurosurg Psychiatry. 1981;44(5):407–13.
85. Rutkove SB, Kothari MJ, Shefner JM. Nerve, muscle, and neuromuscular junction electrophysiology at high temperature. Muscle Nerve. 1997;20(4):431–6.
86. Racinais S, Girard O. Neuromuscular failure is unlikely to explain the early exercise cessation in hot ambient conditions. Psychophysiology. 2012;49(6):853–65.
87. Périard JD, Caillaud C, Thompson MW. Central and peripheral fatigue during passive and exercise-induced hyperthermia. Med Sci Sports Exerc. 2011;43(9):1657–65.
88. Stevens CJ, Mauger AR, Hassmèn P, Taylor L. Endurance performance is influenced by perceptions of pain and temperature: theory, applications and safety considerations. Sports Med. 2018;48(3):525–37.
89. Marks LE, Gonzalez RR. Skin temperature modifies the pleasantness of thermal stimuli. Nature. 1974;247(5441):473–5.
90. Gaoua N, Grantham J, Racinais S, Massioui El F. Sensory displeasure reduces complex cognitive performance in the heat. J Environ Psychol. 2012;32:158–63.
91. Tyler CJ, Reeve T, Hodges GJ, Cheung SS. The effects of heat adaptation on physiology, perception and exercise performance in the heat: a meta-analysis. Sports Med. 2016;46(11):1699–724.
92. Sunderland C, Morris JG, Nevill ME. A heat acclimation protocol for team sports. Br J Sports Med. 2008;42(5):327–33.
93. Kelly M, Gastin PB, Dwyer DB, Sostaric S, Snow RJ. Short duration heat acclimation in Australian football players. J Sports Sci Med. 2016;15:118–25.
94. Walker SM, Dawson B, Ackland TR. Performance enhancement in rally car drivers via heat acclimation and race simulation. Comp Biochem Physiol Part A Mol Integr Physiol. 2001;128(4):701–7.
95. Radakovic SS, Maric J, Surbatovic M, Radjen S, Stefanova E, Stankovic N, et al. Effects of acclimation on cognitive performance in soldiers during exertional heat stress. Mil Med. 2007;172(2):133–6.
96. Racinais S, Wilson MG, Gaoua N, Periard J. Heat acclimation has a protective effect on the central but not peripheral nervous system. J Appl Physiol. 2017;123(4):816–24.
97. Dreosti AO. The results of some investigations into the medical aspect of deep mining on the Witwatersrand. J Chem Metal Mining Soc S Afr. 1935;36:102–29.
98. Périard JD, Racinais S. Self-paced exercise in hot and cool conditions is associated with the maintenance of %VO2peak within a narrow range. J Appl Physiol [Internet]. 2015;118(10):1258–65. Available from: http://eutils.ncbi.nlm.nih.gov/entrez/eutils/elink.fcgi?dbfrom=pubmed&id=25814635&retmode=ref&cmd=prlinks.
99. Sawka MN, Young AJ, Cadarette BS, Levine L, Pandolf KB. Influence of heat stress and acclimation on maximal aerobic power. Eur J Appl Physiol Occup Physiol. 1985;53(4):294–8.

100. Racinais S, Périard JD, Karlsen A, Nybo L. Effect of heat and heat acclimatization on cycling time trial performance and pacing. Med Sci Sports Exerc. 2015;47(3):601–6.
101. Keiser S, Flück D, Hüppin F, Stravs A, Hilty MP, Lundby C. Heat training increases exercise capacity in hot but not in temperate conditions: a mechanistic counter-balanced cross-over study. Am J Physiol Heart Circ Physiol. 2015;309(5):H750–61.
102. Neal RA, Corbett J, Massey HC, Tipton MJ. Effect of short-term heat acclimation with permissive dehydration on thermoregulation and temperate exercise performance. Scand J Med Sci Sports. 2015;26(8):875–84.
103. Karlsen A, Racinais S, Jensen MV, Norgaard SJ, Bonne T, Nybo L. Heat acclimatization does not improve VO2max or cycling performance in a cool climate in trained cyclists. Scand J Med Sci Sports. 2015;25:269–76.
104. Shvartz E, Shapiro Y, Magazanik A, Meroz A, Birnfeld H, Mechtinger A, et al. Heat acclimation, physical fitness, and responses to exercise in temperate and hot environments. J Appl Physiol. 1977;43(4):678–83.
105. Sawka MN, Toner MM, Francesconi RP, Pandolf KB. Hypohydration and exercise: effects of heat acclimation, gender, and environment. J Appl Physiol. 1983;55(4):1147–53.
106. Scoon GSM, Hopkins WG, Mayhew S, Cotter JD. Effect of post-exercise sauna bathing on the endurance performance of competitive male runners. J Sci Med Sport. 2007;10(4):259–62.
107. Buchheit M, Voss SC, Nybo L, Mohr M, Racinais S. Physiological and performance adaptations to an in-season soccer camp in the heat: associations with heart rate and heart rate variability. Scand J Med Sci Sports. 2011;21(6):e477–85.
108. Buchheit M, Racinais S, Bilsborough J, Hocking J, Mendez-Villanueva A, Bourdon PC, et al. Adding heat to the live-high train-low altitude model: a practical insight from professional football. Br J Sports Med. 2013;47(Suppl 1):i59–69.
109. Henschel A, Taylor HL, Keys A. The persistence of heat acclimatization in man. Am J Phys. 1943;140:321–5.
110. Wyndham CH, Jacobs GE. Loss of acclimatization after six days of work in cool conditions on the surface of a mine. J Appl Physiol. 1957;11(2):197–8.
111. Robinson S, Turrell ES, Belding HS, Horvath SM. Rapid acclimatization to work in hot climates. Am J Phys. 1943;140:168–76.
112. Shapiro Y, Moran D, Epstein Y. Acclimatization strategies--preparing for exercise in the heat. Int J Sports Med. 1998;19(Suppl 2):S161–3.
113. Fein JT, Haymes EM, Buskirk ER. Effects of daily and intermittent exposures on heat acclimation of women. Int J Biometeorol. 1975;19(1):41–52.
114. Barnett A, Maughan RJ. Response of unacclimatized males to repeated weekly bouts of exercise in the heat. Br J Sports Med. 1993;27(1):39–44.
115. Chalmers S, Esterman A, Eston R, Bowering KJ, Norton K. Short-term heat acclimation training improves physical performance: a systematic review, and exploration of physiological adaptations and application for team sports. Sports Med. 2014;44(7):971–88.
116. Gill N, Sleivert G. Effect of daily versus intermittent exposure on heat acclimation. Aviat Space Environ Med. 2001;72(4):385–90.
117. Garrett AT, Goosens NG, Rehrer NJ, Rehrer NG, Patterson MJ, Cotter JD. Induction and decay of short-term heat acclimation. Eur J Appl Physiol. 2009;107(6):659–70.
118. Garrett AT, Creasy R, Rehrer NJ, Patterson MJ, Cotter JD. Effectiveness of short-term heat acclimation for highly trained athletes. Eur J Appl Physiol [Internet]. 2012;112(5):1827–37. Available from: http://eutils.ncbi.nlm.nih.gov/entrez/eutils/elink.fcgi?dbfrom=pubmed&id=21915701&retmode=ref&cmd=prlinks.
119. Garrett AT, Rehrer NJ, Patterson MJ. Induction and decay of short-term heat acclimation in moderately and highly trained athletes. Sports Med. 2011;41(9):757–71.
120. Lind AR, Bass DE. Optimal exposure time for development of acclimatization to heat. Fed Proc. 1963;22:704–8.
121. Daanen HAM, Racinais S, Périard JD. Heat acclimation decay and re-induction: a systematic review and meta-analysis. Sports Med. 2018;48(2):409–30.

122. Racinais S, Alonso JM, Coutts AJ, Flouris AD, Girard O, González-Alonso J, et al. Consensus recommendations on training and competing in the heat. Scand J Med Sci Sports. 2015;25:6–19.
123. Brazaitis M, Skurvydas A. Heat acclimation does not reduce the impact of hyperthermia on central fatigue. Eur J Appl Physiol. 2010;109(4):771–8.
124. Racinais S, Wilson MG, Periard J. Passive heat acclimation improves skeletal muscle contractility in humans. Am J Phys Regul Integr Comp Phys. 2017;312:R101–7.
125. Takamata A, Yoshida T, Nishida N, Morimoto T. Relationship of osmotic inhibition in thermoregulatory responses and sweat sodium concentration in humans. Am J Phys Regul Integr Comp Phys. 2001;280(3):R623–9.
126. Beaudin AE, Clegg ME, Walsh ML, White MD. Adaptation of exercise ventilation during an actively-induced hyperthermia following passive heat acclimation. Am J Phys Regul Integr Comp Phys. 2009;297(3):R605–14.
127. Zurawlew MJ, Walsh NP, Fortes MB, Potter C. Post-exercise hot water immersion induces heat acclimation and improves endurance exercise performance in the heat. Scand J Med Sci Sports [Internet]. 2016;26(7):745–54. Available from: http://eutils.ncbi.nlm.nih.gov/entrez/eutils/elink.fcgi?dbfrom=pubmed&id=26661992&retmode=ref&cmd=prlinks.
128. Mee JA, Peters S, Doust JH, Maxwell NS. Sauna exposure immediately prior to short-term heat acclimation accelerates phenotypic adaptation in females. J Sci Med Sport. 2018;21(2):190–5.
129. Nelms JD, Turk J. A self-regulating method for rapid acclimatization to heat. J Physiol. 1972;221(1):2P–3P.
130. Armstrong LE, Hubbard RW, DeLuca JP, Christensen EL. Self-paced heat acclimation procedures. Natick: U.S. Army Research Institute of Environmental Medicine; 1986. p. 1–36.
131. Nielsen B, Strange S, Christensen NJ, Warberg J, Saltin B. Acute and adaptive responses in humans to exercise in a warm, humid environment. Pflugers Arch. 1997;434(1):49–56.
132. Fox RH, Goldsmith R, KIDD DJ, LEWIS HE. Acclimatization to heat in man by controlled elevation of body temperature. J Physiol. 1963;166:530–47.
133. Rowell LB, Kraning KK, Kennedy JW, Evans TO. Central circulatory responses to work in dry heat before and after acclimatization. J Appl Physiol. 1967;22(3):509–18.
134. Taylor NAS. Principles and practices of heat adaptation. J Hum-Environ Syst. 2000;4(1):11–22.
135. Fox RH, Goldsmith R, Hampton IF, Hunt TJ. Heat acclimatization by controlled hyperthermia in hot-dry and hot-wet climates. J Appl Physiol. 1967;22(1):39–46.
136. Regan JM, Macfarlane DJ, Taylor NA. An evaluation of the role of skin temperature during heat adaptation. Acta Physiol Scand. 1996;158(4):365–75.
137. Gibson OR, Mee JA, Taylor L, Tuttle JA, Watt PW, Maxwell NS. Isothermic and fixed-intensity heat acclimation methods elicit equal increases in Hsp72 mRNA. Scand J Med Sci Sports. 2015;25:259–68.
138. Corbett J, Rendell RA, Massey HC, Costello JT, Tipton MJ. Inter-individual variation in the adaptive response to heat acclimation. J Therm Biol. 2018;74:1–35.
139. Williams CG, Wyndham CH, Morrison JF. Rate of loss of acclimatization in summer and winter. J Appl Physiol. 1967;22(1):21–6.
140. Adam JM, Fox RH, Grimby G. Acclimatization to heat and its rate of decay in man. J Physiol J Physiol. 1960;152:26–7.
141. Ashley CD, Ferron J, Bernard TE. loss of heat acclimation and time to re-establish acclimation. J Occup Environ Hyg. 2015;12(5):302–8.

Part III

Application to Specific Sports

The Application of Heat Stress to Team Sports: Football/Soccer, Australian Football and Rugby

9

Katie Slattery and Aaron J. Coutts

9.1 Introduction

Training and competing for team sports in the heat places the body under an increased thermal load, influencing performance capacity and potentially leading to heat-related illness in extreme conditions. While football/soccer, Australian football (AF) and rugby are usually played in the cooler winter months, there are occasions where matches and major competitions are held in thermally challenging conditions [1, 2]. For example, of the 64 matches played in the 2014 FIFA World Cup in Brazil, 16 were played under high environmental heat stress [wet-bulb globe temperature (WBGT) 28–33 °C] [1]. The professional football league in Australia (A-League) is held over the summer months. Similarly, during the Australian domestic AF season, the temperature and humidity can reach levels that place the players at a high to extreme risk of heat illness [2]. Players also train for the upcoming season in warmer temperatures and compete in pre-season competitions under hot and/or humid conditions [3, 4]. It is therefore important to prepare players to tolerate exercise in the heat from both a performance and health perspective. This can be achieved through both adaptive mechanisms (i.e. heat acclimatisation) and by utilising acute strategies to reduce heat strain such as pre-cooling and ensuring adequate hydration. The development and implementation of heat policies are also needed to protect players from severe environmental conditions.

Team sports performance is the product of a complex interplay between tactical (interaction with other individuals), technical (individual skills), physical (physiological capacity) and psychological constructs [5, 6]. Each moment of the match involves a tactical decision and the execution of a technical skill that requires a specific physical movement. The effectiveness of this process is dependent on an

K. Slattery (✉) · A. J. Coutts
Sport and Exercise Discipline Group, University of Technology Sydney (UTS), Moore Park, NSW, Australia
e-mail: Katie.Slattery@uts.edu.au

J. D. Périard, S. Racinais (eds.), *Heat Stress in Sport and Exercise*,
https://doi.org/10.1007/978-3-319-93515-7_9

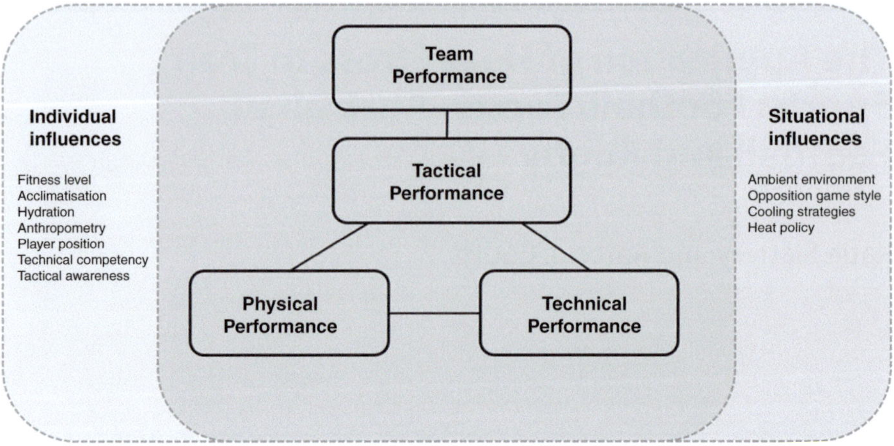

Fig. 9.1 Influences of heat stress on team sport performance. Adapted from Henderson et al. [7]

individual player's readiness to perform (i.e. psychological state, fitness level, health and technical competency) within the context of the match. These constructs are then layered upon situational factors such as the opposition strength, win/loss margin and environmental conditions that combine to influence the outcome of the match. As presented in the conceptual model (Fig. 9.1), any change in an individual construct or situational factor can influence the overall team performance. Moreover, when these changes occur, a whole system adjustment must be made between the tactical, technical and physical performance aspects. This chapter examines how an added heat stress can affect each construct and provides insight into how players adapt their game style or utilise strategies to cope with the thermal load.

9.2 Influence of Heat on Performance and Health

Competitive success in team sports is dependent on a combination of tactical, technical, physical and psychological factors, all of which have been shown to be affected by the heat. In team sport, work demands are stochastic, consisting of explosive multidirectional movement patterns, sport-specific technical skills, running and collisions with varying and unpredictable recovery periods. Due to the self-paced nature, fatigue is characterised by a progressive reduction in work rate throughout the match. It is commonplace for the total distance covered in the second half of football, AF and rugby to be lower than the first [2, 8]. Players may also experience transient fatigue following intense periods of play [9]. As few studies have investigated the impact of heat stress on team sport in an ecological setting, it is difficult to have a complete understanding of how it affects match performance. Below is a summary based on research to date on how hot and/or humid conditions can influence tactical, technical, physical and psychological performance.

9.2.1 Physical Performance

Physical performance during football, AF and rugby matches is highly variable and influenced by situational match-related variables, including the environmental conditions [6, 10–13]. To date, few studies have investigated how an added heat stress affects the physical characteristics of team sport matches [1–3, 14, 15], and changes in activity profiles are still to be determined in rugby league and union. Based on the findings from comparisons between matches played in the heat versus temperate conditions in football and AF, evidence suggests that when in a hot environment, players modify their physical activity patterns to enable the maintenance of key physiological parameters that are critical to match performance [1, 3]. For example, in football total distance covered in a match played at 43 °C was 7% less than a match played at 21 °C [14]. Yet, sprint frequency and sprint distance covered in both matches was similar and in the heat, a more consistent pacing of high-intensity running was observed throughout the entire match. A similar result was found when a match was played in moderately hot (34 °C, 38%RH) compared to hot conditions (36 °C, 61%RH) [15], where no difference in sprinting or high-speed running was found between the matches. To achieve this, players reduced the amount of time spent jogging and increased their time spent walking in the hotter environment. Likewise, AF players competing in the heat have been reported to reduce the total distance covered but sustain high-intensity running performance and increase the number of accelerations [2]. Further evidence of players modifying physical performance in the heat has been reported during the 2014 FIFA World Cup in Brazil [1]. During the competition, players were exposed to a range of environmental conditions. When the physical activity responses to low, moderate and high levels of heat stress were compared, there was no difference in playing time, or total distance covered. Notably however, there was a reduction in high-intensity running distance associated with matches played in the heat. Collectively, these results suggest that team sport athletes modify physical activity patterns to ensure that they are able to perform high-intensity efforts when required. However, this conclusion is based on observational data from a small sample group of matches. Further well-controlled research is needed to establish whether these initial findings are consistent across football codes and different levels of participation.

9.2.2 Physiological Response

When players train and compete in hot and humid conditions, the metabolic heat created by the contracting muscle combined with the greater thermal load from the environment puts a considerable stress on the body's thermoregulatory system. Heat stress can negatively affect many aspects of team sport performance including a player's aerobic capacity [14, 16], cognitive ability [17], perception of effort [3], muscle force generation capacity and ability to perform repeat sprints [18, 19]. As players complete intensive physical training in hot conditions, coaches and support staff should be aware of the effect of heat exposure on the ability to perform high-intensity intermittent exercise. Changes observed in the physiological response of team sport players when training and competing in the heat are summarised in the following sections.

9.2.2.1 Core Temperature

High-intensity intermittent running creates a large thermal load and can rapidly increase core temperature [20]. However, there can be little difference in the core temperature reached when matches are played in temperate or hot environments [2, 3]. For instance, even in cooler conditions, core temperatures above 39 °C have been reported in professional AF players [2]. There is evidence of players altering physical activity patterns or applying pacing strategies to preserve physical capacity during pre-season AF matches in the heat [3]. Where after an initial rise in the first quarter, core temperature plateaued for the remainder of the match. Similarly, in semi-professional football players, core temperature peaked at the end of the first half and did not reach the same level again in the second half when matches were played in hot conditions [15]. Due to this lack of relationship between core and ambient temperature, it has been proposed that the amount of physical activity completed is more closely linked to rises in core temperature during matches than the environmental conditions [2, 3, 14, 21].

However, in extreme conditions body temperature can be significantly higher during team sport matches. When elite players completed a football match in very hot (~43 °C) conditions, both core and muscle temperature were significantly elevated compared to a match in a temperate environment (~21 °C) [14]. As increases in whole-body temperature can lead to severe heat illness and injury (see Chap. 5), precautionary measures should be applied (i.e. additional drink breaks or schedule training for earlier in the day) to minimise the thermal load when training and competing in a hot and/or humid environment.

9.2.2.2 Heart Rate

Changes in heart rate can provide insight into the cardiovascular response to a thermal load. During steady-state exercise at a fixed workload in the heat, heart rate increases to counteract the reduction in stroke volume and to meet the higher demands on cardiac output [22]. This is also evident when high-intensity shuttle runs are completed at a standardised workload [20, 23]. Under heat stress, the heart rate response is higher than the equivalent work undertaken in temperate conditions. However, in team sport matches, where the players can self-regulate their physical activity, the resultant heart rate response in the heat is similar to matches that are played in a cool climate [14, 15]. Despite a ~22 °C difference in playing conditions, a very similar average and maximal heart rate was observed in professional football players [14]. Average heart rate during a match played at ~43 °C (hot) was 158 ± 2 bpm, compared to 160 ± 2 bpm when the temperature was ~21 °C (temperate). Although, total distance (hot: −7%) and the amount of high-intensity running (hot: 883 ± 45 m, temperate: 978 ± 97 m) completed in the heat was significantly reduced. Similar results were shown during a high-intensity intermittent running protocol where university-level football players continued to exercise regardless of whether their prescribed velocities were met [16]. Whether the protocol was completed in a hot (30 °C, 50% relative humidity [RH]) or temperate (18 °C, 50%RH) environment, only a ~2 bpm difference was observed in average heart rate. Like the findings in actual football matches, total distance covered and high-speed distance were reduced by ~4% and ~8%, respectively, in the hotter conditions. Collectively, these findings suggest that rather than continuing to exercise at an unsustainable intensity under hot conditions, players adjust their workload to maintain a similar relative exercise intensity. This is in line with the observation that core temperatures of team sport athletes during matches are often alike, irrespective of the environmental conditions [2, 3].

9.2.2.3 Sweat Rate

Evaporation of sweat is a primary thermoregulatory mechanism to dissipate heat and is dependent on the thermal gradient between the environment (i.e. temperature, humidity and wind speed) and the skin [24]. For team sport players, sweat losses can be considerable due to the high metabolic load of intense exercise, uniform requirements and their typically larger body mass [25]. Even in temperate conditions, team sport athletes lose a significant amount of body water during exercise [26, 27]. This loss of fluid results in a reduction in plasma volume, placing greater strain on the cardiovascular system and can lead to accelerated increases in core temperature [16]. The sweat response can vary greatly between individuals. For instance, sweat rates have been reported to range from 0.3 L/h to 2.5 L/h in football players, 0.4 L/h to 2.0 L/h in rugby players and from 0.9 L/h to 2.1 L/h in AF players in a range of environmental conditions [25]. This variability in sweat loss during training and competition is explained in part by the intensity of exercise [21], fitness level [28] and acclimatisation status of the player [29].

Interestingly, while sweat rates for team sport athletes are higher in a hot environment [14, 20], body mass changes are similar to games or training conducted in temperate conditions when fluid is available [2, 14, 20]. This suggests that players are able to increase their fluid intake accordingly when playing in the heat. For example, when an intermittent running shuttle test designed to replicate the physical demands of a team sport match was completed in a hot (~30 °C) and temperate (~20 °C) environment, the change in body mass was almost identical (hot: 0.56 ± 0.20 kg, temperate: 0.59 ± 0.24 kg) as the players drank almost twice the amount of fluid in the hot trial (hot: 1.18 ± 0.12 L/h, temperate: 0.63 ± 0.07 L/h) [20]. It is important to note that even with restricted access to fluid during game conditions, football players were able to minimise body mass changes while playing in the heat [14, 30]. These findings suggest that players are able to adequately hydrate during training and competition in hot and/or humid conditions.

9.2.2.4 Metabolic Response

Team sports are metabolically demanding, requiring a high rate of energy turnover to complete the multiple intense efforts, maximal sprints and accelerations required for success [31–33]. For players, an additional heat stress may place a greater reliance on anaerobic pathways for ATP regeneration [34]. Indeed, a greater carbohydrate utilisation and higher rate of glycogenolysis have been associated with exercise in the heat which can accelerate the accumulation of metabolic by-products and lead to an earlier onset of peripheral muscle fatigue when exercising at a given workload [35]. However, in a practical setting, studies have not found a relationship between markers of metabolic fatigue and a reduction in team sport running capacity in a hot environment. While an indirect measure of anaerobic metabolism, no difference in the post-exercise blood lactate values have been found following either high-intensity intermittent shuttle runs [16, 20, 23, 36], repeat sprint efforts [18] or during match play [14] in hot compared to temperate conditions. These findings are in line with changes observed at a muscular level during repeated cycling sprints performed under heat stress [9] where an elevated core temperature (39.5 ± 0.2 °C) and reduced power output, accompanied by a lower extracellular potassium, muscle hydrogen ion concentration and lactate accumulation, were observed following 5 × 15-s maximal sprints at 40 °C compared to 20 °C. Likewise, a link between an altered substrate use and high-intensity exercise in the heat has not been found. When prolonged high-intensity exercise was completed in both hot (~30 °C) and temperate conditions (~20 °C), similar levels of plasma ammonia, plasma free fatty acids and blood glucose were measured [20]. In support of this finding, no difference in the amount of muscle glycogen depletion was observed between a football match played at 43 °C or 21 °C [37]. Combined, these results suggest that an added environmental thermal load does not substantially affect the metabolic response to team sports matches as the effect of an increased temperature is negated by a reduced overall physical performance.

9.2.2.5 Neuromuscular Response

Peak speeds and sprint running distances during team sport matches are often maintained [2, 15], or even improved [1, 14] in hot conditions. This may be a physiological effect as increases in muscle temperature accelerate fibre conduction velocity, increase ATP turnover and cross bridge cycling rate to enhance sprint performance [38]. Improved sprint performance in the heat has been shown in isolated bouts of explosive and high-intensity exercise. For example, when rugby union [39] and rugby league players [40] wore passive heat maintenance jackets to maintain core temperature following a warm-up, peak power output in a counter-movement jump and repeat sprint ability (6 × 40 -m shuttle sprints, 20-s passive recovery) were both improved compared to the control condition. Alternatively, as players typically reduce the amount of low-moderate intensity activity when competing in the heat [1, 2], the capacity to perform maximal efforts at crucial time points throughout a match may be preserved. In support, neuromuscular testing prior to and following team sport matches in thermally challenging environments has shown that muscle function is unaffected by the conditions [3, 14, 37, 41, 42]. Players are able to maintain power output in single maximal efforts such as a vertical jump [3], counter-movement jump [41], 30-m sprint [14] or a maximal voluntary contraction torque of the plantar flexors [37] when tested following a match in the heat. Moreover, if reductions in neuromuscular capacity occur, they appear to be no different to the expected decrease after a match played in a temperate environment. For instance, a similar decrease in repeat sprint performance (~2%), reduction in voluntary activation (~1.4%) and lower peak twitch torque of the plantar flexors was observed following football matches played in either 43 °C or 21 °C [37]. These results are contrary to investigations that have assessed the neuromuscular response of team sport athletes during field-based tests of physical performance. Where during prolonged high-intensity shuttle running, a significant heat-induced reduction in sprint performance has been reported [16, 23]. This observation is in line with the reductions in central nervous system drive, impaired neural recruitment and decreased muscle activation that are associated with hyperthermia-induced fatigue during sustained voluntary maximal isometric contractions [22]. These differing findings highlight the importance of assessing any changes in the neuromuscular response associated with the heat within the tactical and technical context of a match.

9.2.3 Tactical and Technical Performance

Skill level, decision-making ability and tactical awareness are key differentiators for success in team sport [43, 44]. While the physical characteristics of a player allow them to perform these actions, matches are not necessarily won by the fittest or fastest team. When competing in hot and/or humid conditions, technical performance parameters appear to be unaffected by the environment [1, 14]. For example, in football matches played under high heat stress, there was a higher success rate for all passes [1, 14] and a similar total number and length of passes and crosses [14]. This may be the result of slower game play, as in these matches a reduction in physical activity was also observed, either by reducing total distance [14] or the amount of high-intensity running completed [1]. In addition, the number of goals scored, actual playing time and the total number of penalty cards given were similar, suggesting that these aspects of game play were unaffected by the environmental conditions [1]. This is in line with observations during professional football matches in temperate conditions where technical actions and ball involvement are maintained despite declines in physical performance throughout the match [45]. However, these findings are based on highly skilful professional players who did not exhibit any hyperthermia-induced signs of fatigue. If hyperthermia-induced fatigue was present during a team sport match, it is possible that skill execution, cognitive function and decision-making ability could be impaired [46].

9.2.4 Cognitive Function

Tactical performance is closely linked to cognitive function as during matches players are continually processing task-specific information, evaluating movement patterns within the context of the match and making behavioural adjustments accordingly [47]. Increases in core temperature during football matches played at

~34 °C and 64% RH have been associated with reductions in cognitive function [17]. Collectively, the results from a range of tests for viso-motor reaction time, fine motor speed and working memory revealed a slower, yet more accurate response after the match. Thermally challenging environments can also exacerbate mental fatigue and lead to further impairments of technical and tactical performance [46]. For example, when mental fatigue levels are high, speed and accuracy during football-specific decision-making skill [48], offensive and defensive technical performance in small-sided games [49] and passing accuracy in a football-specific skill test [48] are all negatively impacted. While these findings provide examples of how an added heat stress can affect technical and tactical performance in football, they may not translate to other competitions, players or team sports. Moreover, a match outcome will only be impacted if one team's technical and tactical capacity is affected more than their opponent. Nonetheless, they do show that increases in body temperature can compromise motor skill performance and cognitive functioning and suggest that players reduce physical workload to preserve technical and tactical parameters in hot conditions.

9.2.5 Health

When team sport training and competition are completed in thermally challenging environments, there is a potential increased risk of heat illness and injury [50, 51]. However, from the available published data, it appears that football, AF and rugby players are able to cope with hot and humid conditions without any adverse effects [1, 3, 14, 52, 53]. Despite players reaching core temperatures of above 40 °C, no signs or symptoms of heat illness were reported during AF matches in hot and/or humid conditions [2, 3]. Similarly, no heat-related illness or injuries were reported during major competitions including the 2003 Rugby World Cup [53] or the 2014 FIFA World Cup [1], which were held in hot environments. These examples are from professional competitions however, where medical staff are well-prepared and educated on prevention strategies to reduce the likelihood of heat illness and injury. At a participation level, the early signs of heat stress may more easily be missed. Without immediate cooling interventions, players can then rapidly progress along the continuum of heat illness and injury to the more severe and potentially fatal, exertional heat stroke. It is also important to note that heat-related illness and injury can occur in cool climates as the environmental heat stress is only one of several predisposing factors [50]. When players are unfit, unacclimatised, hypohydrated, have an infective illness, are wearing non-breathable clothing or have taken stimulants, they are more susceptible to heat illness and injury, regardless of the ambient conditions [50, 54]. It is therefore necessary to have an understanding of a player's health status prior to training and competition. Additionally, all staff, officials and players should be educated on the early warning signs of heat strain such as dizziness, nausea or a headache [54]. Countermeasures such as acclimatisation, hydration and cooling can also help reduce the likelihood of heat illness and injury.

9.3 Countermeasures to Optimise Performance and Health

9.3.1 Heat Acclimation

Football, AF and rugby players can naturally acclimatise by travelling earlier to a hot location before the match or acclimate by completing a block of training in an artificially heated environment [51]. To ensure the best outcome from the training block, players need sufficient exposure to the heat without compromising training quality or causing excessive fatigue. This can be achieved through carefully structuring the training program. Generally, it is recommended that acclimation protocols replicate the demands of team sport performance [55]. This allows players to experience heat in a situation that reflects their competitive environment and has the additional benefit of more easily integrating the heat stimulus into an existing training schedule [51]. One approach is to use natural heat exposure during skills-based training, then complete resistance and high-intensity conditioning sessions in a cool environment [56]. Another is to use alternating days of heat training sessions (i.e. intermittent exposure) to better fit into a player's overall training load [57]. Other example strategies of how heat acclimatisation can be applied in team sport are outlined in Table 9.1. Forms of passive exposure, including saunas [59] or hot-water immersion [65], can also be used to further promote physiological adaptations to the heat.

Both short-term [61, 63, 66] and medium-term (8–14 days) [56, 67] heat acclimation protocols can improve physical performance parameters in team sport athletes. A 7-day training camp in a hot environment (~39.8 °C, 27% RH) has been shown to induce thermoregulatory adaptions including plasma volume expansion, increased sweat rate and reduced sweat sodium concentration in semi-professional football players [51]. Similar signs of heat adaptation were observed in professional AF players living and training in a hot environment (~32 °C and ~39% RH) for 14 days [56, 68]. However, travelling in advance to prepare for a competition is not always a practical option for teams. Short-term acclimation protocols (≤7 days) can provide a time- and cost-efficient alternative [66]. While the thermoregulatory adaptations are not as pronounced after a shorter exposure period, players have been shown to have an improved thermal tolerance following as little as four sessions in the heat [63].

9.3.1.1 Player Monitoring During Heat Acclimation

Heat places an additional stress on the body. Like any training stressor, how a player tolerates and adapts to the heat should be monitored, from both the perspective of assessing the thermoregulatory response and the management of a player's well-being. Table 9.2 provides a summary of common measures used in team sport players.

Changes in exercise heart rate during standardised heat exposure provide valuable insight into how a player is responding to the increased thermal load. However, it can be difficult to distinguish between an adaptive or maladaptive state based on heart rate alone [69]. When assessing heat acclimatisation, it is important to

Table 9.1 Practical application of heat acclimation strategies in football, AF and rugby

Training phase	Objective	Protocol length	Distribution of exposures	Type of heat stress	Ambient temperature	Relative humidity	Total duration of heat exposure (min)	Training prescription	Training sessions	Reference
Early preparatory	Initial preparatory camp to evaluate individual responses to the heat to optimise re-acclimation during future heat protocols	10–14-day	Consecutive	Natural or artificial	30–35 °C	40–60%	600–840	Constant rate	Low-moderate intensity cycling	Castle et al. [58] Racinais et al. [56]
	Promote thermoregulatory adaptions associated with improved aerobic performance.	10-day	Consecutive	Sauna	87 °C	11%	300	Passive	Post-training	Stanley et al. [59]
Late preparatory	Preparation for a major tournament	8–14 days	Consecutive	Natural	Same as competition		600–1200	Self-regulated	Routine training	Buchheit et al. [60] Racinais et al. [61] Racinais et al. [56]
In-season	Preparation for a mid-season one-off match in hot conditions	4–5 sessions	Intermittent	Artificial	30–35 °C	40–60%	150–200	Constant rate or self-regulated	High-intensity intermittent running	Brade et al. [62] Kelly et al. [57] Sunderland et al.[63]
Injury management and rehabilitation	Reduce mechanical workload for an equivalent metabolic response	As required	Intermittent	Artificial	30–35 °C	40–60%	150–200/week	Controlled hyperthermia	Low-moderate intensity cycling	Philp et al. [64]

Table 9.2 Implementation of tools used in team sport to monitor training and adaptive responses to the heat

Monitoring tool	Use	Interpretation / sign of heat acclimation	When to measure	Reference
Health and well-being				
Psychometric questionnaires i.e. rating of perceived fatigue, sleep, mood, soreness and stress scored on a 5-point scale (1 = poor, 5 = very good)	Measure the perceived state of well-being	High ratings of wellness variables: Player is tolerating overall training stress Low ratings of wellness variables: Player is not tolerating the overall training stress	Daily or prior to each key training session. The questionnaire should be completed before, during and after the heat acclimation period	Buchheit et al. [68] Buchheit et al. [67]
Urine-specific gravity	Assess level of hydration	A high concentration of solutes in the urine indicates dehydration	Daily or at regular intervals, upon waking Pre- and post-match or heat-response test	Aughey et al. [2] Kelly et al. [57] Kurdak et al. [30]
Body mass	Assess fluid balance over consecutive days	Large daily fluctuations in body mass can indicate a player has not adequately re-hydrated following the previous session	Daily, upon waking	Racinais et al. [56] Racinais et al. [51]
Sub-maximal running test 5′-5′ test: 5-min run at a fixed speed, followed by 5-min of seated recovery	Provides insight into changes in fitness and fatigue	Fatigue measure: Increased RPE Fitness measure: Improved heart rate recovery	Complete as part of warm-up. 2–3/week, or 3 consecutive days at beginning and 3 consecutive days at the end of protocol	Buchheit et al. [60] Buchheit et al. [68]
Thermoregulatory adaptations				
Heat-response test i.e. 24–30-min walking at 5 km/h, 24–30-min seated rest at 44 °C, 44% RH or standardised prolonged repeated sprint protocol	Lab-based test to directly measure thermoregulatory responses to rest and exercise	Reduced core and skin temperature, lower sub-maximal heart rate at the fixed work rate, higher sweat rate, reduced sodium concentration and reduced haematocrit percentage	Pre- and post-heat acclimation period. Can also re-test 1–2 weeks following to assess decay of adaptations	Brade et al. [62] Kelly et al. [57] Racinais et al. [61] Racinais et al. [56]

(continued)

Table 9.2 (continued)

Monitoring tool	Use	Interpretation / sign of heat acclimation	When to measure	Reference
CO re-breathing test	Indirect measure of plasma volume	Plasma volume expansion	Pre and post-heat acclimation period. Can also re-test 1–2 weeks following to assess decay of adaptations	Buchheit et al. [68] Racinais et al. [56]
Haemoglobin and haematocrit	Relative change in plasma volume	Decreased haemoglobin concentration and haematocrit percentage	Pre- and post-heat acclimation period. Can also re-test 1–2 weeks following to assess decay of adaptations Daily measures at rest Pre- and post-heat-response test	Buchheit et al. [60] Philp et al. [64] Racinais et al. [56]
Sub-maximal running test i.e. 5′-5′ test: 5-min run at a fixed speed, followed by 5-min of seated recovery	Field-based test to estimate heat adaptation	Reduced sub-maximal heart rate and increased heart rate variability	Complete as part of warm-up. 2–3/weeks, or 3 consecutive days at beginning and 3 consecutive days at the end of protocol	Buchheit et al. [60] Buchheit et al. [68] Buchheit et al. [67]
Body mass pre- and post-exercise and account for fluid intake during the session	Estimate sweat loss and sweat rate	A greater amount of body mass loss suggests an increased sweat rate	Pre- and post-heat acclimation period. Can be assessed more regularly to prescribe fluid intake for rehydration.	Aughey et al. [2] Brade et al. [62] Buchheit et al. [67] Kelly et al. [57] Philp et al. [64] Racinais et al. [56]
Performance measures/fitness testing				
YoYo intermittent recovery test	Assess high-intensity intermittent running performance	Increased distance covered during the test	Pre- and post-heat acclimation period	Buchheit et al. [60] Racinais et al. [56]

Table 9.2 (continued)

Monitoring tool	Use	Interpretation / sign of heat acclimation	When to measure	Reference
30–15 intermittent fitness test	Assess high-intensity intermittent running performance	Improved peak velocity (V_{IFT})	Pre- and post-heat acclimation period	Philp et al. [64]
Match analysis	Assess match-running performance	Similar physical activity profile to matches played in temperate conditions	Pre- and post-heat acclimation period	Racinais et al. [61]
Standardised training drills	Assess sport-specific running performance	Increased total distance and greater amount of high-intensity running	Pre-, during and post-heat acclimation period	Racinais et al. [56]
Repeated sprint efforts i.e. 4 × 60-m runs with 30-s passive recovery	Assess neuromuscular efficiency	Improved neuromuscular efficiency index (accelerometer-derived ratio between velocity load and force load)	Pre-, during and post-heat acclimation period	Buchheit et al. [67]
Session monitoring				
Record environmental conditions	Quantify thermal load		Heat training sessions	Buchheit et al. [60] Racinais et al. [61]
Thermal comfort	Assess perceptual response to the ambient environment	Improved thermal comfort	Heat training sessions	Kelly et al. [57]
Thermal sensation	Assess perceptual response to the ambient environment	Reduced heat sensation	Heat training sessions	Brade et al. [62] Philp et al. [64] Racinais et al. [56]
Heart rate	Internal training load measure	Average heart rate or prescribed heart rate for a constant rate training session	Field-based training and conditioning sessions	Buchheit et al. [60] Racinais et al. [61]
Running volume and intensity	External training load measure	Total distance and high-intensity running	Field-based training sessions and running conditioning	Racinais et al. [56]
RPE:min	Contextualise internal training load to external load	Increased ratio suggests increased level of heat strain	Field-based training sessions and running conditioning	Buchheit et al. [67]

(continued)

Table 9.2 (continued)

Monitoring tool	Use	Interpretation / sign of heat acclimation	When to measure	Reference
Session RPE	Quantify overall training load	Training time and associated post-session RPE	Each session	Buchheit et al. [68] Buchheit et al. [67] Philp et al. [64]
HR-based TRIMP	Quantify overall training load	Training time and associated average heart rate	Each session	Buchheit et al. [60] Racinais et al. [61]

contextualise changes in heart rate to the physical work completed or to the perception of effort. For example, standardised sub-maximal running [60, 67, 70] or training drills [56] can be used at regular intervals to track the time course of heat adaptation and ensure that players are adequately coping with the training stress. These field-based monitoring tests are practical to implement, repeatable, relatively non-fatiguing, time efficient and can be completed as part of the warm-up. Heart rate measures and rating of perceived exertion (RPE) during a 5-min run at a fixed speed, followed by 5-min of seated recovery (5′-5′ test) has been used throughout an in-season football [60] and a pre-season AF [70] heat training camp. A trend for average heart rate during the final 30 s of exercise (HRex) to decrease and heart rate variability to increase (HRV; logarithm of the standard deviation of instantaneous beat-to-beat R–R interval variability, log SD1, measured from Poincaré plots during the last 3 -min of recovery) as the camp progressed was associated with heat-induced plasma volume expansion. Heart rate recovery (HRR; absolute difference between HRex the HR recorded after 60-s recovery) and RPE remained stable, indicating a balanced fitness/fatigue status. Similar sub-maximal tests can be used as part of a warm-up before training sessions done in a heat chamber [71]. It is highly recommended to incorporate the test used for monitoring into the training schedule prior to the heat acclimation period. By having a better understanding on how players respond to training in a temperate environment, it will allow for a more meaningful interpretation of the heat-induced changes.

A measure of both internal and external load should be taken to capture the overall training stimulus during heat acclimation. Otherwise, when conditioning sessions, technical practise or competition are completed under hot and/or humid conditions, training stress can be over- or underestimated. For instance, if training load is quantified based on external load parameters alone, such a total distance covered, or time in specific velocity bands, practitioners may conclude that the training stimulus is lower in the heat as a player's running output is reduced. To gain a more accurate understanding, it is also important to consider a measure of internal training stress, such as heart rate, or RPE when quantifying the training response, particularly when players are under a high thermal load. Heat-induced elevations in heart rate for a reduced mechanical workload are indicative of the greater training stress that occurs during exercise in hot compared to cooler environments [64]. Moreover, a reduction in internal training measures as the heat acclimation period progresses can be used as an indicator of thermoregulatory adaptation. For example, the average heart rate of AF players during a standardised training drill was reduced and external load measures (total distance and amount of high-intensity running) improved at the completion of a heat acclimation camp [56]. To a similar effect, RPE has been used to contextualise changes in training volume (m/min) [67]. For the first two sessions of a heat training camp, RPE:m/min ratio of footballers increased substantially, suggesting an increased perceived exertion for a given workload. The RPE:m/min ratio then followed a downward trend and stabilised after 5 days. This coincides with the 5–7 days it typically takes for the initial signs of heat acclimation to appear. These examples show how player monitoring can provide insight into the effectiveness of the heat acclimation period.

9.3.2 Hydration

Adequate hydration can minimise the effects of heat stress during training and competition in football, AF and rugby [17, 50, 51, 72]. It is generally recommended that players drink 5–6 ml of fluid per kg of body mass every 2–3 h prior to training and competition [51]. As sweat rates can vary considerably between each person, it is suggested that hydration practices are individualised [21, 25]. Individualisation of fluid intake can be based on the intensity and duration of the match or training session [21] or by using pre- and post-session changes in body mass to estimate fluid loss (as shown Table 9.1). To ensure that the fluid intake is sufficient, urine specific gravity (<1.020) and fluctuations in body mass (<1%) upon waking are useful measures to track hydration status over a period of days (see Table 9.1) [51]. Alternatively, for youth or players at a participation level, a urine colour chart can be an appropriate tool to assess hydration [73]. The inclusion of additional drink breaks and recommended fluid intakes in heat policies highlights the importance of hydration to reduce the adverse effects of heat exposure [74, 75]. For example, the Australian Football League (AFL) suggests players consume 500–700 ml per quarter [75]. FIFA recommends players replace each kilogram of weight lost during training or matches with 1.2–1.5 L of fluid [74]. Moreover, when replacing sweat losses, replenishment of electrolytes should also be considered [51].

9.3.3 Cooling Strategies

Cooling strategies can be used in football, AF and rugby to improve heat storage capacity (pre-cooling), reduce exercise-induced increases in core temperature (mid-cooling) and accelerate recovery (post-cooling). While the evidence supporting improved performance in a competitive team sport setting has been equivocal, it is prudent to provide access to cooling from a heat-illness prevention perspective, particularly for players who are unacclimatised to hot conditions [58, 62]. To optimise performance, it is also important to balance the effects of pre-cooling and ensuring that players are adequately warmed-up [76], as excessive external cooling may negatively impact muscle function and speed [77]. The summary below outlines examples of cooling interventions that can benefit training and competition in the heat and be incorporated into a player's schedule. More details on cooling are available in Chap. 7.

Pre-cooling: A mixed method approach to cooling with both internal (i.e. ice slurry) and external (i.e. ice vest, cold pack and iced towels) techniques applied for 20–30-min prior to warming up [78, 79]. Provide access to fans, spray bottles, shaded area and/or air conditioning.

Mid-cooling: Drinks should be provided at less than 15 °C to assist with absorption [80]. At half time, reapply ice vest, cold pack and iced towels [78]. Provide access to fans, spray bottles and either a shaded area, a cool room and/or air conditioning [75]. On interchange bench, provide fans, spray bottles and shade [75]. Substitute players should also have access to fluids and ice [80].

Post-cooling: Cold water immersion or cold showers can be used to quickly reduce the core and skin temperature of players [81].

9.4 Heat Policy and Implementation

Players and officials are susceptible to heat illness and injury when training and competing in the heat. By outlining preventative measures in heat policies, the governing bodies of football, AF and rugby aim to reduce this risk. These policies consider both the environmental conditions and each individual player's capacity to tolerate heat stress and then provide strategies to reduce heat exposure, such as the rescheduling of games and countermeasures to minimise the impact of hot and humid environments. For example, the Australian National Rugby League (NRL) employs the Sports Medicine Australia guidelines to determine the appropriate game day intervention [82]. Based on a combination of external factors such as the ambient environment, duration and intensity of the exercise, availability of fluid, playing facilities, access to medical staff or first aid and the individual player characteristics (i.e. age and fitness level), the match may be postponed. Common countermeasures to heat stress across the federations include the provision of additional drink/cooling breaks [74, 80, 82] or other approaches, such as an increased number of on-field water carriers [75] to ensure ample drinking opportunities for players. A strong emphasis on player preparedness and education is also present, where players are encouraged to begin the match or training session well-hydrated, utilise cooling inventions and be aware of, then seek medical help at the first signs of heat distress [74, 75, 80].

Considering the commercial impact of rescheduling or cancelling a professional match, it is important that robust heat policies are in place to protect the welfare of players. World Rugby [80] and the AFL [75] provide the most comprehensive policies that outline the responsibilities of the player, team officials, medical officers and event organisers to manage thermal load during both training and competition. The AFL policy also advocates that clubs should continue to research the effect of heat stress and management strategies for players [75]. In comparison, the FIFA [74] and NRL policies are quite brief and centred on heat management strategies during competition. Another key difference between the policies is how the level of heat risk is assessed. FIFA [74] and the NRL [82] utilise a WBGT of 32 °C and 30 °C, respectively, to consider measures to minimise heat illness and injury. World Rugby uses a heat stress index of 150 to signify when the interventions outlined in the heat policy should be implemented [80]. The AFL does not specify exact criteria that determine when additional heat management strategies are warranted. Instead, it is under the discretion of the event organiser to base the decision on the information provided by the Australian Bureau of Meteorology [75]. It is difficult to comment on the most appropriate approach to assess when the environmental conditions become severe enough to activate a heat policy, as there is little published information on the incidence of heat-related illness and injury during matches. In the future, greater collaboration, surveillance and reporting of all cases of heat illness and injury ranging from mild to severe in these sports will assist in better understanding the potential mechanisms of heat stress. Based on this knowledge, improved prevention and management of strategies can then be developed.

9.5 Conclusion

Football, AF and rugby players can gain a competitive advantage in hot conditions by being able to better tolerate an added heat stress. This can be from the strategic application of cooling interventions or by being well-acclimatised to enable the maintenance of match-running performance in hot environments. Consecutive days of heat exposure during on-field training sessions is an effective acclimation method in team sports. Although, due to the variable nature of these sessions it is important to closely monitor and track individual responses to the increased thermal load. This ensures that players are receiving a sufficient heat stress and adapting appropriately. Moreover, to reduce the likelihood of heat-related illness and injury during both training and competition, players, support staff and coaches should be educated on the early warning signs and have the facilities (i.e. fans, ice, fluids and shaded areas) to quickly cool players if required. In general, it appears that in football, AF and rugby, players can adequately manage heat stress as severe heat illness and injury are rarely reported. This is in line with initial evidence that suggests players reduce their physical activity outputs during matches in hot conditions in response to an increase in thermal strain, which preserves the capacity to execute tactical strategies during key moments of the match. However, further well-controlled, field-based studies are needed to confirm if these findings occur across football codes and in different levels of participation.

References

1. Nassis GP, Brito J, Dvorak J, Chalabi H, Racinais S. The association of environmental heat stress with performance: analysis of the 2014 FIFA world cup Brazil. Br J Sports Med. 2015;49(9):609–13. https://doi.org/10.1136/bjsports-2014-094449.
2. Aughey RJ, Goodman CA, McKenna MJ. Greater chance of high core temperatures with modified pacing strategy during team sport in the heat. J Sci Med Sport. 2014;17(1):113–8. https://doi.org/10.1016/j.jsams.2013.02.013.
3. Duffield R, Coutts AJ, Quinn J. Core temperature responses and match running performance during intermittent-sprint exercise competition in warm conditions. J Strength Cond Res. 2009;23(4):1238–44. https://doi.org/10.1519/JSC.0b013e318194e0b1.
4. Jones TW, Smith A, Macnaughton LS, French DN. Strength and conditioning and concurrent training practices in elite rugby union. J Strength Cond Res. 2016;30(12):3354–66. https://doi.org/10.1519/JSC.0000000000001445.
5. Impellizzeri FM, Marcora SM. Test validation in sport physiology: lessons learned from clinimetrics. Int J Sports Physiol Perform. 2009;4(2):269–77.
6. Carling C. Interpreting physical performance in professional soccer match-play: should we be more pragmatic in our approach? Sports Med. 2013;43(8):655–63. https://doi.org/10.1007/s40279-013-0055-8.
7. Henderson MJ, Harries SK, Poulos N, Fransen J, Coutts AJ. Rugby sevens match demands and measurement of performance: a review. Kinesiology. 2018;50(Suppl. 1):49–59.
8. Mohr M, Krustrup P, Bangsbo J. Match performance of high-standard soccer players with special reference to development of fatigue. J Sports Sci. 2003;21(7):519–28. https://doi.org/10.1080/0264041031000071182.
9. Mohr M, Rasmussen P, Drust B, Nielsen B, Nybo L. Environmental heat stress, hyperammonemia and nucleotide metabolism during intermittent exercise. Eur J Appl Physiol. 2006;97(1):89–95. https://doi.org/10.1007/s00421-006-0152-6.
10. McLaren SJ, Weston M, Smith A, Cramb R, Portas MD. Variability of physical performance and player match loads in professional rugby union. J Sci Med Sport. 2016;19(6):493–7. https://doi.org/10.1016/j.jsams.2015.05.010.
11. Ryan S, Coutts AJ, Hocking J, Kempton T. Factors affecting match running performance in professional Australian football. Int J Sports Physiol Perform. 2017;12(9):1199–204. https://doi.org/10.1123/ijspp.2016-0586.
12. Kempton T, Coutts AJ. Factors affecting exercise intensity in professional rugby league match-play. J Sci Med Sport. 2016;19(6):504–8. https://doi.org/10.1016/j.jsams.2015.06.008.
13. Bradley PS, Noakes TD. Match running performance fluctuations in elite soccer: indicative of fatigue, pacing or situational influences? J Sports Sci. 2013;31(15):1627–38. https://doi.org/10.1080/02640414.2013.796062.
14. Mohr M, Nybo L, Grantham J, Racinais S. Physiological responses and physical performance during football in the heat. PLoS One. 2012;7(6):e39202. https://doi.org/10.1371/journal.pone.0039202.
15. Ozgunen KT, Kurdak SS, Maughan RJ, Zeren C, Korkmaz S, Yazici Z, Ersoz G, Shirreffs SM, Binnet MS, Dvorak J. Effect of hot environmental conditions on physical activity patterns and temperature response of football players. Scand J Med Sci Sports. 2010;20(Suppl 3):140–7. https://doi.org/10.1111/j.1600-0838.2010.01219.x.
16. Aldous JW, Chrismas BC, Akubat I, Dascombe B, Abt G, Taylor L. Hot and hypoxic environments inhibit simulated soccer performance and exacerbate performance decrements when combined. Fronters in Physiology. 2015;6:421. https://doi.org/10.3389/fphys.2015.00421.
17. Bandelow S, Maughan R, Shirreffs S, Ozgunen K, Kurdak S, Ersoz G, Binnet M, Dvorak J. The effects of exercise, heat, cooling and rehydration strategies on cognitive function in football players. Scand J Med Sci Sports. 2010;20(Suppl 3):148–60. https://doi.org/10.1111/j.1600-0838.2010.01220.x.

18. Maxwell NS, Aitchison TC, Nimmo MA. The effect of climatic heat stress on intermittent supramaximal running performance in humans. Exp Physiol. 1996;81(5):833–45.
19. Girard O, Brocherie F, Bishop DJ. Sprint performance under heat stress: a review. Scand J Med Sci Sports. 2015;25(Suppl 1):79–89. https://doi.org/10.1111/sms.12437.
20. Morris JG, Nevill ME, Riddell C, Williams C. Effect of a hot environment on performance of prolonged, intermittent, high-intensity shuttle running. J Sports Sci. 1998;16(7):677–86. https://doi.org/10.1080/026404198366489.
21. Duffield R, McCall A, Coutts AJ, Peiffer JJ. Hydration, sweat and thermoregulatory responses to professional football training in the heat. J Sports Sci. 2012;30(10):957–65. https://doi.org/10.1080/02640414.2012.689432.
22. Nybo L, Rasmussen P, Sawka MN. Performance in the heat-physiological factors of importance for hyperthermia-induced fatigue. Compr Physiol. 2014;4(2):657–89. https://doi.org/10.1002/cphy.c130012.
23. Sunderland C, Nevill ME. High-intensity intermittent running and field hockey skill performance in the heat. J Sports Sci. 2005;23(5):531–40. https://doi.org/10.1080/02640410410001730197.
24. Gagnon D, Jay O, Kenny GP. The evaporative requirement for heat balance determines whole-body sweat rate during exercise under conditions permitting full evaporation. J Physiol. 2013;591(11):2925–35. https://doi.org/10.1113/jphysiol.2012.248823.
25. Nuccio RP, Barnes KA, Carter JM, Baker LB. Fluid balance in team sport athletes and the effect of hypohydration on cognitive, technical, and physical performance. Sports Med. 2017;47(10):1951–82. https://doi.org/10.1007/s40279-017-0738-7.
26. Ali A, Gardiner R, Foskett A, Gant N. Fluid balance, thermoregulation and sprint and passing skill performance in female soccer players. Scand J Med Sci Sports. 2011;21(3):437–45. https://doi.org/10.1111/j.1600-0838.2009.01055.x.
27. Phillips SM, Sykes D, Gibson N. Hydration status and fluid balance of elite european youth soccer players during consecutive training sessions. J Sports Sci Med. 2014;13(4):817–22.
28. Mora-Rodriguez R. Influence of aerobic fitness on thermoregulation during exercise in the heat. Exerc Sport Sci Rev. 2012;40(2):79–87. https://doi.org/10.1097/JES.0b013e318246ee56.
29. Fox RH, Goldsmith R, Hampton IF, Lewis HE. The nature of the increase in sweating capacity produced by heat acclimatization. J Physiol. 1964;171:368–76.
30. Kurdak SS, Shirreffs SM, Maughan RJ, Ozgunen KT, Zeren C, Korkmaz S, Yazici Z, Ersoz G, Binnet MS, Dvorak J. Hydration and sweating responses to hot-weather football competition. Scand J Med Sci Sports. 2010;20(Suppl 3):133–9. https://doi.org/10.1111/j.1600-0838.2010.01218.x.
31. Coutts AJ, Kempton T, Sullivan C, Bilsborough J, Cordy J, Rampinini E. Metabolic power and energetic costs of professional Australian football match-play. J Sci Med Sport. 2015;18(2):219–24. https://doi.org/10.1016/j.jsams.2014.02.003.
32. Kempton T, Sirotic AC, Rampinini E, Coutts AJ. Metabolic power demands of rugby league match play. Int J Sports Physiol Perform. 2015;10(1):23–8. https://doi.org/10.1123/ijspp.2013-0540.
33. Bangsbo J, Iaia FM, Krustrup P. Metabolic response and fatigue in soccer. Int J Sports Physiol Perform. 2007;2(2):111–27.
34. Dimri GP, Malhotra MS, Sen Gupta J, Kumar TS, Arora BS. Alterations in aerobic-anaerobic proportions of metabolism during work in heat. Eur J Appl Physiol Occup Physiol. 1980;45(1):43–50.
35. Febbraio MA, Snow RJ, Stathis CG, Hargreaves M, Carey MF. Effect of heat stress on muscle energy metabolism during exercise. J Appl Physiol. 1994;77(6):2827–31. https://doi.org/10.1152/jappl.1994.77.6.2827.
36. Morris JG, Nevill ME, Williams C. Physiological and metabolic responses of female games and endurance athletes to prolonged, intermittent, high-intensity running at 30 degrees and 16 degrees C ambient temperatures. Eur J Appl Physiol. 2000;81(1–2):84–92. https://doi.org/10.1007/PL00013801.

37. Nybo L, Girard O, Mohr M, Knez W, Voss S, Racinais S. Markers of muscle damage and performance recovery after exercise in the heat. Med Sci Sports Exerc. 2013;45(5):860–8. https://doi.org/10.1249/MSS.0b013e31827ded04.
38. Gray SR, De Vito G, Nimmo MA, Farina D, Ferguson RA. Skeletal muscle ATP turnover and muscle fiber conduction velocity are elevated at higher muscle temperatures during maximal power output development in humans. Am J Physiol Regul Integr Comp Physiol. 2006;290(2):R376–82. https://doi.org/10.1152/ajpregu.00291.2005.
39. Russell M, West DJ, Briggs MA, Bracken RM, Cook CJ, Giroud T, Gill N, Kilduff LP. A passive heat maintenance strategy implemented during a simulated half-time improves lower body power output and repeated sprint ability in professional Rugby union players. PLoS One. 2015;10(3):e0119374. https://doi.org/10.1371/journal.pone.0119374.
40. Kilduff LP, West DJ, Williams N, Cook CJ. The influence of passive heat maintenance on lower body power output and repeated sprint performance in professional rugby league players. J Sci Med Sport. 2013;16(5):482–6. https://doi.org/10.1016/j.jsams.2012.11.889.
41. Mohr M, Krustrup P. Heat stress impairs repeated jump ability after competitive elite soccer games. J Strength Cond Res. 2013;27(3):683–9. https://doi.org/10.1097/JSC.0b013e31825c3266.
42. Girard O, Nybo L, Mohr M, Racinais S. Plantar flexor neuromuscular adjustments following match-play football in hot and cool conditions. Scand J Med Sci Sports. 2015;25(Suppl 1):154–63. https://doi.org/10.1111/sms.12371.
43. Kempton T, Sirotic AC, Coutts AJ. A comparison of physical and technical performance profiles between successful and less-successful professional rugby league teams. Int J Sports Physiol Perform. 2017;12(4):520–6. https://doi.org/10.1123/ijspp.2016-0003.
44. Russell M, Kingsley M. Influence of exercise on skill proficiency in soccer. Sports Med. 2011;41(7):523–39. https://doi.org/10.2165/11589130-000000000-00000.
45. Carling C, Dupont G. Are declines in physical performance associated with a reduction in skill-related performance during professional soccer match-play? J Sports Sci. 2011;29(1):63–71. https://doi.org/10.1080/02640414.2010.521945.
46. Schmit C, Hausswirth C, Le Meur Y, Duffield R. Cognitive functioning and heat strain: performance responses and protective strategies. Sports Med. 2017;47(7):1289–302. https://doi.org/10.1007/s40279-016-0657-z.
47. Lex H, Essig K, Knoblauch A, Schack T. Cognitive representations and cognitive processing of team-specific tactics in soccer. PLoS One. 2015;10(2):e0118219. https://doi.org/10.1371/journal.pone.0118219.
48. Smith MR, Zeuwts L, Lenoir M, Hens N, De Jong LM, Coutts AJ. Mental fatigue impairs soccer-specific decision-making skill. J Sports Sci. 2016;34(14):1297–304. https://doi.org/10.1080/02640414.2016.1156241.
49. Badin OO, Smith MR, Conte D, Coutts AJ. Mental fatigue: impairment of technical performance in small-sided soccer games. Int J Sports Physiol Perform. 2016;11(8):1100–5. https://doi.org/10.1123/ijspp.2015-0710.
50. American College of Sports Medicine, Armstrong LE, Casa DJ, Millard-Stafford M, Moran DS, Pyne SW, Roberts WO. American College of Sports Medicine position stand. Exertional heat illness during training and competition. Med Sci Sports Exerc. 2007;39(3):556–72. https://doi.org/10.1249/MSS.0b013e31802fa199.
51. Racinais S, Alonso JM, Coutts AJ, Flouris AD, Girard O, Gonzalez-Alonso J, Hausswirth C, Jay O, Lee JK, Mitchell N, Nassis GP, Nybo L, Pluim BM, Roelands B, Sawka MN, Wingo JE, Periard JD. Consensus recommendations on training and competing in the heat. Scand J Med Sci Sports. 2015;25(Suppl 1):6–19. https://doi.org/10.1111/sms.12467.
52. Borges NR, Doering TM, Reaburn P, Scanlan AT. Hydration status of rugby union players in hot and humid conditions: a comparative team case study of day and night training sessions. S Afr J Res Sport Phys Educ Recreat. 2017;39(2):21–31.
53. Best JP, McIntosh AS, Savage TN. Rugby world cup 2003 injury surveillance project. Br J Sports Med. 2005;39(11):812–7. https://doi.org/10.1136/bjsm.2004.016402.

54. Nichols AW. Heat-related illness in sports and exercise. Curr Rev Muscoskelet Med. 2014;7(4):355–65. https://doi.org/10.1007/s12178-014-9240-0.
55. Casadio JR, Kilding AE, Cotter JD, Laursen PB. From lab to real world: heat acclimation considerations for elite athletes. Sports Med. 2017;47(8):1467–76. https://doi.org/10.1007/s40279-016-0668-9.
56. Racinais S, Buchheit M, Bilsborough J, Bourdon PC, Cordy J, Coutts AJ. Physiological and performance responses to a training camp in the heat in professional Australian football players. Int J Sports Physiol Perform. 2014;9(4):598–603. https://doi.org/10.1123/ijspp.2013-0284.
57. Kelly M, Gastin PB, Dwyer DB, Sostaric S, Snow RJ. Short duration heat acclimation in Australian football players. J Sports Sci Med. 2016;15(1):118–25.
58. Castle P, Mackenzie RW, Maxwell N, Webborn AD, Watt PW. Heat acclimation improves intermittent sprinting in the heat but additional pre-cooling offers no further ergogenic effect. J Sports Sci. 2011;29(11):1125–34. https://doi.org/10.1080/02640414.2011.583673.
59. Stanley J, Halliday A, D'Auria S, Buchheit M, Leicht AS. Effect of sauna-based heat acclimation on plasma volume and heart rate variability. Eur J Appl Physiol. 2015;115(4):785–94. https://doi.org/10.1007/s00421-014-3060-1.
60. Buchheit M, Voss SC, Nybo L, Mohr M, Racinais S. Physiological and performance adaptations to an in-season soccer camp in the heat: associations with heart rate and heart rate variability. Scand J Med Sci Sports. 2011;21(6):e477–85. https://doi.org/10.1111/j.1600-0838.2011.01378.x.
61. Racinais S, Mohr M, Buchheit M, Voss SC, Gaoua N, Grantham J, Nybo L. Individual responses to short-term heat acclimatisation as predictors of football performance in a hot, dry environment. Br J Sports Med. 2012;46(11):810–5. https://doi.org/10.1136/bjsports-2012-091227.
62. Brade CJ, Dawson BT, Wallman KE. Effect of pre-cooling on repeat-sprint performance in seasonally acclimatised males during an outdoor simulated team-sport protocol in warm conditions. J Sports Sci Med. 2013;12(3):565–70.
63. Sunderland C, Morris JG, Nevill ME. A heat acclimation protocol for team sports. Br J Sports Med. 2008;42(5):327–33. https://doi.org/10.1136/bjsm.2007.034207.
64. Philp CP, Buchheit M, Kitic CM, Minson CT, Fell JW. Does short-duration heat exposure at a matched cardiovascular intensity improve intermittent-running performance in a cool environment? Int J Sports Physiol Perform. 2017;12(6):812–8. https://doi.org/10.1123/ijspp.2016-0072.
65. Zurawlew MJ, Walsh NP, Fortes MB, Potter C. Post-exercise hot water immersion induces heat acclimation and improves endurance exercise performance in the heat. Scand J Med Sci Sports. 2016;26(7):745–54. https://doi.org/10.1111/sms.12638.
66. Brade C, Dawson B, Wallman K. Effect of precooling and acclimation on repeat-sprint performance in heat. J Sports Sci. 2013;31(7):779–86. https://doi.org/10.1080/02640414.2012.750006.
67. Buchheit M, Cholley Y, Lambert P. Psychometric and physiological responses to a preseason competitive camp in the heat with a 6-hour time difference in elite soccer players. Int J Sports Physiol Perform. 2016;11(2):176–81. https://doi.org/10.1123/ijspp.2015-0135.
68. Buchheit M, Racinais S, Bilsborough J, Hocking J, Mendez-Villanueva A, Bourdon PC, Voss S, Livingston S, Christian R, Periard J, Cordy J, Coutts AJ. Adding heat to the live-high train-low altitude model: a practical insight from professional football. Br J Sports Med. 2013;47(Suppl 1):i59–69. https://doi.org/10.1136/bjsports-2013-092559.
69. Le Meur Y, Buchheit M, Aubry A, Coutts AJ, Hausswirth C. Assessing overreaching with heart-rate recovery: what is the minimal exercise intensity required? Int J Sports Physiol Perform. 2017;12(4):569–73. https://doi.org/10.1123/ijspp.2015-0675.
70. Buchheit M, Racinais S, Bilsborough JC, Bourdon PC, Voss SC, Hocking J, Cordy J, Mendez-Villanueva A, Coutts AJ. Monitoring fitness, fatigue and running performance during a pre-season training camp in elite football players. J Sci Med Sport. 2013;16(6):550–5. https://doi.org/10.1016/j.jsams.2012.12.003.

71. Crowcroft S, Duffield R, McCleave E, Slattery K, Wallace LK, Coutts AJ. Monitoring training to assess changes in fitness and fatigue: the effects of training in heat and hypoxia. Scand J Med Sci Sports. 2015;25(Suppl 1):287–95. https://doi.org/10.1111/sms.12364.
72. Cheuvront SN, Kenefick RW, Montain SJ, Sawka MN. Mechanisms of aerobic performance impairment with heat stress and dehydration. J Appl Physiol. 2010;109(6):1989–95. https://doi.org/10.1152/japplphysiol.00367.2010.
73. Guelinckx I, Fremont-Marquis AS, Eon E, Kavouras SA, Armstrong LE. Assessing hydration in children: from science to practice. Ann Nutr Metab. 2015;66(Suppl 3):5–9. https://doi.org/10.1159/000381814.
74. FIFA Playing in the heat. 2018. http://www.fifa.com/development/medical/players-health/minimising-risks/heat.html. Accessed 08 Jan 2018.
75. AFL Football in extreme conditions: guidelines for prevention of heat injury. 2018. http://www.aflcommunity.com.au/fileadmin/user_upload/Manage_Your_Club/3._Club_Management_Program/1._Club_Policies___Guidelines/heat_policy.PDF. Accessed 08 Jan 2018.
76. Racinais S, Cocking S, Periard JD. Sports and environmental temperature: from warming-up to heating-up. Temperature. 2017;4(3):227–57. https://doi.org/10.1080/23328940.2017.1356427.
77. Bergh U, Ekblom B. Physical performance and peak aerobic power at different body temperatures. J Appl Physiol Respir Environ Exerc Physiol. 1979;46(5):885–9. https://doi.org/10.1152/jappl.1979.46.5.885.
78. Duffield R, Coutts A, McCall A, Burgess D. Pre-cooling for football training and competition in hot and humid conditions. Eur J Sport Sci. 2013;13(1):58–67. https://doi.org/10.1080/17461391.2011.589474.
79. Brade C, Dawson B, Wallman K. Effects of different precooling techniques on repeat sprint ability in team sport athletes. Eur J Sport Sci. 2014;14(Suppl 1):S84–91. https://doi.org/10.1080/17461391.2011.651491.
80. Rugby W. World rugby heat guideline. 2018. http://playerwelfare.worldrugby.org/?subsection=6. Accessed 08 Jan 2018.
81. Bongers CC, Hopman MT, Eijsvogels TM. Cooling interventions for athletes: an overview of effectiveness, physiological mechanisms, and practical considerations. Temperature. 2017;4(1):60–78. https://doi.org/10.1080/23328940.2016.1277003.
82. NRL heat guidelines. 2018. https://playnrl.com/media/1936/heat-guidelines_with-changes1.pdf. Accessed 08 Jan 2018.

Heat Stress During American Football

10

William M. Adams, Luke N. Belval, Yuri Hosokawa,
Andrew J. Grundstein, and Douglas J. Casa

10.1 Introduction

American football is the most popular televised sport in the USA. Teams of 11 players perform sets of intermittent activity, called "downs," attempting to advance the football down of a 100-yard (91.44 m) field. Each down is culminated by either a team scoring, an incomplete pass to a receiver, or a runner with the ball being tackled to the ground. Most downs are completed within a matter of 4–6 s, resulting in a high velocity game [1]. The construct of gameplay varies depending on the level of competition and consists of four quarters that are 12–15 min in length with a 12–20 min half-time rest period separating quarters two and three.

American children begin playing the sport as early as 5 years old and can advance all the way to professional American football in the National Football League. Most American secondary schools and colleges also field interscholastic teams. Being a Northern hemisphere fall sport, most teams begin training during the warmer summer months. Many teams have historically utilized this time of year to conduct the hardest training sessions and to practice multiple times in 1 day [2].

W. M. Adams (✉)
Department of Kinesiology, University of North Carolina at Greensboro, Greensboro, NC, USA
e-mail: wmadams@uncg.edu

L. N. Belval · D. J. Casa
Korey Stringer Institute, Department of Kinesiology, University of Connecticut, Storrs, CT, USA

Y. Hosokawa
Faculty of Sport Sciences, Waseda University, Tokorozawa, Japan

A. J. Grundstein
Department of Geography, University of Georgia, Athens, GA, USA

J. D. Périard, S. Racinais (eds.), *Heat Stress in Sport and Exercise*,
https://doi.org/10.1007/978-3-319-93515-7_10

The combination of athletes of large body mass, warm environmental conditions, heavy protective equipment, and intense exercise has created situations where heat illnesses are a serious concern for medical staffs and coaches. The danger of these combined circumstances can be highlighted by the high-profile death of National Football League player Korey Stringer [3]. Similarly, athletes at both the secondary school and collegiate levels of American football die every year from exertional heat stroke, typically during summer pre-season training [4]. Epidemiological data shows that 54 American football players (42 secondary school, nine collegiate, two professional, and one sandlot) have died from EHS between the years of 1995–2014 [5]. Since nearly all heat related problems that American Football players experience occur during training, activity modifications and other preventative manners can greatly decrease risk if they are understood and implemented by coaching, training, and medical staffs.

Given the unique characteristics of American football (i.e., physical characteristics of these athletes, timing of their competitive season, and physiological demands of the sport), environmental heat stress can have detrimental effects on both athlete safety and performance. Acknowledging the factors that predispose these athletes to heat stress is essential for mitigating risk and optimizing performance in the heat. This chapter will focus on the physical characteristics of the American football athlete and discuss the influence that environmental-based activity modifications, heat acclimatization, hydration, body cooling, sleep, and protective equipment have on safety and performance during participation in the heat. At the conclusion of this chapter, the authors will also discuss some strategies that can be utilized in American football to keep athletes safe and performing at an optimal level (Fig. 10.1).

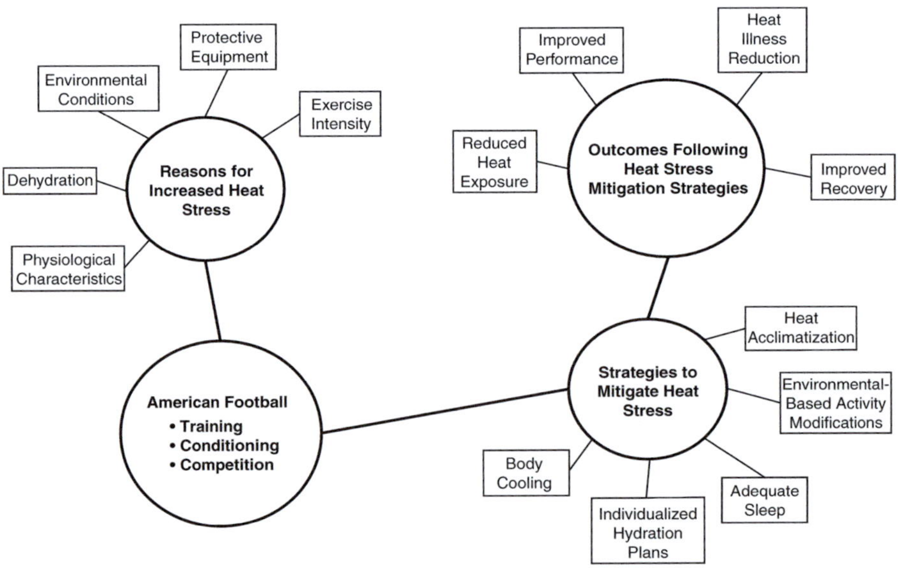

Fig 10.1 Factors responsible for increased heat stress and subsequent strategies to mitigate associated risk factors in American football

10.2 Physical Characteristics of Football Athletes

American football athlete's anthropometrics vary based upon the positions they play, with most American football players playing only one position. Example anthropometrics by position are presented in Fig. 10.2. Skilled position players like

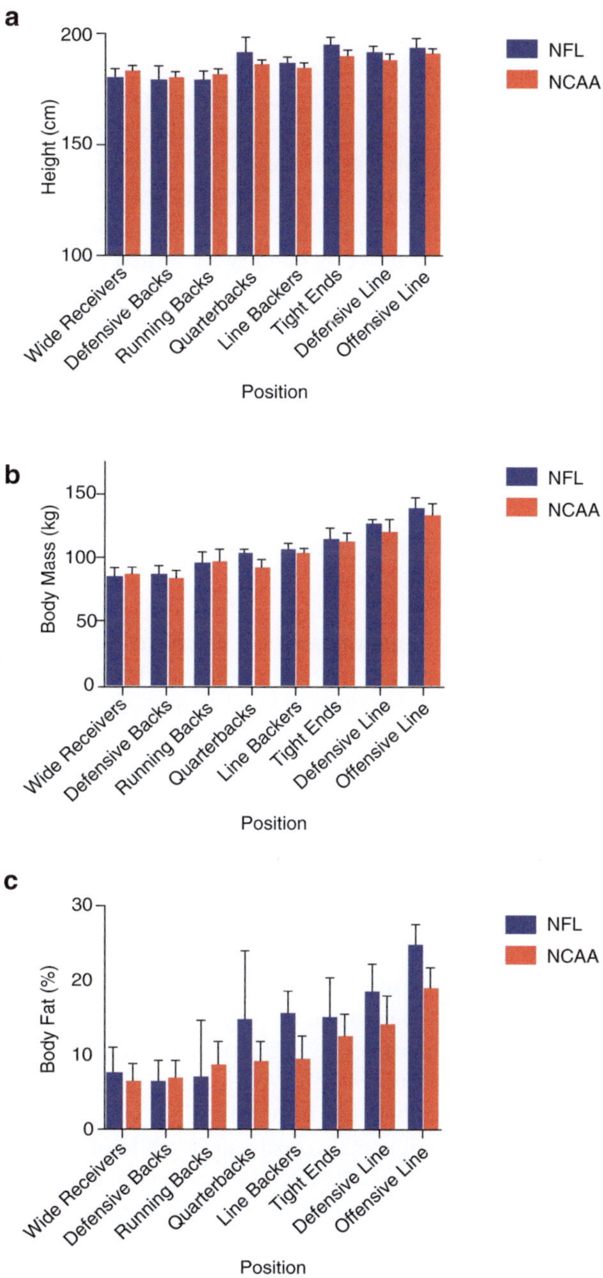

Fig. 10.2 Average (**a**) height, (**b**) body mass, and (**c**) body fat percentage among professional and collegiate American football athletes by position. *NFL*, National Football League; *NCAA*, National Collegiate Athletics Association

the quarterback, kicker, and punter are typically leaner as their positions require more technique. Wide receivers and the defensive backs who try to prevent them from catching the ball are typically the tallest and fastest players on the field [6]. Running backs and line backers typically have stockier builds but remain highly athletic [6]. Finally, offensive and defensive linemen are the largest players on the field, relying on strength, power and technique more than pure athleticism.

While all American football athletes face the challenges of training in warm environmental conditions, linemen face a unique challenge as a result of their size. American football lineman have been observed to have a high internal body temperature during exercise [7] and are at an increased risk of exertional heat illness [4], which can be attributed to several factors. Linemen in comparison to other players possess increased body fat, lower surface area to body mass ratio, and lower aerobic capacities [8]. This increases the metabolic demands of a given activity while concomitantly decreasing the ability to dissipate heat. Furthermore, since they do not move as quickly as skilled positions, a lower convective airflow leads to lower heat loss potential despite an increased sweat production and skin wettedness [9, 10]. The anthropometric features of linemen can also affect hydration. Linemen have been found to have higher sweat rates independent of metabolic heat production [10]. Interestingly this still typically only yields mild dehydration with the frequency of water breaks that has become commonplace during practices [8].

10.3 Environmental Considerations and Activity Modifications

Meteorological conditions are an extrinsic factor that can affect an athlete's thermal environment and thermal response. There are multiple approaches for measuring environmental heat stress but the wet bulb globe temperature (WBGT) is the most widely endorsed for use in an athletics setting, including American football [11, 12]. This index integrates the influences of key weather variables affecting the human heat balance including ambient air temperature, humidity, radiant heating, and wind speed via a weighted average of the wet bulb temperature (WB), globe temperature (GT), and the dry bulb temperature (DB) in the following equation:

$$\text{WBGT} = 0.7\,\text{WB} + 0.2\text{GT} + 0.1\text{DB}$$

The WBGT is often coupled with guidelines that adjust other factors such as length/intensity of physical activity, modification of work-to-rest ratios, and alteration of clothing/protective equipment worn during activity. Activity modification guidelines developed by the American College of Sports Medicine (ACSM) are frequently used to guide heat safety policies for many sports in the USA [11]. American football specific heat safety guidelines have also been developed by the Georgia High School Athletic Association (GHSA) that incorporate activity modification and equipment adjustments with changes in WBGT [13] (Table 10.1).

Several epidemiological case studies of American football players show strong associations between WBGT and variations in heat injury occurrence and provide

Table 10.1 Environmental-based activity modification policy developed by the Georgia High School Association (GHSA) with American football specific guidelines [12]

WBGT (°C)	Activity guidelines and rest break guidelines
<27.78	Normal activities—Provide at least three separate rest breaks each hour with a minimum duration of 3 min each during the workout
27.78–30.50	Use discretion for intense or prolonged exercise; watch at-risk players carefully Provide at least three separate rest breaks each hour with a minimum duration of 4 min each
30.50–32.17	Maximum practice time is 2 h. *For football:* players are restricted to helmet, shoulder pads, and shorts during practice and all protective equipment must be removed during conditioning activities. If the WBGT rises to this level during practice, players may continue to work out wearing football pants without changing of shorts. *For all sports:* Provide at least four separate rest breaks each hour with a minimum duration of 4 min each
32.17–33.33	Maximum practice time is 1 h. *For football:* no protective equipment may be worn during practice, and there may be no conditioning activities. *For all sports*: There must be 20 min of rest breaks distributed throughout the hour of practice
>33.33	No outdoor workouts. Delay practices until a cooler WBGT level is reached

WBGT wet bulb globe temperature

support for the use of WBGT as a useful heat exposure metric. The most extensive research has examined exertional heat injuries (EHIs) among collegiate-level American football players. A study of five Division I universities in the Southeastern United States during the late summer and early autumn 2003 collected data on 128 EHIs (e.g., exercise-associated muscle cramps, heat syncope, and heat exhaustion; no exertional heat strokes) from 26,993 athlete exposures with associated WBGT measurements [14]. Incidence rates per 1000 athlete exposures (AEs) increased markedly from approximately 1 in the "moderate" (18–23 °C) to near or over 6 for "high" (23–28 °C) or "extreme" (>28 °C) risk categories. Further, all of the heat exhaustion and heat syncope incidents occurred in the "high" or "extreme" WBGT categories. Building upon this work, Cooper et al. developed a more expansive 4-year study of 60 US colleges and universities (553 EHIs over 365,810 athlete exposures) spread across five geographic regions [14]. They observed that injury rates for the more serious heat illnesses (i.e., heat syncope and heat exhaustion) increased with higher WBGT risk category. EHI rates per 1000 AEs were 0.23 for WBGTs <27.8 °C, 1.34 for WBGTs between 27.9–30.0 °C and 30.1–32.2 °C, and over 4 for WBGTs ≥32.3 °C [14].

There is more limited research on environmental conditions and heat injuries among secondary school and youth athletes. A 3-year study (2009–2011) investigated EHIs among secondary school football players in the state of Georgia, USA, found that EHI rates increased for WBGTs over 27.8 °C with a peak in the 30.1–32.2 °C range; EHI rates, however, decreased for WBGTs ≥32.3 °C as many schools limited or cancelled practice activities under these conditions [15]. Furthermore, in a 3-month study of American football players at 12 Florida high schools, Tripp et al. found that almost three-quarters of EHSs (73.7%) occurred with WBGTs considered "high" (23–28 °C) or "extreme" (>28 °C) risk [16]. There are no similar studies of youth football players but Yeargin et al. noted that almost half of youth football events occurred under conditions that would warrant an EHI concern based on ACSM WBGT activity modification criteria [17].

Finally, Grundstein et al. performed a retrospective analysis of 58 fatal exertional heat stroke (EHS) cases of American football players, including youth, collegiate, and professional levels. Median exposure WBGTs were high during these incidents at 27.7 °C for morning practices and 30.2 °C for afternoon practices [4]. In a follow-up study, Grundstein et al. found that relative environmental conditions were also important, reflecting the role of local acclimatization. In mild climates (e.g., Northern portions of the USA), 80% of fatal EHSs had above average WBGTs and 50% had WBGTs that were one standard deviation from the local long-term mean. In comparison, half of the incidents in hotter climates (e.g., Southern United States) occurred with WBGTs that were near or below normal, as even typical conditions in these areas are sufficiently hot and humid to cause heat stress [18]. Given this evidence, proper policies should be adopted and implemented across all levels of American football that utilizes region specific environmental conditions to establish graded levels of activity modifications to enhance athlete safety and performance in hot conditions.

10.4 Heat Acclimatization

Heat acclimatization describes the process of obtaining physiological adaptations to heat stress that results in decreased exercising heart rate and internal body temperature, increased plasma volume, increased sweat efficiency, and increased work capacity [19, 20]. Induction of heat acclimatization requires one to exercise for approximately 7–14 consecutive days in a hot environment and under a moderate to vigorous intensity training load that would induce an elevation of internal body temperature (≈38.5–39.5 °C), skin temperature, and sweat rate for ≥60 min [19, 20]. Superior outcomes in protocols longer than 14 days have been reported, which is required to achieve 95% of the physiological adaptations [19, 20].

The physiological adaptations observed following induction of heat acclimatization permit performance benefits primarily via cardiovascular and thermoregulatory adaptations. These are characterized by plasma volume expansion, decreased exercising heart rate, the leftward-shift in sweat onset response (i.e. the onset of sweating begins at a lower internal body temperature) and increased sweat rate that will enhance evaporative heat losses, if the extent of dehydration during activity is minimized [20]. These responses also lead to reduced propensity to exertional heat illness by providing cardiovascular and thermoregulatory stability during exercise [20]. For further details regarding heat acclimatization, please refer to Chap. 8 in this text.

In the USA, the governing bodies overseeing American football at the secondary school, collegiate, and the professional levels have implemented heat acclimatization guidelines [21–23]. Although the extent of implementation is currently limited in the secondary school setting, a roughly 50% reduction in overall exertional heat illness incidence was observed in eight states that mandated the guidelines in the secondary school setting [24, 25]. While there are a series of progressive guidelines established at the secondary and collegiate setting (Table 10.2) to facilitate heat acclimatization, the professional league in the USA (National Football League) only recommends altering the schedule of practices to avoid the hottest times of the

Table 10.2 American football heat acclimatization protocols in secondary school and collegiate athletics

	National Collegiate Athletics Association [22, 23]		Secondary school recommendation [21]		
Days	Practice considerations	Equipment considerations	Practice considerations		Equipment considerations
1–5	• Single 3-h on-field practice or • One 2-h on-field practice and one 1-h field testing (i.e., speed, conditioning, agility) session • One 1-h walk-through consisting of positional skills • 3-h of continuous rest between practice, testing, walk-through	• Only helmets permitted (Days 1–2) • Only helmets and shoulder pads permitted (Days 3–5)	• No more than 3-h of total practice time and • One 1-h maximum walk-through with at least 3-h recovery period between the practice and the walk-through	• Contact with blocking sleds and tackling dummies	• Only helmets permitted (Days 1–2) • Only helmets and shoulder pads permitted (Days 3–5)
6–14	• No more than one 3-h on-field practice permitted per day • One 1-h walk-through permitted • 3 h of continuous rest between on-field practice and walk-through	• Full pads permitted	• No more than 3-h of total practice time per practice • Two a day practices permitted, however, double-practice days must be followed by a single-practice day • On double-practice days, total practice time (summation of both practice sessions) must not exceed 5 h	• On single-practice days, one 1-h walk-through is permitted, separated from the practice by at least 3 h of continuous rest • Live, full contact drills permitted	• Full pads permitted

Adapted from [26]

day, providing more rest between practices and altering the activities occurring during practice during the first 2–3 days of preseason training.

10.5 Hydration

The maintenance of a normal level of hydration is essential for the physiological processes that sustain life. Throughout the day, the regulation of total body water is tightly controlled within the body via neural, hormonal, and osmotic responses, which allows the body to remain in a state of euhydration. However, during exercise, particularly in hot environmental conditions, fluid balance and homeostasis is often disrupted, which can adversely affect athletic performance and safety in exercising individuals. Body water losses during exercise, particularly when losses exceed 2% of body mass, exacerbates cardiovascular and thermoregulatory strain [27–30], impairs musculoskeletal and cognitive performance [31–33], and increases the risk of exertional heat illness [11, 12, 34]. Refer to Chap. 6 for further details on hydration in sport and exercise.

The sport of American Football poses unique challenges surrounding minimizing fluid losses during activity. Factors such as the size of the athlete (i.e., body surface area and overall mass of the athlete), protective equipment worn during participation, the commencement of the competitive season starting in the summer (Northern hemisphere) months which exposes athletes to potentially severe heat stress, and position specific responsibilities of the athletes all contribute to the ability of the athlete to mitigate the dehydration-mediated performance and safety risks [9, 35–38]. Prior literature shows that average sweat rates vary from 1.6 L/h to 2.3 L/h and are dependent upon player position, with linemen exhibiting greater sweat rates than backs [39]. In addition, research examining the hydration practices of secondary school, collegiate, and professional American football players' hydration strategies has found that these athletes are persistently in a state of hypohydration and are unable to replace the fluids necessary to return to baseline hydration levels [37–40]. Due to the propensity of these athletes remaining in a state of hypohydration during and following participation, this evidence suggests that American football athletes may be at greater risk for performance deficits and onset of exertional heat illness.

Establishing hydration strategies and guidelines for American football athletes should take an individualistic approach as generalized guidelines [41] are not appropriate and may inadvertently place athletes at risk for other potentially serious medical conditions such as exertional hyponatremia [42]. Calculation of individual sweat rates is an efficient method for determining the volume of fluid that athletes should consume during activity. Coupled with the unlimited access to fluids during activity creates an optimal environment for athletes to optimize both safety and performance while competing in hot environmental conditions.

Current recommendations indicate that athletes should begin activity in a euhydrated state, minimize fluid losses during physical activity to prevent fluid losses that exceed 2% of body mass loss, and replace the remaining losses following activity in preparation of the next bout of exercise [43]. With the variability in sweat rates in American football athletes, it may not be feasible to keep fluid losses under 2% of body mass loss during activity. For example, high sweat rates (i.e., >2–2.5 L/h) that are unmatched to the fluid absorption capacity of the gastrointestinal tract will result in dehydration. In these instances, it is still imperative that athletes minimize fluid losses during exercise to reduce physiological strain during activity and complete activity with a lesser degree of dehydration. Establishing procedures where American football athletes track pre- and post-activity body mass allows for the determination of fluid needs for the subsequent bout of training or competition.

10.6 Body Cooling

Prior research examining the effect of protective equipment, body mass size, and player position has found that American football athletes exercising in hot and humid environmental conditions with full protective equipment were more likely to experience uncompensable heat stress, an earlier onset of fatigue and exaggerated perception of effort [35, 44]. Furthermore, athletes with a greater body mass, particularly linemen, are at greater risk of exertional heat illness that results from a combination of greater metabolic heat production and reduction of convective heat losses [8, 9]. From this evidence, the utility of body cooling during American football is of particular interest for enhancing safety and optimizing performance. Refer to Chap. 7 for further details on heat stress, exercise, and body cooling.

Body cooling has been extensively studied from both a performance and safety perspective during exercise in hot conditions. The utilization of body cooling prior to and during exercise has relayed convincing evidence supporting its performance enhancing effects [45–47]. Furthermore, from a clinical perspective, the treatment of EHS requires immediate, aggressive, whole-body cooling using a cooling modality with a cooling rate >0.155 °C/min to ensure survivability from this medical emergency [9, 11, 12, 48, 49]. When considering body cooling options for optimizing performance, Adams et al. provide three basic parameters that must be examined: what sport and type of activity (i.e., training or competition) is taking place, when is it most feasible to utilize body cooling (i.e., prior to competition, during competition, or during defined breaks such as half-time in competition), and what is the best cooling modality to use to elicit maximal benefit [50].

Body cooling during training and conditioning sessions (as compared to competitions) is logistically more feasible to manage as body cooling procedures do not have to fall within the confines of game flow logistics, space, and feasibility. The flexibility allotted during training and conditioning allows for a multi-faceted approach for the attenuation of body temperature during activity. Designing training and conditioning sessions to include body cooling breaks in designated shaded body cooling stations, unlimited access to cold fluids for hydration and modification of work-to-rest ratios based on environmental conditions allows an optimal environment focused on athlete safety. Furthermore, inclusion of active cooling modalities such as rotating cold/wet towels, cold water immersion (CWI), immersion of hands/

forearms and elbows in ice water, cooling vests and misting fans can supplement the aforementioned strategies to further enhance body cooling capabilities. While CWI should be onsite to treat EHS due to its unparalleled effectiveness [51, 52], the other body cooling options identified above might be more conducive depending on the flow and logistics of the training and conditioning session when cooling is being used for preventative reasons or to enhance performance.

Body cooling considerations for American football competition are also very unique. The evidence from Bongers et al. [47, 53] and Wegmann et al. [54] has shown the benefit of pre-cooling, as long as the pre-cooling option is an effective cooling modality (e.g., cold water ingestion, mixed method cooling, cooling vests, etc.) (Fig. 10.3). Lopez et al. and DeMartini et al. have shown that not all modalities have the same level of cooling potential [55, 56], thus requiring careful consideration as to what cooling modalities are (1) most logistically feasible for use during an American football competition and (2) which cooling modalities are going to offer the best cooling potential. During competition, body cooling can take place prior to the start of the competition, during competition, and during the half-time period of competition. Pre-cooling can occur during warm-ups (e.g., linemen using cooling vests) or during the time period post warm-up when the teams are back in the locker room prior to the start of the competition (e.g., cooling fans, cooling towels, etc.). If a quality warm-up can be performed with minimal influence on the athlete's body temperature, then the body cooling has been effective. Cooling options during the competition will have to be confined to items that can be used on the sideline. Beyond the obvious methods such as removal of equipment (i.e., helmets, gloves, etc.), hydrating with cold beverages, moving to shaded area, prior research has shown the potential benefits for active cooling (e.g., cooling fans, cooling towels) during the flow of the game (i.e., when offense is on the sideline and defense is on the field) (Fig. 10.3) [47, 53]. The half-time of American football (~12–20 min) offers a unique opportunity for body cooling. Athletes may cool the body by equipment removal, CWI, and other cooling options (e.g., cooling towels)

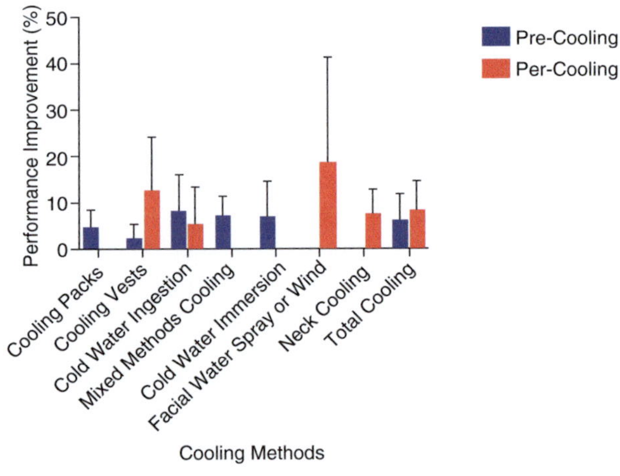

Fig. 10.3 Ergogenic effects of cooling modalities when applied prior to (pre-cooling) or during (per-cooling) exercise in the heat [51, 52, 55]

that would be of benefit from the opportunity to having access to a greater body surface area while the athlete is without equipment. Supporting evidence has also shown benefit from an endurance performance perspective when cooling was utilized during designated breaks in activity [57].

10.7 Sleep

Sleep is a primal necessity for life and is responsible for regeneration of energy stores, regulation of stress-based hormones, and improvements in neural plasticity [58–60]. In the context of sport and physical activity, sleep is important for the optimization of performance and safety [61, 62]. Evidence suggests that partial (sleep ≥4 h <8 h) and long-term sleep deprivation (46–72 h) are indicative of motor- and cognitive-related performance deficits [61, 63]. Sleep loss has also been identified as a predisposing factor for exertional heat illness [11, 12], which may be due to the altered thermoregulatory effector responses (i.e., changes in the threshold and sensitivity of local sweating and cutaneous vasodilation) that occur with sleep loss [64–67]. However, it must be noted that the findings on the effects of sleep deprivation on whole-body temperature responses remain equivocal [68, 69]. For American football athletes, it is recommended that 7–9 h of sleep is achieved on a nightly basis to allow for restoration of daily and exercise-related stressors to optimize performance and reduce risk for alteration of body temperature regulation during exercise in the heat.

10.8 Equipment

Due to the high-speed collisions of the sport, substantial protective equipment is required to minimize the risk of serious injuries. All players wear helmets, shoulder pads, thigh pads, knee pads, and hip pads both during games and most practice sessions. These pads can significantly alter the microenvironment experienced by players and lead to uncompensable heat stress. Typical ensembles cover a majority of the skin surface and decrease convective and evaporative heat loss [70]. Individuals wearing football uniforms and equipment during exercise in the heat had significantly decreased time to exhaustion and increased rates of rise of rectal and skin temperature, even when only wearing a partial ensemble of equipment [35, 44]. A key strategy to decrease the strain of football in the heat is to minimize the amount of equipment required for the given training activity.

10.9 Educational/Clinical

Effectiveness of Policy Change Specific to Heat Illness While it is not possible to prevent 100% cases of exertional heat illness occurring during sport or physical activity, especially in American football, there are a multitude of steps that can be taken to mitigate risk. Developing policies that are grounded in scientific evidence have proven successful in reducing the risk of conditions such as EHS. One such example is the adoption and implementation of heat acclimatization policies for preseason training within American football. Given the start of the season occurring in August in the Northern hemisphere, athletes subjected to potentially extreme heat stress and the benefits allotted from heat acclimatization can be protective for the athlete [71]. Specifically, the implementation of heat acclimatization guidelines at the secondary school level has shown a ~55% reduction in the incidence of EHS [25], which further supports the effectiveness of heat acclimatization and reducing risk.

On-field Heat Illness Prevention Strategies Prevention of exertional heat illness starts with identification of modifiable risk factors that contribute to increased thermal strain, excess heat gain, impede heat dissipation, or hamper adequate recovery. The following section will list four strategies that American football teams can implement to prevent exertional heat illness and optimize performance in the heat.

1. *Priming physical fitness.* Physical fitness, particularly cardiovascular fitness, contributes greatly to one's thermal-tolerance during exercise [20]. Priming physical fitness well before the warm season has been shown to adequately support the heat acclimatization response in runners, which can also be applied in American football players [72]. As physical characteristics of American football athletes greatly differ by position, special considerations must be placed when forming the training regimen (i.e., linemen should not be expected to complete similar training as wide receivers).
2. *Active body cooling.* Active body cooling is a direct method to attenuate excess heat gain during exercise. The protective equipment required for participation is a unique aspect of American football that warrants special consideration in choosing body cooling modalities [50]. Establishing body cooling procedures for American football athletes should consider the timing of body cooling both training (e.g., having designated cooling breaks during training) and competition (e.g., prior to the start of competition, during half-time and sideline cooling during offensive/defensive sides), and type of cooling modality that is most effective (as determined by the modalities overall cooling rate) for athlete safety and performance optimization [50].
3. *Environmental monitoring.* Avoidance of continuous exercise during extreme heat can be achieved through environmental monitoring. Epidemiological study suggests high correlation between the incidence of exertional heat illness when the WBGT exceeds 28 °C [14]. Measurement of the environmental conditions should be taken at the location of activity since the values collected at the nearest weather station are likely to differ from the microenvironment of the athletic venue [73].
4. *Hydration.* Dehydration directly affects thermal-tolerance by elevating the internal body temperature [74]. Typical sweat rates among American football players have been shown to be ≈2 L/h, predisposing athletes to a potential to sustain high levels of dehydration if fluid losses are not minimized [7]. Since the intermittent nature of typical American football training and competition offers frequent

opportunities to rehydrate, each athlete should monitor the change in body weight before and after practice to recognize the adequate and appropriate amount of fluid needs during and following activity.

The success of aforementioned strategies relies on organizational support. For example, identification of at-risk individual through physical examination and fitness testing (e.g., maximal oxygen consumption test) will help identify individuals with previous history of exertional heat illness and low physical fitness, which is vital information to be shared among the coaching and medical staff to make systematic decisions in providing individualized interventions in the team sport setting to minimize risk. Likewise, predetermined policies and procedures for training during extreme heat regarding heat acclimatization and environmental monitoring will allow safe and efficient scheduling of practices that allows adequate heat exposure to induce heat acclimatization while averting performance decrements from excess heat exposure.

10.10 Summary

The physical demands, athlete demographics, protective equipment requirements, and the timing of year in which the sport is played elevate the risk profile for American football in terms of susceptibility of heat illness. Implementation of prevention strategies such as heat acclimatization, environmental-based activity modifications, individualized hydration strategies, and body cooling optimizes the health and performance of the athletes competing in this sport.

References

1. Edwards T, Spiteri T, Piggott B, Haff GG, Joyce C. A narrative review of the physical demands and injury incidence in American football: application of current knowledge and practices in workload management. Sports Med. 2018;48(1):45–55.
2. Wellman AD, Coad SC, Flynn PJ, Siam TK, McLellan CP. A comparison of pre-season and in-season practice and game loads in NCAA division I football players. J Strength Cond Res. 2017. In Press.
3. Grundstein A, Knox JA, Vanos J, Cooper ER, Casa DJ. American football and fatal exertional heat stroke: a case study of Korey stringer. Int J Biometeorol. 2017;61(8):1471–80.
4. Grundstein AJ, Ramseyer C, Zhao F, Pesses JL, Akers P, Qureshi A, et al. A retrospective analysis of American football hyperthermia deaths in the United States. Int J Biometeorol. 2012;56(1):11–20.
5. Kucera KL, Klossner D, Colgate B, Cantu RC. Annual survey of football injury research. Chapel Hill: National Center for Catastrophic Sport Injury Research; 2015. Report No.: 2015–01.
6. Pincivero DM, Bompa TO. A physiological review of American football. Sports Med Auckl NZ. 1997;23(4):247–60.
7. Godek SF, Bartolozzi AR, Burkholder R, Sugarman E, Dorshimer G. Core temperature and percentage of dehydration in professional football linemen and backs during preseason practices. J Athl Train. 2006;41(1):8–14.

8. Davis JK, Baker LB, Barnes K, Ungaro C, Thermoregulation SJ. Fluid balance, and sweat losses in American football players. Sports Med Auckl NZ. 2016;46(10):1391–405.
9. Demartini JK, Casa DJ, Stearns R, et al. Effectiveness of cold water immersion in the treatment of exertional heat stroke at the falmouth road race. Med Sci Sports Exerc. 2015;47(2):240–5.
10. Deren TM, Coris EE, Bain AR, Walz SM, Jay O. Sweating is greater in NCAA football linemen independently of heat production. Med Sci Sports Exerc. 2012;44(2):244–52.
11. Armstrong LE, Casa DJ, Millard-Stafford M, Moran DS, Pyne SW, Roberts WO. American College of Sports Medicine position stand. Exertional heat illness during training and competition. Med Sci Sports Exerc. 2007;39(3):556–72.
12. Casa DJ, DeMartini JK, Bergeron MF, Csillan D, Eichner ER, Lopez RM, et al. National Athletic Trainers' Association position statement: exertional heat illnesses. J Athl Train. 2015;50(9):986–1000.
13. Georgia High School Association. Practice Policy for Heat and Humidity | GHSA.net [Internet]. [cited 2018 Feb 11]. Available from: https://www.ghsa.net/practice-policy-heat-and-humidity.
14. Cooper ER, Ferrara MS, Casa DJ, Powell JW, Broglio SP, Resch JE, et al. Exertional heat illness in American football players: when is the risk greatest? J Athl Train. 2016;51(8):593–600.
15. Ferrara MS, Cooper ER, Miles JD, Curry PR, Grundstein AJ, Powell JP. The Georgia study: risk of exertional heat illnesses in high school players. J Athl Train. 2013;48(3):83.
16. Tripp BL, Eberman LE, Smith MS. Exertional heat illnesses and environmental conditions during high school football practices. Am J Sports Med. 2015;43(10):2490–5.
17. Yeargin SW, Cahoon E, Hosokawa Y, Mensch JM, Dompier TP, Kerr ZY. Environmental conditions and seasonal variables in American youth football leagues. Clin Pediatr (Phila). 2017;56(13):1209–18.
18. Grundstein AJ, Hosokawa Y, Casa DJ. Fatal exertional heat stroke and American football players: the need for regional heat-safety guidelines. J Athl Train. 2018;53(1):43–50.
19. Tyler CJ, Reeve T, Hodges GJ, Cheung SS. The effects of heat adaptation on physiology, perception and exercise performance in the heat: a meta-analysis. Sports Med Auckl NZ. 2016;46(11):1699–724.
20. Armstrong LE, Maresh CM. The induction and decay of heat acclimatisation in trained athletes. Sports Med Auckl NZ. 1991;12(5):302–12.
21. Adams WM, Scarneo SE, Casa DJ. State-level implementation of health and safety policies to prevent sudden death and catastrophic injuries within secondary school athletics. Orthop J Sports Med. 2017;5(9):2325967117727262.
22. National Collegiate Athletics Association. Playing and practice seasons [Internet]. Bylaws, Article 17.10.2.4 Apr 26, 2017. Available from: https://web3.ncaa.org/lsdbi/search/bylawView?id=8823.
23. National Collegiate Athletics Association. Playing and practice seasons [Internet]. Bylaws, Article 17.10.2.5 Apr 14, 2017. Available from: https://web3.ncaa.org/lsdbi/search/bylawView?id=8823.
24. Attanasio SM, Adams WM, Stearns RL, Huggins RA, Casa DJ. Occurrence of exertional heat stroke in high school football athletes before and after implementation of evidence-based heat acclimatization guidelines. J Athl Train. 2016;51(6):S168.
25. Kerr ZY, Register-Mihalik JK, Pryor RR, et al. The effect of the National Athletic Trainers Association Inter-Association Task Force (NATA-IATF) preseason heat acclimatization guidelines on high school football preseason exertional heat illness rates. J Athl Train. 2018;53(6 Supplement):S-72.
26. Hosokawa Y, Adams WM. Heat risks in athletics. In: Hosokawa Y, editor. Human health and physical activity during heat exposure. Cham: Springer; 2018. p. 73–84.
27. González-Alonso J, Mora-Rodríguez R, Below PR, Coyle EF. Dehydration markedly impairs cardiovascular function in hyperthermic endurance athletes during exercise. J Appl Physiol Bethesda Md 1985. 1997;82(4):1229–36.
28. Adams WM, Ferraro EM, Huggins RA, Casa DJ. Influence of body mass loss on changes in heart rate during exercise in the heat: a systematic review. J Strength Cond Res. 2014;28(8): 2380–9.

29. Montain SJ, Coyle EF. Influence of graded dehydration on hyperthermia and cardiovascular drift during exercise. J Appl Physiol Bethesda Md 1985. 1992;73(4):1340–50.
30. Sawka MN, Young AJ, Francesconi RP, Muza SR, Pandolf KB. Thermoregulatory and blood responses during exercise at graded hypohydration levels. J Appl Physiol Bethesda Md 1985. 1985;59(5):1394–401.
31. Judelson DA, Maresh CM, Anderson JM, Armstrong LE, Casa DJ, Kraemer WJ, et al. Hydration and muscular performance: does fluid balance affect strength, power and high-intensity endurance? Sports Med Auckl NZ. 2007;37(10):907–21.
32. Grandjean AC, Grandjean NR. Dehydration and cognitive performance. J Am Coll Nutr. 2007;26(5 Suppl):549S–54S.
33. Cheuvront SN, Carter R 3rd, Sawka MN. Fluid balance and endurance exercise performance. Curr Sports Med Rep. 2003;2(4):202–8.
34. Casa DJ. Exercise in the heat. II. Critical concepts in rehydration, exertional heat illnesses, and maximizing athletic performance. J Athl Train. 1999;34(3):253–62.
35. Armstrong LE, Johnson EC, Casa DJ, Ganio MS, McDermott BP, Yamamoto LM, et al. The American football uniform: uncompensable heat stress and hyperthermic exhaustion. J Athl Train. 2010;45(2):117–27.
36. McCullough EA, Kenney WL. Thermal insulation and evaporative resistance of football uniforms. Med Sci Sports Exerc. 2003;35(5):832–7.
37. Godek SF, Bartolozzi AR, Godek JJ. Sweat rate and fluid turnover in American football players compared with runners in a hot and humid environment. Br J Sports Med. 2005;39(4):205–11. discussion 205-211.
38. Godek SF, Godek JJ, Bartolozzi AR. Hydration status in college football players during consecutive days of twice-a-day preseason practices. Am J Sports Med. 2005;33(6):843–51.
39. Godek SF, Bartolozzi AR, Burkholder R, Sugarman E, Peduzzi C. Sweat rates and fluid turnover in professional football players: a comparison of National Football League linemen and backs. J Athl Train. 2008;43(2):184–9.
40. Stover EA, Zachwieja J, Stofan J, Murray R, Horswill CA. Consistently high urine specific gravity in adolescent American football players and the impact of an acute drinking strategy. Int J Sports Med. 2006;27(4):330–5.
41. Sawka MN, Burke LM, Eichner ER, Maughan RJ, Montain SJ, Stachenfeld NS. American College of Sports Medicine position stand. Exercise and fluid replacement. Med Sci Sports Exerc. 2007;39(2):377–90.
42. Almond CSD, Shin AY, Fortescue EB, Mannix RC, Wypij D, Binstadt BA, et al. Hyponatremia among runners in the Boston Marathon. N Engl J Med. 2005;352(15):1550–6.
43. McDermott BP, Anderson SA, Armstrong LE, Casa DJ, Cheuvront SN, Cooper L, et al. National Athletic Trainers' association position statement: fluid replacement for the physically active. J Athl Train. 2017;52(9):877–95.
44. Johnson EC, Ganio MS, Lee EC, Lopez RM, McDermott BP, Casa DJ, et al. Perceptual responses while wearing an American football uniform in the heat. J Athl Train. 2010;45(2):107–16.
45. Tyler CJ, Sunderland C, Cheung SS. The effect of cooling prior to and during exercise on exercise performance and capacity in the heat: a meta-analysis. Br J Sports Med. 2015;49(1):7–13.
46. Ranalli GF, Demartini JK, Casa DJ, McDermott BP, Armstrong LE, Maresh CM. Effect of body cooling on subsequent aerobic and anaerobic exercise performance: a systematic review. J Strength Cond Res. 2010;24(12):3488–96.
47. Bongers CCWG, Thijssen DHJ, Veltmeijer MTW, Hopman MTE, Eijsvogels TMH. Precooling and percooling (cooling during exercise) both improve performance in the heat: a meta-analytical review. Br J Sports Med. 2014;*49*(6):377–84.
48. McDermott BP, Casa DJ, Ganio MS, Lopez RM, Yeargin SW, Armstrong LE, et al. Acute whole-body cooling for exercise-induced hyperthermia: a systematic review. J Athl Train. 2009;44(1):84–93.
49. Adams WM, Hosokawa Y, Casa DJ. The timing of exertional heat stroke survival starts prior to collapse. Curr Sports Med Rep. 2015;14(4):273–4.
50. Adams WM, Hosokawa Y, Casa DJ. Body-cooling paradigm in sport: maximizing safety and performance during competition. J Sport Rehabil. 2016;25(4):382–94.

51. Zhang Y, Davis J-K, Casa DJ, Bishop PA. Optimizing cold water immersion for exercise-induced hyperthermia: a meta-analysis. Med Sci Sports Exerc. 2015;47(11):2464–72.
52. Proulx CI, Ducharme MB, Kenny GP. Effect of water temperature on cooling efficiency during hyperthermia in humans. J Appl Physiol Bethesda Md 1985. 2003;94(4):1317–23.
53. Bongers CCWG, Hopman MTE, Eijsvogels TMH. Cooling interventions for athletes: an overview of effectiveness, physiological mechanisms, and practical considerations. Temp Austin Tex. 2017;4(1):60–78.
54. Wegmann M, Faude O, Poppendieck W, Hecksteden A, Fröhlich M, Meyer T. Pre-cooling and sports performance: a meta-analytical review. Sports Med Auckl NZ. 2012;42(7):545–64.
55. Lopez RM, Eberman LE, Cleary MA. Superficial cooling does not decrease core body temperature before, during, or after exercise in an American football uniform. J Strength Cond Res. 2012;26(12):3432–40.
56. DeMartini JK, Ranalli GF, Casa DJ, Lopez RM, Ganio MS, Stearns RL, et al. Comparison of body cooling methods on physiological and perceptual measures of mildly hyperthermic athletes. J Strength Cond Res. 2011;25(8):2065–74.
57. Yeargin SW, Casa DJ, McClung JM, Knight JC, Healey JC, Goss PJ, et al. Body cooling between two bouts of exercise in the heat enhances subsequent performance. J Strength Cond Res. 2006;20(2):383–9.
58. Adam K. Sleep as a restorative process and a theory to explain why. Prog Brain Res. 1980;53:289–305.
59. Alvarez GG, Ayas NT. The impact of daily sleep duration on health: a review of the literature. Prog Cardiovasc Nurs. 2004;19(2):56–9.
60. Banks S, Dinges DF. Behavioral and physiological consequences of sleep restriction. J Clin Sleep Med. 2007;3(5):519–28.
61. Pilcher JJ, Huffcutt AI. Effects of sleep deprivation on performance: a meta-analysis. Sleep. 1996;19(4):318–26.
62. Himashree G, Banerjee PK, Selvamurthy W. Sleep and performance--recent trends. Indian J Physiol Pharmacol. 2002;46(1):6–24.
63. VanHelder T, Radomski MW. Sleep deprivation and the effect on exercise performance. Sports Med Auckl NZ. 1989;7(4):235–47.
64. Sawka MN, Leon LR, Montain SJ, Sonna LA. Integrated physiological mechanisms of exercise performance, adaptation, and maladaptation to heat stress. Compr Physiol. 2011;1(4):1883–928.
65. Kolka MA, Stephenson LA. Exercise thermoregulation after prolonged wakefulness. J Appl Physiol Bethesda Md 1985. 1988;64(4):1575–9.
66. Dewasmes G, Bothorel B, Hoeft A, Candas V. Regulation of local sweating in sleep-deprived exercising humans. Eur J Appl Physiol. 1993;66(6):542–6.
67. Sawka MN, Gonzalez RR, Pandolf KB. Effects of sleep deprivation on thermoregulation during exercise. Am J Phys. 1984;246(1 Pt 2):R72–7.
68. Moore JP, Harper Smith AD, Di Felice U, Walsh NP. Three nights of sleep deprivation does not alter thermal strain during exercise in the heat. Eur J Appl Physiol. 2013;113(9):2353–60.
69. Tokizawa K, Sawada S-I, Tai T, Lu J, Oka T, Yasuda A, et al. Effects of partial sleep restriction and subsequent daytime napping on prolonged exertional heat strain. Occup Environ Med. 2015;72(7):521–8.
70. Kulka TJ, Kenney WL. Heat balance limits in football uniforms how different uniform ensembles alter the equation. Phys Sportsmed. 2002;30(7):29–39.
71. Adams WM, Casa DJ, Drezner JA. Sport safety policy changes: saving lives and protecting athletes. J Athl Train. 2016;51(4):358–60.
72. Armstrong LE, Hubbard RW, DeLuca JP, Christensen EL. Heat acclimatization during summer running in the northeastern United States. Med Sci Sports Exerc. 1987;19(2):131–6.
73. Pryor JL, Pryor RR, Grundstein A, Casa DJ. The heat strain of various athletic surfaces: a comparison between observed and modeled wet-bulb globe temperatures. J Athl Train. 2017;52(11):1056–64. Available from: http://natajournals.org/doi/abs/10.4085/1062-6050-52.11.15.
74. Huggins RA, Martschinske JL, Applegate K, Armstrong LE, Casa DJ. Influence of dehydration on internal body temperature changes during exercise in the heat: a meta-analysis. Med Sci Sports Exerc. 2012;44(S5):791.

Tennis in the Heat

11

Michael F. Bergeron

11.1 Introduction

The heat is a formidable opponent for tennis players at all levels of competition. Extensive sweating prompting measurable body water and electrolyte deficits is noticeably prevalent during extended play in hot and/or humid weather. As environmental conditions become more challenging, even the fittest of players are recognizably affected, as a concomitant level of thermal and cardiovascular strain increasingly challenges one's physiology, perception of effort, and on-court well-being and performance. This is visibly evident in recreational tennis, but the challenge is typically greatest in tournament competition where the intensity of play is high and persistent, and the duration of each match often extends to several hours or more [1–4]. Even with just one match a day, repeated daily exposure to the demanding environmental conditions can progressively take its toll as players advance through an increasingly difficult draw. However, the challenge of cumulative tennis and heat exposure is particularly notable and difficult when players are required to compete more than once on the same day across successive days in a single event.

This chapter highlights the relevant evidence and perspective on challenges facing junior and adult tennis players in the heat specific to characteristic thermoregulatory responses. This is followed by practical considerations and steps in mitigating undue on-court thermal strain and clinical risk. The last section emphasizes the need for more inclusive real-world scenarios in research and the informed utility of technology and advanced analytics. More detailed explanations and supporting evidence are presented in the earlier chapters, regarding the physiological mechanisms and consequent changes in related systems functionality that are

M. F. Bergeron (✉)
SIVOTEC Analytics, Boca Raton, FL, USA
e-mail: mbergeron@sivotecanalytics.com

J. D. Périard, S. Racinais (eds.), *Heat Stress in Sport and Exercise*,
https://doi.org/10.1007/978-3-319-93515-7_11

fundamental to these responses and challenges confronting tennis players in the heat. Thorough discussion of exertional heat illness is also covered earlier, as well as applicable recommended prevention strategies and clinical management.

11.2 Thermoregulatory Responses and Challenges in Tennis

11.2.1 Sweat Loss During Tennis

Sweat loss during tennis in the heat can be noticeably extensive. While the format of tennis competition provides regular opportunities to consume fluids, sufficiently effective rehydration during and after play often involves more than simply ample fluid volume intake. Notably, on-court sweat sodium and chloride losses can be substantial, even for players who are well-acclimatized to the heat [5, 6]. Accordingly, deliberate additional salt intake (by way of daily fluid and/or food consumption) is often warranted to ensure complete rehydration and optimal body water retention and distribution in all fluid compartments [7–11]. As with body water loss via sweating, sweat sodium and chloride loss rates are highly variable and must be addressed specific to individual needs [5–7, 12].

11.2.1.1 Junior Tennis

With increasing environmental heat stress (air temperature, humidity, and solar radiation) and intensity and duration of play, young adolescent tennis players are readily capable of on-court sweating rates that are near or exceed 1.0 L/h during practice [13] and tournament singles and doubles competition [14]. Parallel to physical and physiological development and maturation, an increasingly greater muscle mass produces more heat; thus, there is a greater need for evaporative cooling. And with more mature sweat glands and a larger capacity for sweat production [15], sweating rate in older adolescents during intense practice and competitive play in challenging hot conditions can often reach 2.5 L/h or more [5, 6].

11.2.1.2 Adult Tennis

Comparable to older adolescents, and often to a much greater extent, on-court sweating rates in adult tennis players can be extensive during intense practice and competitive play [6, 16–21]. Among the still limited reported examples of on-court sweating rates during tennis in adults, Bergeron et al. [16] evaluated male and female tennis players from two Division I university tennis teams during three successive days of competitive round-robin play in the heat; Périard et al. [20, 21] monitored heat non-acclimatized men during two simulated matches; Hornery et al. [17] examined male professional tennis players during international tournament competition; and Tippet et al. [19] examined professional women players during outdoor tournament play in similar hot environmental conditions. While on-court sweating rate was extensive in many instances (i.e., 2.0 L or more per hour), post-play body weight deficits were only modest (generally around 1% or a little more/less).

11.2.2 Heat Strain

As ongoing metabolic heat production and storage continues, a tennis player's body core temperature can be expected to progressively increase as play persists. This signature of evolving heat strain is especially intensified during a hard practice session or competitive intense match in the heat. Moreover, the repeated complex, intermittent activity patterns, with varying workloads and short recovery periods between points, that are characteristic of tennis can further exacerbate on-court thermal strain and related clinical risk [22]. Notably, however, reports of players (at any level of competition) incurring exertional heatstroke on-court or immediately post-play are rare, whereas heat exhaustion is likely far more common in tennis when collectively prompted by the aggregate of extended strenuous play, a measurable total body water and energy deficit, and central fatigue [23, 24].

Measuring body core and skin temperature and other relevant metrics during tennis competition is not trivial; however, advances in sensor and integrated platform technologies and methods have made it more feasible [25–28]. Nonetheless, the logistics and hesitancy by elite players to alter their routines during sanctioned competitive play are practical barriers. Accordingly, researchers have more commonly utilized simulated tennis protocols to gain helpful insights to expected exertional heat strain incurred on-court during competition. Visible indications of thermal strain in players are routinely observed during hot-weather competition; and more detailed emerging perspective on heat-related court calls and post-match clinical consults has been described at select professional events [3, 4]. However, the specific extent and confirmed prevalence of undue body core temperature and exertional heat illness, for example, as well as a clear appreciation of modifiable contributing factors, across a practical range of heat and humidity are largely not well-described in sanctioned tournament play.

11.2.2.1 Junior Tennis

Bergeron et al. [13] examined thermal strain in highly skilled, fit junior tennis players (on average, 15 years old) during intense on-court training (rallying using a variety of strokes and patterns and playing competitive points) in a very warm environment (WBGT: 26.6 °C in the water trial; 26.3 °C in the carbohydrate-electrolyte commercial sport drink trial). Their findings did not reveal an association of pre-practice hydration status (as indicated by urine specific gravity) with body core temperature during the 2-h practice sessions. However, a lower mean core temperature was observed during the randomized carbohydrate-electrolyte drink trial (38.0 ± 0.2 °C) compared to the water trial (38.2 ± 0.3 °C). Notably, while body core temperature approached or reached 39 °C for some players, none of these young athletes had any visible indications of excessive thermal strain.

To examine thermal strain in junior tennis players during actual sanctioned tournament competition, Bergeron et al. [14] observed eight elite-level young boys during the first round of singles and doubles play in a United States Tennis Association 14 s age group national championships event during early August. In contrast to the 2-h practice sessions described above, pre-play hydration status (urine specific gravity) was increasingly associated with on-court body core temperature (final reading: 38.7 ± 0.3 °C) during singles play as the matches advanced (WBGT: mean 29.6 °C; range 28.5–30.1 °C). This underscores the contrast of metabolic heat production and storage during meaningful competition versus practice or simulated competition. These elite players characteristically still maintained a strong effort and very high intensity, though it was only the first round of competition. This study also highlighted how even doubles play in junior tournament-level tennis can elicit appreciable thermal strain in hot conditions (WBGT: 31.3 °C).

11.2.2.2 Changes Across Adolescence

The long-standing belief that youth athletes have greater difficulty tolerating sports or other physical activity in the heat because of a purported biological maturation-related disadvantage has been distinctly countered by more current research and position stands [24, 29, 30]. Cardiovascular capacity and thermoregulation are generally adequate, if hydration is amply maintained [31–34]. However, there are progressively greater challenges during vigorous tennis in the heat for boys and girls as they physically and physiologically develop and mature, more heat is produced during play from a greater muscle mass, and more mature sweat glands yield increased sweat production [15]. This can result in greater on-court thermal loads and an increase in sweat fluid and electrolyte (primarily sodium) losses. Accordingly, total body water and exchangeable sodium deficits from extensive sweating can be comparatively more substantial in mid- to late-teen tennis players. Moreover, longer and physically demanding workouts and matches are characteristic of more physically developed, fit, and skilled older adolescent players. Thus, there is a potential for greater levels of muscle damage and various physiological carry-over effects that could further increase thermal strain and other clinical risks during the next training session or match [35].

11.2.2.3 Adult Tennis

Périard et al. [21] reported only moderate thermal strain in the heat, as did Hornery et al. [17] with similar peak body core temperatures (38.5–39.4 °C) reached during hard and clay court professional tournament play. And Tippet et al. [19] also observed only moderately high mean (38.7 ± 0.20 °C) and peak (39.1 ± 0.3 °C) body core temperatures in female professional tennis players throughout match play. However, even without conditions prompting excessive thermal strain, Périard et al. [20] found that on-court body core temperature was reduced during the early stages of play when better pre-play hydration achieved with specified individualized water volume intake and a target sodium content.

11.2.3 Repeated Same-Day and Successive-Day Match Play

Arguably the most recognized challenge for a tennis player to maintain hydration, minimize on-court thermal strain, and perform optimally is during a hot-weather tournament when multiple singles matches are played on the same day and the scheduled rest and recovery periods between contests are inappropriately brief. This is particularly characteristic of junior tennis, especially during tournaments at the local and regional levels. Though, in many same-day scenarios or across successive days with older adolescent players *and* adults, on-court and consequent daily accumulated body water and sodium losses can be significant from repeated extensive sweating. Thus, adequately rehydrating and minimizing day-to-day between-match total body water and exchangeable sodium deficits can be particularly difficult for certain individuals [6, 12].

The recognized impact of previous competition-related physical activity and heat exposure on subsequent same-day physiological strain and performance in tennis is very clear to those participating and closely observing. In contrast, this undue burden is evidently not fully appreciated by many event organizers. Moreover, the repeated-bout effect in tennis has not been adequately empirically examined. The only reported measured performance impact of prior same-day tennis-heat exposure (degree-minutes) during official competitive match play was that assessed by Coyle [36] with boys during a 14's national championships event over a 7-year period. However, field and laboratory studies on repeated-bout non-tennis exercise in adults and youth clearly demonstrate a negative impact on the next bout of activity, because of various physiological "carry-over" effects from previous same-day strenuous physical activity and heat exposure. Even with ample hydration and body core temperature returning to baseline before starting a second bout of exercise, a marked carry-over effect on physiological and perceptual strain remained [37–42]. In more stressful environmental conditions and a more typical uncontrolled real-world setting, following a long intense competitive tennis match and measurably incomplete rehydration and body cooling before going on court again, an expected greater impact would be likely. Nevertheless, these limited relevant findings provide important insight that should help guide more practically appropriate tournament scheduling, as heat and humidity and the concomitant threat to health and performance increase.

11.2.4 Clinical Conditions, Medications, and Caffeine

Standard precautions should be taken with tennis players who are currently (or were recently) ill, which may dictate significantly reducing the practice and/or playing load in the heat. A recent bout of exertional heat illness may be a good reason to do the same. While the conservative approach minimally dictates a deliberate focus on implementing additional offsetting measures specific to readily modifiable risk factors in both instances, following best practices may warrant avoiding play and practice altogether.

Using certain medications (e.g., a non-steroidal anti-inflammatory drug, dopamine reuptake inhibitor to treat attention-deficit/hyperactivity disorder, or diuretic) should be explicitly discussed with one's own primary healthcare provider, in advance of training or competing in a hot environment, especially if there is a history of excessive thermal strain or exertional heat illness [23, 29, 43,

44]. Caffeine consumption, despite its routine use by many players [45] and demonstrated ergogenic effects [46, 47], is often discouraged in tennis, especially during hot-weather training and competition. However, caffeine does not appear to be appreciably thermogenic or interfere with heat dissipation during exercise-heat stress; thus, concerns over measurable effects of caffeine intake on thermoregulation, fluid-electrolyte balance, and exercise-heat tolerance do not appear to be warranted [48, 49]. Accordingly, up to moderate caffeine consumption prior to play is not likely to augment thermal strain and exertional heat illness risk or impede or lessen on-court performance.

11.3 Mitigating Heat Strain and Clinical Risk

11.3.1 What Can Tennis Players, Coaches, and Administrators Do?

Any time junior and adult tennis players compete or train in the heat, the potential risk of incurring a significant total body water deficit, undue thermal strain, and/or exertional heat illness should be appreciated by everyone involved. This is especially the case when participating in multiple same-day matches or practice sessions over several days or more in a row. As noted earlier, players should avoid or significantly limit practice, training, and competition in hot and/or humid conditions, if they are currently ill or are recovering from an illness, especially those illnesses involving gastrointestinal distress (e.g., vomiting, diarrhea) and/or fever [23, 29]. These and other individualized appropriate modifications to tennis practice and training are essential for affected players, especially in the heat.

Administratively, for any tennis tournament where warm-to-hot environmental conditions are anticipated, written protocols following "best practices" standard of care should be in place and practiced, and trained personnel with readily available facilities should be on-site, for effectively treating all forms of exertional heat illness [23]. An increasing need for medical personnel to be promptly responsive to an escalating frequency of heat-related on-court calls and post-match clinical consults and treatments should be anticipated as environmental conditions become more challenging [3, 4]. Moreover, all tennis players (and those coaches and administrators overseeing training and competition in the heat) should also focus on other readily modifiable risk factors beyond just hydration management, such as heat acclimatization, cooling strategies, and scheduling of play [20, 29, 50–53].

Fundamental factors contributing to exertional heat illness risk and primary practical offsetting measures during tournament tennis in the heat are highlighted in Table 11.1.

Table 11.1 Fundamental factors contributing to exertional heat illness (EHI) risk and primary practical offsetting measures during tournament tennis in the heat

EHI risk factors	Preventive or offsetting measures
• A hot and/or humid climate (including no or very little breeze and cloud cover) • Players who are traveling and likely not sufficiently acclimatized to playing in the heat/humidity • Extensive competitive play or training in the immediate period leading up to the event • Players who inherently sweat considerably • Long matches • Multiple same-day matches—notable increased risk after the first match owing to carry-over effects • Little or no on-court shade	• Ensure adequate daily water and sodium intake based on *individual* on-court sweat losses – Prior to each match, drink appropriate fluids regularly throughout the day – During play, consume fluid regularly to appropriately offset body water and sodium losses from sweating – After play, drink *up to* ~120% of any remaining body water deficit (indicated by a change in body weight) over the rest of the day and before the next day's match – Consume additional sodium daily via foods and fluids, especially when sweating extensively and repeatedly • Minimize match warm-up time and unnecessary heat exposure • Consider on-court and between-match cooling strategies • Administrators should ensure – Adequate (i.e., extended) rest and recovery time between same-day matches – Close monitoring of (and prompt response to) participants for signs and symptoms of developing heat illness – Readily available (on-site) trained personnel and facilities for effectively treating exertional heat illness

Every player (and, as applicable, reinforced by respective responsible coaches and/or parents) always has the responsibility to begin play healthy, well-hydrated, well-nourished, well-rested, adequately fit, and otherwise prepared for the demands of competition (accordingly, these potential risk factors are not listed). These recommendations apply as well to practice in the same conditions

11.3.2 Pre-Play Preparation

Players and coaches should follow a graduated introduction and exposure to practice and play in a hot and/or humid environment, while progressing similarly with the accompanying intensity and duration of practice/training and competition. Progressive acclimatization to the environment, activity, and physiological demands is essential to minimize on-court performance decrements and the concomitant risk of incurring exertional heat illness. This is not always practically feasible when traveling between locations to an event in a more stressful (greater heat and/or humidity) environment. However, even arriving just an additional 2–3 days early to acclimatize to the new weather conditions can measurably help [54].

Players, coaches, and parents must also be prepared to appropriately adjust for changing weather conditions, recognizing that tolerance to physical activity decreases and exertional heat illness risk increases, as the heat and/or humidity rise [24, 29, 53]. For practice, the duration and overall intensity of activities and each session should be eased, and the frequency and duration of breaks should be increased, to maintain safety, attention, and performance. During competitive scenarios, duration of pre-match warm-ups in hot and/or humid weather should be minimized. To further minimize on-court cardiovascular and thermal strain, all players should be well-hydrated and well-nourished prior to going on court to compete [20, 55–57].

11.3.3 During Play

Players should drink appropriate and appealing fluids regularly on each changeover during every match [5, 6, 13, 14, 16, 43]. While optimal fluid intake (volume and beverage characteristics) during tennis match play is indeed situation- and individual-specific, drinking to thirst is not always adequate to sufficiently maintain hydration status or optimal performance, especially for those with a high sweating rate during extended play [58]. In contrast to earlier findings with non-tennis youth [59, 60], unflavored water has more recently been shown to be equally effective as a carbohydrate-electrolyte sports drink (CHO-E) in minimizing total body water deficits in physically active young girls during intermittent exercise in the heat [61]. The same was demonstrated with high-level, fit junior tennis players during intense on-court training in outdoor hot conditions [13]. However, many older adolescent and adult players who sweat extensively may benefit from an individualized level of regular concomitant sodium intake on-court during breaks. This will enhance rehydration and partially offset a rapidly evolving sweat-induced exchangeable sodium deficit [5–7, 16].

Readily available access to and ample regular intake of appropriate beverages is a priority. In addition, taking advantage of shaded space and a variety of short-term cooling practices (e.g., fans, umbrellas, iced and cold damp towels), along with the full allotted time on changeovers and between points, can improve perceptual and physiological strain [1, 21, 62, 63]. If it is within the scope of viable options, players can select to construct shorter points by modifying playing style as a player's capability allows, thus reducing overall exertion and metabolic heat production and storage to some extent [64]. Nonetheless, all tennis players should be closely monitored in the heat, and a prompt and appropriate response, including immediately stopping participation and seeking appropriate medical attention and treatment, should be implemented at the earliest signs of developing exertional heat illness.

11.3.4 Post-Play Recovery

Effective post-play rehydration involves more than simply ample water intake. Even with frequent breaks and opportunities to rehydrate during practice or with regularly consuming water or other beverages on each changeover during a long match, a post-play body water and/or exchangeable sodium deficit can be significant. Accordingly, a more deliberate individualized rehydration plan between sessions and matches is often essential. Concomitant with appropriate energy and other nutrient intake, electrolytes (especially sodium) need to be sufficiently consumed as well to offset potentially extensive sweat-related electrolyte losses and to better retain and distribute ingested water [5, 7, 12, 14, 16]. For same-day repeated matches, players should stay in the shade and/or airconditioned places to rest and recover. It is also fitting to minimize or eliminate (when the between-match time is brief) any additional pre-match warm-up period.

11.3.5 Evidence-Informed Tennis in the Heat Guidelines

The applicable knowledge base, awareness, and concern for player safety and well-being in the heat have noticeably improved in the past decade or so. Unfortunately, most current hot-weather preparation, on-court, recovery, and scheduling recommendations and guidelines promoted by tennis governing bodies for effectively managing hydration, reducing thermal strain, optimizing performance, and minimizing exertional heat illness risk on-court are not sufficiently sport-specific and evidence-based. For example, practical on-court cooling methods [1, 62, 63] need to be more closely and comprehensively assessed in tennis; and various established guidelines for stoppage or modifying play in the heat should be more thoroughly and objectively scrutinized [3, 4, 65]. Tippet et al. [19] specifically examined of the effectiveness of the Women's Tennis Association Extreme Weather Conditions Rule in reducing thermal strain. This rule (in place since 1992) currently allows for a break in play between the second and third sets of the match when the on-site wet-bulb globe temperature (WBGT) meets or exceeds 30.1 °C [65]. After the permitted 10-min break, body core temperature decreased by 0.25 (0.20) °C. However, core temperature remained slightly higher than it was at the start of the previous second set. By the end of play, peak thermal strain was similar as observed at the end of the second set. But without many more similar observations in competitive (including men's) tennis, as well as a comparison to extended-set vs 3-set scenarios in extreme weather (as defined by this rule) where players do not take the between-set break, full appreciation of the extreme weather rule effectiveness on player well-being in reducing thermal strain would be premature.

Recognizing the insufficient tennis-specific on-court and recovery data and knowledge to better guide players and coaches, certain recommendations of best practices should still be emphasized to minimize exertional heat illness risk and improve player safety, well-being, and on-court performance. This is particularly relevant with the notably greater exertional heat illness risk and potential performance challenges resulting from multiple competitive matches in hot weather on the same day. Tennis governing bodies and tournament administrators should accordingly provide longer rest and recovery periods between same-day matches as environmental heat stress increases. And while coaches and event administrators should always provide adequate between-session/match recovery time, especially in the heat, a greater accommodation for recovery should be anticipated and liberally applied, as young tennis players get older, and train and compete repeatedly and at a higher level.

11.3.6 Players with Disabilities: Wheelchair Tennis

Wheelchair tennis continues to grow in participation and popularity, reinforced and showcased by the Paralympics and International Tennis Federation (ITF) through their widespread schedule of global events, development funds and programs, and high-profile ambassadors [66]. Not surprisingly, wheelchair competitors are exposed to the same extreme heat conditions and related health risks and performance barriers as able-bodied players. Accordingly, many of the same risk mitigation strategies (heat acclimatization, reduced pre-match heat exposure, modification of play style, hydration, etc.) as with able-bodied players are appropriate and effective. However, there are also numerous additional inherent modulating factors and related clinical risks with spinal cord injury (SCI), based on the variance of spinal lesion level and consequent degree of physiological impairment. For example, this population demonstrates proportionate impaired thermoregulatory function and consequent increase in body core temperature and heat storage during vigorous exercise in the heat, compared to able-bodied peers [67]. There is also greater heat retention during exercise recovery. Thus, traditional between-game cooling strategies on changeovers during wheelchair tennis are not as effective. The aggregate potential effect is an intrinsic higher risk (proportional to the injury level) of developing exertional heat illnesses during tennis competition in the heat [68]. While the ITF provides guidelines to allow for a 10-min break between second and third sets when the WBGT meets or exceeds 30.1 °C or suspension of play (32.2 °C) in extreme weather conditions [69], research to determine more evidence-informed heat safety thresholds and guidelines specific to wheelchair tennis is prominently lacking.

11.4 Future Directions

The published empirical evidence of heat-related physiological responses and metrics in youth and adult tennis players during unrestricted sanctioned competition is very limited. As a result, this narrow selection of observations does not fully reveal or represent the full scope of practical hydration and thermal strain challenges, exertional heat illness risks, and barriers to performance that many tennis players face when competing in the heat. Accordingly, currently established general recommendations for maintaining hydration and minimizing on-court thermal strain and exertional heat illness risk in youth and adults [24, 29, 70, 71] should be recognized as having a limited field-supported, evidence base specific to tennis.

11.4.1 Real-World Scenarios in Tennis

Much more research needs to be done to better appreciate the full scope and extent of the physiological demands and hydration and thermal strain challenges confronting junior and adult tennis players in various environments, competition formats, and venues. New investigations should include profiling players during longer, more intense, 3-set to 5-set matches played in the afternoon during later rounds of sanctioned tournaments. This is where the risk is likely to be appreciably greater for significant body water deficits and other physiological carry-over effects from previous rounds of play, as well as undue thermal strain. Reported incidents of heat-related court calls, requests for cooling devices, and post-play medical consults across environmental heat stress zones at the Australian Open have, however, provided some important preliminary practical insights to the escalating parallel heat strain incurred by players during competitive elite tennis [3, 4].

Examining the impact of environmental heat and thermal strain on hyperthermic fatigue, neuromuscular function, and musculoskeletal or other injury risk [72–74] should also be clinical, player management, and research priorities. These effects should be specifically profiled in response to extended and repeated bouts of intense tennis training and sanctioned competition across multiple successive days and weeks, as this reflects practical tennis schedules. Current perspective based on selected discrete test measures during and after only simulated match play bouts [20, 75, 76] in the heat is insufficient. Moreover, current risk-reduction strategies for pre-play, on-court, and between-set cooling, enhanced hydration, and between-match recovery need to be better clarified and validated for practical efficacy in real-world scenarios with able-bodied and wheelchair tennis [77–82]. Research on the interacting effects of schedule, travel, and heat-acclimatization status of players on clinical risk and performance is also needed. Accordingly, it is premature to claim a full appreciation of the real-world challenges these athletes face during sanctioned competition.

11.4.2 Technology and Advanced Analytics in Tennis Heat Safety Research

The experimental setting (i.e., laboratory or on-court with simulated match play) often utilized in evaluating cardiovascular and thermal strain and other physiological responses during tennis in the heat has significant practical translation limitations. Many of the observations, conclusions, and purported applications are, in certain instances, too far removed from a more real-world competitive tennis context. Moreover, the physiological and psychological demands and individual responses of a tennis player's complex biological system while on-court and during tournament play cannot be fully interpreted by evaluating only a minimal number of discrete measures. This is especially evident with measurements limited to only pre- and post-play, or when the players under observation are in a controlled setting detached from the natural environment of daily multiple stressors and other wide-ranging influencing factors. Also, generalized heat-safety guidelines based on environmental conditions alone do not adequately address the practical individual on-court or recovery differences and needs.

There has been a noticeable recent explosion of advanced technology for physiological monitoring in Sport, including a vast array of wearable and remote smart sensors for real-time data collection, supported by integrated platforms with advanced algorithms and analytics and practical easy user interface [25, 83, 84]. Efficiently and effectively collecting, managing, and analyzing massive amounts of structured and unstructured information from multiple disparate domains requiring high-performance computer systems is, moreover, no longer the limiting factor. Today, we have the capacity to address all the variables outlined in this chapter and other influencing factors in a widely inclusive, multi-domain, and real-world knowledgeable and innovative way. This is critical to developing and optimizing more effective and practical, evidence-based recommendations and guidelines for tennis players who are training and competing in the heat. Artificial intelligence-driven machine learning and advanced analytics can also help reveal what has been previously unachievable (or even conceived) in more precisely defining player response/risk signatures and patterns that will assist further in determining definitive evidence-informed heat safety guidelines and protocols. The collective effect of these new tools and approaches allows scientists to consider multiple applicable domains of information, much larger data sets, and many more relevant variables and influences. This will no doubt lead to developing valid new predictive models and applications that can better assist in anticipating clinical care needs, mitigating modifiable risk in advance, and clinical decision support [85, 86]. The compelling impact will be to practically and more optimally mitigate exertional heat illness risk while optimizing performance for all tennis-playing populations. Indeed, this timely approach will help to close the gap between research and practical acceptance and uptake of related heat safety recommendations and guidelines for tennis and all of Sport.

11.5 Key Points

- Sweating during tennis can be extensive, leading to significant total body water and exchangeable sodium deficits.
- Thermal strain can be significantly elevated during tennis; however, the inherent nature and format of tennis may be the principal mitigating factors in players incurring excessive body core temperature and loss of thermoregulatory control leading to exertional heatstroke.
- Much more real-world research of tennis competition is needed to fully appreciate thermoregulatory challenges confronting players training and competing in the heat, as well as to design the most practically efficacious, evidence-informed heat safety guidelines.

References

1. Ellenbecker TS, Stroia KA. Heat research guides current practices in professional tennis. Br J Sports Med. 2014;48(Suppl 1):i5–6. https://doi.org/10.1136/bjsports-2013-093272.
2. Périard JD, Bergeron MF. Competitive match-play tennis under heat stress: a challenge for all players. Br J Sports Med. 2014;48(Suppl 1):i1–3. https://doi.org/10.1136/bjsports-2014-093496.

3. Smith MT, Reid M, Kovalchik S, Wood T, Duffield R. Heat stress incidence and matchplay characteristics in women's Grand Slam tennis. J Sci Med Sport. 2018;21(7):666–70. https://doi.org/10.1016/j.jsams.2017.11.006.
4. Smith MT, Reid M, Kovalchik S, Woods TO, Duffield R. Heat stress incident prevalence and tennis matchplay performance at the Australian Open. J Sci Med Sport. 2018;21(5):467–72. https://doi.org/10.1016/j.jsams.2017.08.019.
5. Bergeron MF. Heat cramps during tennis: a case report. Int J Sport Nutr. 1996;6(1):62–8.
6. Bergeron MF. Heat cramps: fluid and electrolyte challenges during tennis in the heat. J Sci Med Sport. 2003;6(1):19–27.
7. Bergeron MF. Muscle cramps during exercise: is it fatigue or electrolyte deficit? Curr Sports Med Rep. 2008;7(4):S50–S5.
8. Mitchell JB, Phillips MD, Mercer SP, Baylies HL, Pizza FX. Postexercise rehydration: effect of Na^+ and volume on restoration of fluid spaces and cardiovascular function. J Appl Physiol. 2000;89:1302–9.
9. Sanders B, Noakes TD, Dennis SC. Sodium replacement and fluid shifts during prolonged exercise in humans. Eur J Appl Physiol. 2001;84(5):419–25.
10. Sanders B, Noakes TD, Dennis SC. Water and electrolyte shifts with partial fluid replacement during exercise. Eur J Appl Physiol Occup Physiol. 1999;80(4):318–23.
11. Shirreffs SM, Maughan RJ. Volume repletion after exercise-induced volume depletion in humans: replacement of water and sodium losses. Am J Phys. 1998;274(5 Pt 2):F868–75.
12. Bergeron MF. Exertional heat cramps: recovery and return to play. J Sport Rehabil. 2007;16(3):190–6.
13. Bergeron MF, Waller JL, Marinik EL. Voluntary fluid intake and core temperature responses in adolescent tennis players: sports beverage versus water. Br J Sports Med. 2006;40(5):406–10.
14. Bergeron MF, McLeod KS, Coyle JF. Core body temperature during competition in the heat: National Boys' 14's Junior Tennis Championships. Br J Sports Med. 2007;41:779–83.
15. Falk B, Bar-Or O, Calvert R, MacDougall JD. Sweat gland response to exercise in the heat among pre-, mid-, and late-pubertal boys. Med Sci Sports Exerc. 1992;24(3):313–9.
16. Bergeron MF, Maresh CM, Armstrong LE, Signorile JF, Castellani JW, Kenefick RW, et al. Fluid-electrolyte balance associated with tennis match play in a hot environment. Int J Sport Nutr. 1995;5:180–93.
17. Hornery DJ, Farrow D, Mujika I, Young W. An integrated physiological and performance profile of professional tennis. Br J Sports Med. 2007;41(8):531–6; discussion 6. https://doi.org/10.1136/bjsm.2006.031351.
18. Morante SM, Brotherhood JR. Air temperature and physiological and subjective responses during competitive singles tennis. Br J Sports Med. 2007;41(11):773–8.
19. Tippet ML, Stofan JR, Lacambra M, Horswill CA. Core temperature and sweat responses in professional women's tennis players during tournament play in the heat. J Athl Train. 2011;46(1):55–60. https://doi.org/10.4085/1062-6050-46.1.55.
20. Périard JD, Racinais S, Knez WL, Herrera CP, Christian RJ, Girard O. Coping with heat stress during match-play tennis: does an individualised hydration regimen enhance performance and recovery? Br J Sports Med. 2014;48(Suppl 1):i64–70. https://doi.org/10.1136/bjsports-2013-093242.
21. Périard JD, Racinais S, Knez WL, Herrera CP, Christian RJ, Girard O. Thermal, physiological and perceptual strain mediate alterations in match-play tennis under heat stress. Br J Sports Med. 2014;48(Suppl 1):i32–i8. https://doi.org/10.1136/bjsports-2013-093063.
22. Kraning KK 2nd, Gonzalez RR. Physiological consequences of intermittent exercise during compensable and uncompensable heat stress. J Appl Physiol. 1991;71(6):2138–45.
23. American College of Sports Medicine, Armstrong LE, Casa DJ, Millard-Stafford M, Moran DS, Pyne SW, et al. American College of Sports Medicine position stand. Exertional heat illness during training and competition. Med Sci Sports Exerc. 2007;39(3):556–72. https://doi.org/10.1249/MSS.0b013e31802fa199.
24. Bergeron MF. Reducing sports heat illness risk. Pediatr Rev. 2013;34(6):270–9. https://doi.org/10.1542/pir.34-6-270.

25. Buller MJ, Welles AP, Friedl KE. Wearable physiological monitoring for human thermal-work strain optimization. J Appl Physiol (1985). 2018;124(2):432–41. https://doi.org/10.1152/japplphysiol.00353.2017.
26. Byrne C, Lim CL. The ingestible telemetric body core temperature sensor: a review of validity and exercise applications. Br J Sports Med. 2007;41(3):126–33.
27. Epstein Y, Mazgaoker S, Gruber D, Moran DS, Yanovich R, Ketko I, et al. Measuring body core temperature using a novel non-invasive sensor. Extrem Physiol Med. 2015;4(Suppl 1):A28.
28. MacRae BA, Annaheim S, Spengler CM, Rossi RM. Skin temperature measurement using contact thermometry: a systematic review of setup variables and their effects on measured values. Front Physiol. 2018;9:29. https://doi.org/10.3389/fphys.2018.00029.
29. Bergeron MF, Devore C, Rice SG. American Academy of Pediatrics Council on sports medicine and fitness and council on school health. Policy statement - climatic heat stress and exercising children and adolescents. Pediatrics. 2011;128(3):e741–7. https://doi.org/10.1542/peds.2011-1664.
30. Rowland T. Thermoregulation during exercise in the heat in children: old concepts revisited. J Appl Physiol. 2008;105(2):718–24. https://doi.org/10.1152/japplphysiol.01196.2007.
31. Inbar O, Morris N, Epstein Y, Gass G. Comparison of thermoregulatory responses to exercise in dry heat among prepubertal boys, young adults and older males. Exp Physiol. 2004;89(6):691–700.
32. Rivera-Brown AM, Rowland TW, Ramirez-Marrero FA, Santacana G, Vann A. Exercise tolerance in a hot and humid climate in heat-acclimatized girls and women. Int J Sports Med. 2006;27(12):943–50.
33. Rowland T, Garrison A, Pober D. Determinants of endurance exercise capacity in the heat in prepubertal boys. Int J Sports Med. 2007;28(1):26–32.
34. Rowland T, Hagenbuch S, Pober D, Garrison A. Exercise tolerance and thermoregulatory responses during cycling in boys and men. Med Sci Sports Exerc. 2008;40(2):282–7.
35. Fortes MB, Di Felice U, Dolci A, Junglee NA, Crockford MJ, West L, et al. Muscle-damaging exercise increases heat strain during subsequent exercise heat stress. Med Sci Sports Exerc. 2013;45(10):1915–24. https://doi.org/10.1249/MSS.0b013e318294b0f8.
36. Coyle J. Cumulative heat stress appears to affect match outcome in a junior tennis championship. Med Sci Sports Exerc. 2006;38(5):S110.. (Abstract)
37. Bergeron MF, Laird MD, Marinik EL, Brenner JS, Waller JL. Repeated-bout exercise in the heat in young athletes: physiological strain and perceptual responses. J Appl Physiol. 2009;106:476–85. https://doi.org/10.1152/japplphysiol.00122.2008.
38. Brenner IK, Zamecnik J, Shek PN, Shephard RJ. The impact of heat exposure and repeated exercise on circulating stress hormones. Eur J Appl Physiol Occup Physiol. 1997;76(5):445–54.
39. Kruk B, Szczypaczewska M, Opaszowski B, Kaciuba-Uscilko H, Nazar K. Thermoregulatory and metabolic responses to repeated bouts of prolonged cycle-ergometer exercise in man. Acta Physiol Pol. 1990;41(7):22–31.
40. Ronsen O, Haugen O, Hallen J, Bahr R. Residual effects of prior exercise and recovery on subsequent exercise-induced metabolic responses. Eur J Appl Physiol. 2004;92:498–507.
41. Sawka MN, Knowlton RG, Critz JB. Thermal and circulatory responses to repeated bouts of prolonged running. Med Sci Sports. 1979;11(2):177–80.
42. Yamada PM, Golding LA. Physiological responses of female fire fighters to repeated work bouts. AMAA J. 2004;17(1):5–7.
43. Bergeron MF. Hydration and thermal strain during tennis in the heat. Br J Sports Med. 2014;48(Suppl 1):i12–7. https://doi.org/10.1136/bjsports-2013-093256.
44. American Academy of Pediatrics Committee on Sports Medicine and Fitness. Medical conditions affecting sports participation. Pediatrics. 2001;107(5):1205–9.
45. Lopez-Samanes A, Moreno-Perez V, Kovacs MS, Pallares JG, Mora-Rodriguez R, Ortega JF. Use of nutritional supplements and ergogenic aids in professional tennis players. Nutr Hosp. 2017;34(5):1463–8. https://doi.org/10.20960/nh.1404.

46. Hornery DJ, Farrow D, Mujika I, Young WB. Caffeine, carbohydrate, and cooling use during prolonged simulated tennis. Int J Sports Physiol Perform. 2007;2(4):423–38.
47. Ranchordas MK, Rogersion D, Ruddock A, Killer SC, Winter EM. Nutrition for tennis: practical recommendations. J Sports Sci Med. 2013;12(2):211–24.
48. Del Coso J, Estevez E, Mora-Rodriguez R. Caffeine during exercise in the heat: thermoregulation and fluid-electrolyte balance. Med Sci Sports Exerc. 2009;41(1):164–73. https://doi.org/10.1249/MSS.0b013e318184f45e.
49. Ely BR, Ely MR, Cheuvront SN. Marginal effects of a large caffeine dose on heat balance during exercise-heat stress. Int J Sport Nutr Exerc Metab. 2011;21(1):65–70.
50. Bergeron MF. Youth sports - making it safe in the heat. ACSM Sports Medicine Basics. Indianapolis: American College of Sports Medicine; 2017.
51. Périard JD, Racinais S, Sawka MN. Adaptations and mechanisms of human heat acclimation: applications for competitive athletes and sports. Scand J Med Sci Sports. 2015;25(Suppl 1):20–38. https://doi.org/10.1111/sms.12408.
52. Périard JD, Racinais S, Timpka T, Dahlstrom O, Spreco A, Jacobsson J, et al. Strategies and factors associated with preparing for competing in the heat: a cohort study at the 2015 IAAF World Athletics Championships. Br J Sports Med. 2017;51(4):264–70. https://doi.org/10.1136/bjsports-2016-096579.
53. Racinais S, Alonso JM, Coutts AJ, Flouris AD, Girard O, Gonzalez-Alonso J, et al. Consensus recommendations on training and competing in the heat. Sports Med. 2015;45(7):925–38. https://doi.org/10.1007/s40279-015-0343-6.
54. Périard JD, Travers GJS, Racinais S, Sawka MN. Cardiovascular adaptations supporting human exercise-heat acclimation. Auton Neurosci. 2016;196:52–62. https://doi.org/10.1016/j.autneu.2016.02.002.
55. Goto M, Okazaki K, Kamijo Y, Ikegawa S, Masuki S, Miyagawa K, et al. Protein and carbohydrate supplementation during 5-day aerobic training enhanced plasma volume expansion and thermoregulatory adaptation in young men. J Appl Physiol (1985). 2010;109(4):1247–55. https://doi.org/10.1152/japplphysiol.00577.2010.
56. Sawka MN, Montain SJ, Latzka WA. Hydration effects on thermoregulation and performance in the heat. Comp Biochem Physiol A Mol Integr Physiol. 2001;128:679–90.
57. Bergeron MF. Hydration in the pediatric athlete - how to guide your patients. Curr Sports Med Rep. 2015;14(4):288–93. https://doi.org/10.1249/JSR.0000000000000179.
58. Armstrong LE, Johnson EC, Bergeron MF. COUNTERVIEW: is drinking to thirst adequate to appropriately maintain hydration status during prolonged endurance exercise? No. Wilderness Environ Med. 2016;27(2):195–8. https://doi.org/10.1016/j.wem.2016.03.002.
59. Rivera-Brown AM, Gutierrez R, Gutierrez JC, Frontera WR, Bar-Or O. Drink composition, voluntary drinking, and fluid balance in exercising, trained, heat-acclimatized boys. J Appl Physiol. 1999;86(1):78–84.
60. Wilk B, Bar-Or O. Effect of drink flavor and NaCL on voluntary drinking and hydration in boys exercising in the heat. J Appl Physiol. 1996;80(4):1112–7.
61. Wilk B, Rivera-Brown AM, Bar-Or O. Voluntary drinking and hydration in non-acclimatized girls exercising in the heat. Eur J Appl Physiol. 2007;101(6):727–34.
62. Lynch GP, Périard JD, Pluim BM, Brotherhood JR, Jay O. Optimal cooling strategies for players in Australian Tennis Open conditions. J Sci Med Sport. 2018;21(3):232–7. https://doi.org/10.1016/j.jsams.2017.05.017.
63. Schranner D, Scherer L, Lynch GP, Korder S, Brotherhood JR, Pluim BM, et al. In-play cooling interventions for simulated match-play tennis in hot/humid conditions. Med Sci Sports Exerc. 2017;49(5):991–8. https://doi.org/10.1249/MSS.0000000000001183.
64. Reid M, Duffield R. The development of fatigue during match-play tennis. Br J Sports Med. 2014;48(Suppl 1):i7–11. https://doi.org/10.1136/bjsports-2013-093196.
65. 2018 WTA Official Rulebook. Women's Tennis Association, St. Petersburg, Florida, USA. http://wtafiles.wtatennis.com/pdf/publications/2018WTARulebook.pdf. Accessed 29 Apr 2018.

66. Wheelchair ITF. International Tennis Federation, London, UK. http://www.itftennis.com/wheelchair/home.aspx. Accessed 29 Apr 2018.
67. Price MJ. Thermoregulation during exercise in individuals with spinal cord injuries. Sports Med. 2006;36(10):863–79.
68. Girard O. Thermoregulation in wheelchair tennis-how to manage heat stress? Front Physiol. 2015;6:175. https://doi.org/10.3389/fphys.2015.00175.
69. ITF Wheelchair Tennis Regulations. International Tennis Federation, London, UK. 2018. http://www.itftennis.com/media/278520/278520.pdf. Accessed 29 Apr 2018.
70. Armstrong LE, Casa DJ, Millard-Stafford M, Moran DS, Pyne SW, Roberts WO. American College of Sports Medicine position stand. Exertional heat illness during training and competition. Med Sci Sports Exerc. 2007;39(3):556–72.
71. Casa DJ, Armstrong LE, Hillman SK, Montain SJ, Reiff RV, Rich BSE, et al. National athletic trainers' association position statement: fluid replacement for athletes. J Athl Train. 2000;35(2):212–24.
72. Cheung SS, Sleivert GG. Multiple triggers for hyperthermic fatigue and exhaustion. Exerc Sport Sci Rev. 2004;32(3):100–6.
73. Nybo L. Hyperthermia and fatigue. J Appl Physiol. 2008;104(3):871–8. https://doi.org/10.1152/japplphysiol.00910.2007.
74. Périard JD, Caillaud C, Thompson MW. Central and peripheral fatigue during passive and exercise-induced hyperthermia. Med Sci Sports Exerc. 2011;43(9):1657–65. https://doi.org/10.1249/MSS.0b013e3182148a9a.
75. Girard O, Christian RJ, Racinais S, Périard JD. Heat stress does not exacerbate tennis-induced alterations in physical performance. Br J Sports Med. 2014;48(Suppl 1):i39–44. https://doi.org/10.1136/bjsports-2013-093165.
76. Périard JD, Girard O, Racinais S. Neuromuscular adjustments of the knee extensors and plantar flexors following match-play tennis in the heat. Br J Sports Med. 2014;48(Suppl 1):i45–51. https://doi.org/10.1136/bjsports-2013-093160.
77. Diaper NJ, Goosey-Tolfrey VL. A physiological case study of a paralympic wheelchair tennis player: reflective practise. J Sports Sci Med. 2009;8(2):300–7.
78. Duffield R, Bird SP, Ballard RJ. Field-based pre-cooling for on-court tennis conditioning training in the heat. J Sports Sci Med. 2011;10(2):376–84.
79. Goosey-Tolfrey VL, Diaper NJ, Crosland J, Tolfrey K. Fluid intake during wheelchair exercise in the heat: effects of localized cooling garments. Int J Sports Physiol Perform. 2008;3(2):145–56.
80. Magal M, Webster MJ, Sistrunk LE, Whitehead MT, Evans RK, Boyd JC. Comparison of glycerol and water hydration regimens on tennis-related performance. Med Sci Sports Exerc. 2003;35(1):150–6. https://doi.org/10.1249/01.MSS.0000043600.40490.7E.
81. Ranalli GF, Demartini JK, Casa DJ, McDermott BP, Armstrong LE, Maresh CM. Effect of body cooling on subsequent aerobic and anaerobic exercise performance: a systematic review. J Strength Cond Res. 2010;24(12):3488–96. https://doi.org/10.1519/JSC.0b013e3181fb3e15.
82. Siegel R, Mate J, Watson G, Nosaka K, Laursen PB. Pre-cooling with ice slurry ingestion leads to similar run times to exhaustion in the heat as cold water immersion. J Sports Sci. 2012;30(2):155–65. https://doi.org/10.1080/02640414.2011.625968.
83. Mendes JJ Jr, Vieira ME, Pires MB, Stevan SL Jr. Sensor fusion and smart sensor in sports and biomedical applications. Sensors (Basel). 2016;16(10):E1569. https://doi.org/10.3390/s16101569.
84. Sawka MN, Friedl KE. Emerging wearable physiological monitoring technologies and decision aids for health and performance. J Appl Physiol. 2018;124(2):430–1. https://doi.org/10.1152/japplphysiol.00964.2017.
85. Millington B, Millington R. The datafication of everything: toward a sociology of sport and big data. Sociol Sport J. 2015;32(2):140–60.
86. Fry E, Mukherjee S. Big data meets biology. Fortune. 2018:44–56.

Athletics: Track and Field in the Heat 12

Juan-Manuel Alonso and Jordan Santos

12.1 Influence of Heat on Performance

Hot environmental conditions have a major impact on Athletics—Track and Field performance, with the effects on running performance being related to distance. While endurance performance is typically impaired in the heat, single short sprint performance may be improved [1]. This stems from an increase in muscle temperature improving explosive power during sprinting or jumping by enhancing muscle contractile properties [2, 3]. Interestingly, Guy et al. [4] analysed seven consecutive International Association of Athletics Federations (IAAF) World Championships (1999–2011) top ten performances in endurance events (5000, 10,000 m and Marathon) and top six performances in sprint events (100 and 200 m). The authors found that in temperate conditions (<25 °C) athletes participating in endurance events were 2% faster than in the heat (≥25 °C; Fig. 12.1). Marathon runners experienced the largest performance impairment in the heat with a 3.1% and a 2.7% reduction in males and females, respectively, whereas middle distance events were less affected by environmental conditions [4]. In contrast, sprint performance was improved with athletes performing 2% better in hot compared with temperate conditions.

The results reported by Guy et al. [4] are in line with previous research [5–8]. For example, in an analysis of performance from seven marathons over a period of 6–36 years, Ely et al. [9] showed a progressive impairment in performance in relation to a rise in wet-bulb-globe-temperature (WBGT) from 5 to 25 °C. This decrement in marathon performance held true for both men and women of wide ranging

J.-M. Alonso (✉)
Sports Medicine Department, ASPETAR Hospital, Doha, Qatar
e-mail: Juan-Manuel.Alonso@aspetar.com

J. Santos
Department of Physical Education and Sport, University of the Basque Country UPV/EHU, Vitoria-Gasteiz, Spain

J. D. Périard, S. Racinais (eds.), *Heat Stress in Sport and Exercise*,
https://doi.org/10.1007/978-3-319-93515-7_12

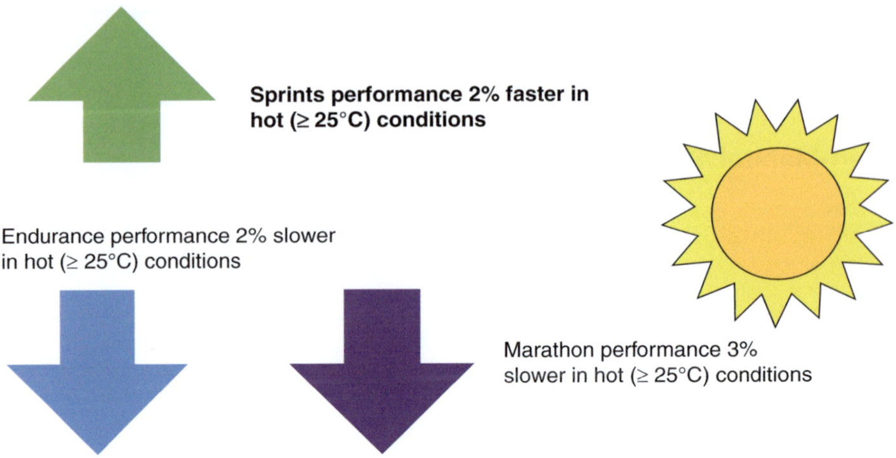

Fig. 12.1 Effects of heat on running performance. Adapted from Jeukendrup [1] and Guy et al. [4]

abilities, although with slower runners being more negatively affected [9]. This could be partly due to the hot microclimate generated by large groups of runners close to each other [10, 11]. El Helou et al. [12] also analysed 1,791,972 performances from three European (Paris, London and Berlin) and three American (Boston, Chicago and New York) marathon races between 2001 and 2010. The authors concluded that the main environmental factor influencing performance was temperature: the more the temperature increased, the more running speed declined and a greater percentage of runners withdrew from the race [12]. This is supported by the findings of Vihma [13], who reported that the percentage of non-finishers in the annual Stockholm Marathon from 1980 to 2008 was negatively affected by the air temperature and relative humidity.

The aforementioned negative effects of heat and humidity are likely enhanced in race walking events when compared to long-distance running events. For example, race walking is characterized by a worse economy of locomotion that ultimately increases internal heat gain relative to running at the same speed. In addition, race walking usually occurs at a lower speed, which limits self-generated wind velocity and convective cooling. Together, the lower economy of movement and reduced heat dissipation capacity associated with race walking in hot environments may lead to an increased risk of hyperthermia [14].

There are several factors playing a fundamental role on the decrement in endurance performance in warm and hot conditions during Athletics events [15]. For example, exercise-heat stress with or without dehydration may impair aerobic performance via one or more mechanisms related to central nervous system, skeletal muscle (i.e. metabolism) and/or cardiovascular (i.e. blood flow) function [5]. While a high core body temperature (T_c) per se was previously suggested as a main factor explaining impaired aerobic performance in the heat [5, 6, 15], it is currently believed that high skin temperatures (T_{sk}), or more specifically the core-to-skin temperature difference, impairs aerobic exercise performance in warm-hot environments due to a rise in cardiovascular stress [5]. Hence, the reductions in running speed noted in longer track and field events are likely mediated by the development of thermal strain, along with sensations of thermal discomfort, which suggests that performance in the heat is influenced by both systemic physiological and behavioural factors [5, 6, 16].

In contrast, sprinting and jumping may be improved in the heat as muscle contractile function is optimized at muscle temperatures slightly higher than resting [8]. Current evidence in team sport athletes also suggests that passive heat maintenance may improve sprinting and jumping performance [17]. Although team sports is not analogous to jumping/throwing field events, there is a vast body of literature converging to suggest that a rise in core and more specifically muscle temperature may be beneficial to the performance of jumpers and throwers [2, 3].

12.2 Influence of Heat on Health

When exercising in hot and humid ambient conditions, the water content in the air acts as a barrier for evaporative cooling [7, 15]. In such conditions, sweating is much less effective for removing heat from the body and in those with high sweat rates may result in dehydration, which impairs thermoregulation [15]. The increased body-water deficit, associated with progressive increase of core temperature, exacerbates the decrement in performance, but also increases the risk of exertional heat illness [3, 7]. The severity of exertional heat illness can escalate from relatively benign symptoms such as exertional muscle cramps to heat exhaustion, exertional associated collapse [18], heat injury and ultimately, exertional heat stroke [3]. It is important to acknowledge that some individuals may be more prone to collapse from exhaustion in the heat than others (i.e. those who are not acclimatized, use certain medications, are dehydrated or have recently been ill). Importantly, whilst core temperature is used in the diagnosis of EHI, along with central nervous system dysfunction, it is common for athletes to reach elevated temperature and remain asymptomatic. For example, Byrne et al. [19] reported that in the 18 participants in which core temperature was measured during a 21-km running race in hot (27 °C) and humid (87% RH) conditions, all reached a temperature of 39 °C, with 56% reaching 40 °C and 11% reaching 41 °C. However, none of these athletes were treated for heat illness.

In the context of Track and Field, exertional heat illness can affect athletes during short, high-intensity or long-duration events. In the best-case scenario, this may result in having to abort a training session in the heat, or withdraw from a competition, and in the worst-case scenario may lead to collapse during or soon after an event [20]. For example, exertional heat illness was a common condition (11–17%) among athletes participating in the World Athletics Championships in Berlin 2009, Daegu 2011 and Moscow 2013, especially those participating in race walking events [21–23]. In a cohort study at the 2015 IAAF World Athletics Championships in Beijing, it was determined that 48% of athletes had previously experienced exertional heat illness symptoms (e.g. cramping, vomiting and nausea) while training and competing in the heat [24]. The authors further reported that 8.5% had been diagnosed with exertional heat illness, with dehydration and heat exhaustion being most prevalent.

12.3 Countermeasures to Optimize Athletics Performance and Health

12.3.1 Heat Acclimatization

Recent reviews have documented the effectiveness of heat acclimatization in alleviating the physiological and perceptual effects of heat stress [25–31]. The most consistent findings are a decreased core temperature and heart rate during exercise in the heat, along with increased total body water and sweat rate. While heat acclimatization can take up to 10 days of consecutive heat exposure to develop, rapid induction of most of the adjustments can be seen in about 7 days (see Chap. 8). Therefore, short-term heat acclimatization may be a practical consideration for track and field athletes faced with congested training and competition schedules. This could involve arriving to the competition venue 1–2 weeks in advance to acclimatize to the setting. Adaptations may also be induced via post-training water immersion [32], thus maintaining the training stimulus and providing an impulse for acclimatization.

12.3.2 Warm-up Strategies and Heat Stress

One of the primary goals of the warm-up is to increase muscle temperature. However, since warming-up in hot and humid conditions increases thermal and circulatory strain, which could be detrimental for prolonged exercise capacity, the warm-up routine should be adapted to the environmental conditions [3]. In endurance events, athletes should approach the warm-up with the aim to optimize physiological readiness through activation of the main energy systems and movement patterns while avoiding too large an increase in core temperature via significant metabolic heat production. Hence, shorter warm-up durations are recommended before prolonged track and field events in hot environments. Sprinting and jumping athletes may wish to maintain or only slightly adjust the warm-up routine.

12.3.3 Pre-cooling and Per-cooling

Given that skin cooling can reduce cardiovascular strain and that cold water or ice-slurry ingestion may reduce core temperature, it has been suggested that cooling interventions may be used to improve prolonged exercise capacity in the heat. In this regard, it has been reported that the combined use of internal (i.e. cold drinks and ice-slurry beverages) and external (i.e. cooling garments, towels and fanning) cooling strategies can lead to greater cooling capacity [3, 25, 33]. Thus, mixed

Table 12.1 Recommendations for track and field athletes [3, 25–29, 33, 34]

• Endurance athletes (marathon and race walking) competing in hot conditions should acclimatize
• Warm-up routines should be shortened when competing in hot conditions
• Using a mixture of internal (cold drinks and ice-slurry beverages) and external (cooling garments, towels and fanning) cooling strategies is encouraged when competing in the heat
• An intake of 5–6 mL of water per kg of body mass, at a frequency of every 2–3 h, prior to training or competing in the heat is advised

cooling methods should be part of the warm-up in the heat for endurance athletes. Athletes involved in sprinting, jumping, throwing and combined events (i.e. decathlon and heptathlon) could also benefit from mixed cooling strategies, and when implemented during rest periods might improve performance [3, 25, 34]. These methods would require evaluation and familiarization, so as not to disrupt the routine of the athlete or lead to too great a decrease in muscle temperature.

12.3.4 Hydration

Sweat rate during exercise in the heat varies depending on environmental conditions, the metabolic rate and acclimatization status of the athlete. Values ranging from 1 L/h to 1.5 L/h are common for athletes performing vigorous exercise in hot environments, although heavy sweaters can exceed 2.5 L/h [25]. Interestingly, Alberto Salazar reportedly had a sweat rate of 3.7 L/h during the 1984 Olympic Marathon [35]. As it is not possible to replenish that amount of fluid during exercise, it is particularly important to start the event euhydrated as the in-competition strategy would likely aim to minimize the progressive fluid loss, not to totally offset it. For most individuals, an intake of 5–6 mL of water per kg of body mass, at a frequency of every 3–4 h in the days leading up to competition as well as 2–3 h prior to either training or competing in the heat is advisable to replenish sweat losses and minimize the risk of undertaking an event in a hypohydrated state. Cold drinks and ice-slurry beverages of crushed ice of ≤4 °C are also recommended [3]. The main electrolyte lost in sweat is sodium and supplementation during longer events may be required. During events lasting longer than 1 h (e.g. marathon and race walking), athletes should aim to consume a solution containing 0.5–0.7 g/L of sodium. In those experiencing muscle cramping, it is recommended to increase the sodium supplementation to 1.5 g/L of fluid [25] (Table 12.1).

12.4 Heat Policy and Implementation

The IAAF has shown awareness in the past on the health risks posed to athletes by environmental factors and has implemented measures to help protect the athletes. The on-field assessment of the WBGT has been adopted in World IAAF Athletics Championships since Osaka 2007. The IAAF employed the flag system based on

Table 12.2 Wet bulb globe temperature (WBGT) index flag coding system, risks and recommendations (adapted from: [20, 36–38])

	Roberts 1998/NATA 2002/IAAF 2013			ACSM 2007 Continuous activity and competition	
	Risk	WGBT	Recommendation	WGBT	Recommendation
Black flag	Extreme	>28 °C (82 °F)	Consider rescheduling or delaying the event until safer conditions prevail; if the event must take place, be on high alert	27.9–30.0 °C (82.1–86 °F)	Cancel level for EHS risk
Red flag	High	23–28 °C (73–82 °F)	Everyone should be aware of injury potential; individuals at risk should not compete	25.7–27.8 °C (78.1–82 °F)	Risk for unfit, non-acclimatized individuals is high
Yellow flag	Moderate risk	18–23 °C (65–73 °F)	Risk increases as event progresses through the day	22.3–25.6 °C (72.1–78 °F)	Risk for all competitors is increased
Green flag	Low	<18 °C (65 °F)	Risk low but still exists based on risk factors	18.4–22.2 °C (65.1–72 °F)	Risk of EHS and other heat illness begins to rise; high-risk individuals should be monitored or not compete
White flag	Low	<10 °C (50 °F)	Lower risk for hyperthermia but increasing risk for hypothermia	10.1–18.3 °C (50.1–65 °F)	Generally safe; EHS can occur

the WBGT developed by the US Navy, made popular by the American College of Sports Medicine [20] and National Association of Athletics Trainers (see Table 12.2) [36, 37], and which is used by various sporting federations. Although it has some limitations [25, 39, 40], WBGT measurements in warm-up areas at the competing track and on road race courses along with appropriate announcements of the readings were implemented in Daegu 2011 and Beijing 2015 World Athletics Championships by the local organizing committees [38]. In Beijing, WGBT readings remained extreme (black) or high (red) as per the flag system [24], representing high to extreme risk for thermal injury.

The WBGT is an environmental heat stress index and not an accurate representation of the actual heat strain faced by competing athletes. This is due to the rate of heat production experienced in individual athletes, based on the event they partake in. The thermal strain experienced by the athletes is also depending on his acclimation status. For example, it has been reported that the amount of heat-related collapse on a set course with similar ambient conditions was reduced if those conditions were seasonal [41]. It is therefore difficult to establish absolute participation cut-off values across sports and is rather recommended to implement preventive countermeasures, including appropriate scheduling of the competition throughout the calendar year and the most suitable start times in championship timetables. In warm and humid climatic geographical areas, events should be rescheduled to the late evening or early morning to avoid the hottest hours of the day and to minimize exposure to solar radiation. Event cancellation may be necessary if the environment poses a serious health hazard to the athletes [25]. However, the decision to cancel an event should be based not only on the prevailing conditions but also on the amount of heat produced by the athlete, with methods such as ingestible thermometer pills used to evaluate this parameter in competition. A host of other factors such as time spent in the heat and the nature of the athletic population (i.e. elite or amateur athletes, youth or master athletes and males or females) should be considered at the time of taking event cancelling decisions, which should never be made by the medical team alone, but together with other technical officials [42].

An adequate medical response protocol for the medical teams of the local organizing committee is of paramount importance. The possibility to adapt rules or implement specific arrangements at certain competition sites to facilitate larger cooling facilities [25] and an extra supply of drinks are further provisions complementing the preventive approach in hot and humid conditions.

12.5 Conclusion

The IAAF plays an integral role in promoting and supporting the scientific advancement of knowledge specific to the physiological challenges related to hyperthermia in elite athletes and the prevention of exertional heat illness and injury that is not necessarily limited to Track and Field Athletics. As such, the Athletics governing body should consider adopting a more comprehensive policy for awarding host cities. This should include policies to avoid hot and humid climates and ensure appropriate event scheduling to hold the higher-risk, longer-lasting events during the coolest time of the day, both fundamental principles of any heat-related illness prevention strategy. The implementation of educational opportunities to outline adequate acclimatization, hydration and cooling strategies for athletes is essential. Finally, high-standard medical coverage is a need that should be guaranteed in every major Athletics Championships. The combination of these actions will hopefully contribute to reduce the risk and incidence of heat-related illness in Track and Field athletes [43] (Table 12.3).

Table 12.3 Recommendations for track and field event organizers [24, 25, 38, 42, 43]

• Appropriate awarding hosting cities to avoid hot and humid climates
• Ensure adequate event scheduling to hold the higher-risk, long-lasting events during the best time of the day
• Implementation of effective educational actions with athletes, coaches and organizers
• Perform on-field assessment of the wet bulb globe temperature (WBGT)
• Contingency plans for delaying or cancelling the event should be in place
• Deploy a high-standard medical coverage

References

1. Jeukendrup A. Effects of heat on running performance. In: Jeukendrup - trusted sports nutrition advice & exercise science news. http://www.mysportscience.com/single-post/2015/03/06/Effects-of-heat-on-running-performance. Accessed 8 Oct 2017.
2. Racinais S, Oksa J. Temperature and neuromuscular function. Scand J Med Sci Sports. 2010;20:1–18.
3. Racinais S, Cocking S, Périard JD. Sports and environmental temperature: from warming-up to heating-up. Temperature. 2017;4:227–57.
4. Guy JH, Deakin GB, Edwards AM, Miller CM, Pyne DB. Adaptation to hot environmental conditions: an exploration of the performance basis, procedures and future directions to optimise opportunities for elite athletes. Sports Med. 2015;45:303–11.
5. Cheuvront SN, Kenefick RW, Montain SJ, Sawka MN. Mechanisms of aerobic performance impairment with heat stress and dehydration. J Appl Physiol. 2010;109:1989–95.
6. James CA, Hayes M, Willmott AGB, Gibson OR, Flouris AD, Schlader ZJ, Maxwell NS. Defining the determinants of endurance running performance in the heat. Temperature. 2017;4:314–29.
7. Bergeron MF. Heat stress and thermal strain challenges in running. J Orthop Sports Phys Ther. 2014;44:831–8.
8. Girard O, Brocherie F, Bishop DJ. Sprint performance under heat stress: a review: sprinting in hot ambient temperature. Scand J Med Sci Sports. 2015;25:79–89.
9. Ely MR, Cheuvront SN, Roberts WO, Montain SJ. Impact of weather on marathon-running performance. Med Sci Sports Exerc. 2007;39:487–93.
10. Dawson NJ, De Freitas CR, Mackey WJ, Young AA. The stressful microclimate created by massed fun runners. Trans Menzies Found. 1987;14:41–4.
11. De Freitas CR, Dawson NJ, Young AA, Mackey WJ. Microclimate and heat stress of runners in mass participation events. J Clim Appl Meteorol. 1985;24:184–91.
12. El Helou N, Tafflet M, Berthelot G, Tolaini J, Marc A, Guillaume M, Hausswirth C, Toussaint J-F. Impact of environmental parameters on marathon running performance. PLoS One. 2012;7:e37407.
13. Vihma T. Effects of weather on the performance of marathon runners. Int J Biometeorol. 2010;54:297–306.
14. Mora-Rodriguez R, Ortega JF, Hamouti N. In a hot–dry environment racewalking increases the risk of hyperthermia in comparison to when running at a similar velocity. Eur J Appl Physiol. 2011;111:1073–80.
15. Jeukendrup A. Why is performance in the heat decreased? In: Jeukendrup - trusted sports nutrition advice & exercise science news. http://www.mysportscience.com/single-post/2017/08/06/Why-is-performance-in-the-heat-decreased?_amp_=true. Accessed 8 Oct 2017.
16. Flouris AD, Schlader ZJ. Human behavioral thermoregulation during exercise in the heat. Scand J Med Sci Sports. 2015;25:52–64.
17. Kilduff LP, West DJ, Williams N, Cook CJ. The influence of passive heat maintenance on lower body power output and repeated sprint performance in professional rugby league players. J Sci Med Sport. 2013;16:482–6.
18. Asplund CA, O'Connor FG, Noakes TD. Exercise-associated collapse: an evidence-based review and primer for clinicians. Br J Sports Med. 2011;45:1157–62.
19. Byrne BC, Lee JKW, Chew S, Lim CL, Tan E. Continuous thermoregulatory responses to mass-participation distance running in heat. Med Sci Sports Exerc. 2006;38:803–10.
20. American College of Sports Medicine, Armstrong LE, Casa DJ, Millard-Stafford M, Moran DS, Pyne SW, Roberts WO. American College of Sports Medicine position stand. Exertional heat illness during training and competition. Med Sci Sports Exerc. 2007;39:556–72.
21. Alonso J-M, Tscholl PM, Engebretsen L, Mountjoy M, Dvorak J, Junge A. Occurrence of injuries and illnesses during the 2009 IAAF World Athletics Championships. Br J Sports Med. 2010;44:1100–5.
22. Alonso J-M, Edouard P, Fischetto G, Adams B, Depiesse F, Mountjoy M. Determination of future prevention strategies in elite track and field: analysis of Daegu 2011 IAAF Championships injuries and illnesses surveillance. Br J Sports Med. 2012;46:505–14.

23. Alonso J-M, Jacobsson J, Timpka T, Ronsen O, Kajenienne A, Dahlström Ö, Spreco A, Edouard P. Preparticipation injury complaint is a risk factor for injury: a prospective study of the Moscow 2013 IAAF Championships. Br J Sports Med. 2015;49:1118–24.
24. Périard JD, Racinais S, Timpka T, Dahlström Ö, Spreco A, Jacobsson J, Bargoria V, Halje K, Alonso J-M. Strategies and factors associated with preparing for competing in the heat: a cohort study at the 2015 IAAF World Athletics Championships. Br J Sports Med. 2017;51(4):264–70.
25. Racinais S, Alonso J-M, Coutts AJ, et al. Consensus recommendations on training and competing in the heat. Br J Sports Med. 2015;49:1164–73.
26. Racinais S, Périard JD, Karlsen A, Nybo L. Effect of heat and heat acclimatization on cycling time trial performance and pacing. Med Sci Sports Exerc. 2015;47:601–6.
27. Daanen HAM, Racinais S, Périard JD. Heat acclimation decay and re-induction: a systematic review and meta-analysis. Sports Med. 2018;48:409–30.
28. Racinais S, Wilson MG, Gaoua N, Périard JD. Heat acclimation has a protective effect on the central but not peripheral nervous system. J Appl Physiol. 2017;123:816–24.
29. Tyler CJ, Reeve T, Hodges GJ, Cheung SS. The effects of heat adaptation on physiology, perception and exercise performance in the heat: a meta-analysis. Sports Med. 2016;46:1699–724.
30. Périard JD, Racinais S, Sawka MN. Adaptations and mechanisms of human heat acclimation: applications for competitive athletes and sports. Scand J Med Sci Sports. 2015;25:20–38.
31. Périard JD, Travers GJS, Racinais S, Sawka MN. Cardiovascular adaptations supporting human exercise-heat acclimation. Auton Neurosci. 2016;196:52–62.
32. Zurawlew MJ, Walsh NP, Fortes MB, Potter C. Post-exercise hot water immersion induces heat acclimation and improves endurance exercise performance in the heat. Scand J Med Sci Sports. 2016;26:745–54.
33. Bongers CCWG, Hopman MTE, Eijsvogels TMH. Cooling interventions for athletes: an overview of effectiveness, physiological mechanisms, and practical considerations. Temperature. 2017;4:60–78.
34. Naito T, Sagayama H, Akazawa N, Haramura M, Tasaki M, Takahashi H. Ice slurry ingestion during break times attenuates the increase of core temperature in a simulation of physical demand of match-play tennis in the heat. Temperature. 2018:1–9. https://doi.org/10.1080/23328940.2018.1475989.
35. Armstrong LE, Hubbard RW, Jones BH, Daniels JT. Preparing Alberto Salazar for the heat of the 1984 olympic marathon. Phys Sportsmed. 1986;14:73–81.
36. Binkley HM, Beckett J, Casa DJ, Kleiner DM, Plummer PE. National Athletic Trainers' Association position statement: exertional heat illnesses. J Athl Train. 2002;37:329.
37. Roberts WO. Medical management and administration manual for long distance road racing. In: Brown CH, Gudjónsson B, editors. IAAF medical manual for athletics and road racing competitions: a practical guide. Monte-Carlo: International Amateur Athletic Federation Publications; 1998. p. 39–75.
38. IAAF. IAAF competition medical guidelines. 2013. https://www.iaaf.org/download/download?filename=a92551d5-35d1-410b-90f8-5354765b875d.pdf&urlslug=IAAF%20Competition%20Medical%20Guidelines%20-%20January%202013%20First%20Edition. Accessed 3 Feb 2018.
39. Budd GM. Wet-bulb globe temperature (WBGT)--its history and its limitations. J Sci Med Sport. 2008;11:20–32.
40. Brocherie F, Millet GP. Is the wet-bulb globe temperature (WBGT) index relevant for exercise in the heat? Sports Med. 2015;45:1619–21.
41. Gosling CM, Gabbe BJ, McGivern J, Forbes AB. The incidence of heat casualties in sprint triathlon: the tale of two Melbourne race events. J Sci Med Sport. 2008;11:52–7.
42. Périard JD, Jay O, Alonso J-M, et al. Author's reply to Brocherie and Millet: 'is the wet-bulb globe temperature (WGBT) index relevant for exercise in the heat?'. Sports Med. 2015;45:1623–4.
43. Alonso J-M. Bringing the games to the world II - environmental challanges to athlete health and performance in the heat. A track and field perspective. In: Aspetar Sports Med. J. 2016. http://www.aspetar.com/journal/upload/PDF/2016310145135.pdf. Accessed 3 Feb 2018.

Cycling in the Heat

13

Julien D. Périard and Sébastien Racinais

13.1 Introduction

Exercise per se, even in temperate conditions, increases core body temperature [1]. When undertaken in the heat, the rise in temperature, including skin, muscle, and tissue temperature, increases to a great extent. This increased level of thermal strain results in performance impairments during endurance events, and can lead to development of exertional heat illness, especially when racing or training at a high intensity [2]. Within the spectrum of cycling events and sub-disciplines, performance on the road and mountain is particularly susceptible to being influenced by hot climatic conditions. This stems from road cycling incorporating prolonged single day and multi-stage races, as well as individual and team time trials. Performance and health in mountain biking events, particularly cross-country, marathon, and enduro races, are also susceptible to the effects of environmental heat stress.

The impact of hot and humid conditions on endurance cycling performance is well documented, with impairments occurring in conjunction with the development of thermal strain [3–5]. Despite performance in prolonged events declining in hot weather, the ability to produce high power outputs during sprinting may be improved in the heat when the level of hyperthermia is not excessively severe [6, 7]. In conditions of elevated thermal strain however, repeated-sprint performance is negatively impacted [8], with the impairment attributable to similar fatigue pathways to that of endurance cycling [9]. Countermeasures have been developed however, not only to optimise performance in the heat, but also to minimise the potential risk of heat illness. Currently, epidemiological data regarding heat illness incidence rates is sparse in cycling. An unpublished cohort study at the 2016 UCI Road World Cycling

J. D. Périard (✉)
University of Canberra Research Institute for Sport and Exercise, Bruce, ACT, Australia
e-mail: julien.periard@canberra.edu.au

S. Racinais
Aspetar Sports Medicine Hospital, Doha, Qatar

J. D. Périard, S. Racinais (eds.), *Heat Stress in Sport and Exercise*,
https://doi.org/10.1007/978-3-319-93515-7_13

Championships indicated that over 50% of cyclists responding to a questionnaire reported previously experiencing heat illness symptoms (e.g. nausea, headache, cramping, fainting) during training or competition in the heat [10]. Approximately 38% of respondents indicated having conducted at least 5 days of heat acclimation prior to the championships, which were held in Qatar in ~37 °C conditions. Interestingly, the athletes also reported arriving in Qatar ~5 days before competing in their event, affording most of them a short-term heat acclimatisation stimulus.

This chapter will focus on the impact of heat stress on cycling performance (i.e. endurance and sprinting), along with the pathways via which performance is altered. These pathways include the development of thermal and cardiovascular strain, as well as central and peripheral fatigue, and adjustments in perception and behaviour. A comparison will also be made of cycling research conducted in the laboratory and in the field to allow for an appreciation of each setting. An outline of countermeasures to optimise cycling performance and minimise heat-related health issues will be presented, along with future research directions to enhance the safety of cyclists competing in the heat.

13.2 Influence of Heat Stress on Prolonged Cycling Performance

It is well documented that endurance capacity is impaired when cycling in warm/hot conditions [5, 11], as well as when air velocity is low [12], and solar radiation [13, 14] and relative humidity are high [15]. For example, Galloway and Maughan [5] reported that time to exhaustion during constant rate cycling (70% maximal oxygen uptake: VO_{2max}) followed an inverted U relationship with exercise capacity optimised at 11 °C (93.5 min), relative to 4 °C (81.4 min), 21 °C (81.2 min), and 31 °C (51.6 min) conditions with 70% relative humidity. Similarly, time to exhaustion was progressively reduced when relative humidity was increased from 24 to 40, 60, and 80% during ergometer cycling at 70% VO_{2max} in 30 °C [15]. These performance impairments relate to various environmental factors contributing independently and synergistically to exacerbate the development of thermal strain, and decrease exercise capacity. Although climatic characteristics strongly influence exercise capacity and the rate of rise in whole-body temperature, a primary factor mediating the development of hyperthermia during exercise is the production or release of metabolically generated heat. Indeed, it has been shown that time to exhaustion in trained individuals is essentially halved when cycling at 75% VO_{2max} (~30 min) compared with 60% (~60 min) in 40 °C and 50% relative humidity conditions [16]. This relates in part to the larger imbalance between the rate of evaporation required for heat balance (E_{req}) and the maximum environmental capacity for evaporation (E_{max}) when exercising at a higher power output, which renders the environment less compensable and increases the rate of rise in core temperature (see Chap. 2).

During time trial events in which power output is self-selected, cycling performance is also impaired under heat stress, even in elite cyclists (e.g. Fig. 13.1). For

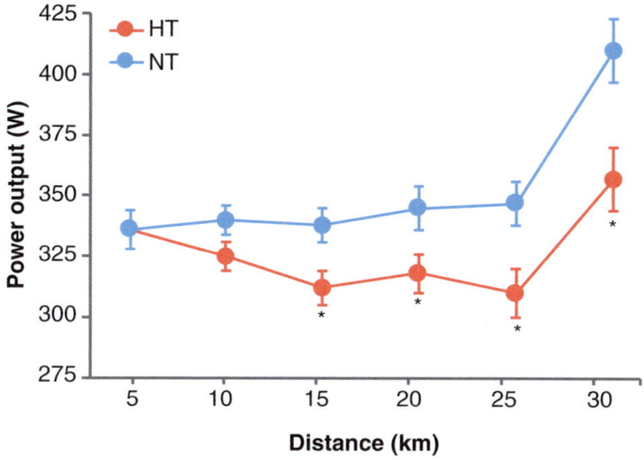

Fig. 13.1 Power output during a 30-min cycling time trial at 32 °C (HT) and 23 °C (NT) in elite road cyclists. *Significantly different from NT ($P < 0.05$). Reproduced with permission from Tatterson et al. [3]

instance, a rise in ambient temperature to 32 °C decreased 40 km time trial performance (309 W) relative to 27 °C (322 W), 22 °C (324 W), and 17 °C (329 W) [11]. This holds true for radiant heat as well, with 15-km time trial performance decreasing when exposed to a radiant heat of 1100 W m^{-2}, even in 18 °C conditions [14]. These performance impairments stem from the rise in whole-body temperature during prolonged cycling in the heat, and in large part to an exacerbated cardiovascular response (see Chap. 3).

13.2.1 Cardiovascular Strain

The impairment in prolonged exercise performance in the heat has historically been related to a thermoregulatory redistribution of blood towards the periphery and a temperature-mediated increase in intrinsic heart rate, which contribute to compromise central blood volume and the maintenance of cardiac output [17–19]. Although cutaneous blood flow reaches a virtual plateau, or upper limit when a core temperature of ~38 °C is attained during exercise [20], a significant volume of blood perfuses peripheral vascular beds [17, 21]. Moreover, for each 1.0 °C elevation in core (i.e. atrial) temperature, a ~7 beats min^{-1} increase in heart rate occurs during passive heat stress [22]. This suggests that the greater increase in heart rate during exercise in the heat may stem in part from the direct effects of temperature on cardiac nodal cells, along with adjustments in autonomic nervous system activity (i.e. sympathetic activation and parasympathetic withdrawal) [23]. Thus, the metabolic and thermoregulatory requirements associated with cycling under heat stress appear to interact and alter cardiac function, modify the distribution of cardiac output, and/or compromise the ability to sustain adequate blood pressure [2, 24]. An example of this is presented in Fig. 13.2 where a thermoregulatory-mediated increase in

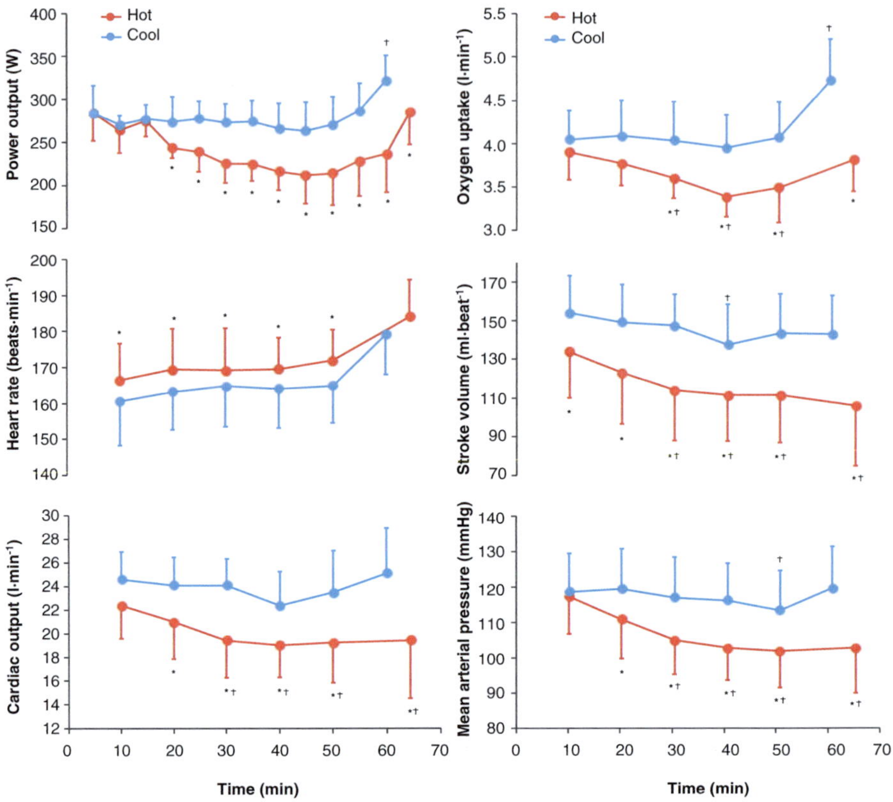

Fig. 13.2 Power output, oxygen uptake, heart rate, stroke volume, cardiac output, and mean arterial pressure during a 40-km cycling time trial in hot (35 °C and 60% relative humidity) and cool (20 °C and 40% relative humidity) conditions. *Significantly different from cool ($P < 0.05$). †Significantly different from 10 min ($P < 0.05$). Reprinted with permission from Périard et al. [4]

cardiovascular strain contributes to the inability to maintain power output during a 40 km self-paced time trial in the heat [4].

The cardiovascular adjustments associated with heat exposure have been shown to decrease VO_{2max} during exhaustive exercise [25–27], as well as during the end-spurt of prolonged time trial efforts (e.g. Fig. 13.2) [4, 28]. During graded exercise on a treadmill, Arngrimsson et al. [29] showed that VO_{2max} was reduced by 4, 9, and 18% in both men and women in 35, 40, and 45 °C conditions following preheating to core and skin temperatures of 37.0 and 35.1 °C, 37.3 and 36.2 °C, and 37.8 and 37.6 °C, respectively. The reduction in VO_{2max} at the completion of self-paced exercise under heat stress lasting 50–60 min is purported to be ~17%, which is larger than that observed at the end of a time trial in cool conditions (~5%) [4, 28, 30]. Interestingly, the reduction in VO_{2max} occurs progressively during time trial efforts in the heat, which leads to a progressive decline in absolute work rate (i.e. power output); however, relative exercise intensity (i.e. the fractional utilisation of VO_{2max}: $\%VO_{2max}$) is maintained within a relatively narrow range, similar to that of a time trial performed in cool conditions [30]. This range widens under heat stress when exercise becomes protracted however, as a disassociation develops between %VO_{2max} and heart rate, and RPE. Typically, the %VO_{2max} sustained during self-paced cycling is related to the duration or distance of the event, with time trials of 40 km conducted at ~85% VO_{2max} and shorter efforts (e.g. 4–20 km) performed at a greater fraction of VO_{2max} [30–34]. The progressive decrease in VO_{2max} appears from a performance perspective, to be the primary determinant resulting in power output being maintained at a similar level at the onset of self-paced cycling in cool and hot conditions, but then progressively causing its decline in the heat.

Perhaps not surprisingly, fitness is an important factor contributing to enhance performance when cycling in the heat. For instance, trained individuals exercising in the heat at the same relative intensity, but at a higher absolute metabolic rate, experience a greater rate of heat storage than untrained individuals [35] and fatigue at a similar or higher core temperature [16, 36–38]. This suggests that training and aerobic fitness may allow for greater rates of body heat accumulation before a reduction in work rate occurs [39].

13.2.2 Central and Peripheral Fatigue

Although an increase in circulatory and cardiovascular strain represents the foremost determinant impairing performance during prolonged cycling in the heat, neuromuscular alterations also influence the ability to exercise optimally under heat stress (see Chap. 4). These alterations include the development of both central and peripheral fatigue, which correspond to a reduction in voluntary muscle activation and a loss of force production capacity, respectively [40–42].

Early studies suggested that the attainment of a high core temperature (39.2–39.7 °C) during prolonged cycling might influence central nervous system function by reducing the mental drive (i.e. motivation) for motor performance [43, 44]. This notion was reinforced by data showing a progressive impairment in force production capacity during a sustained (120 s) maximum voluntary isometric contraction of the knee extensors following exhaustive cycling in the heat, relative to a contraction performed after 60 min of steady state cycling in cool conditions [41]. The impairment was attributed to a hyperthermia-induced reduction in voluntary muscle activation, in response to a 2 °C difference in core temperature between conditions. A more recent study however demonstrated that voluntary activation and force production capacity were similarly reduced during a 20 s maximum voluntary isometric contraction following a 40 km time trial in hot and cool conditions [45]. The reduction in voluntary activation accounted for only ~20% of the decrease in total force production in both conditions, indicating that the higher core temperature (0.8 °C) noted following time trial exercise under heat stress did not exacerbate central fatigue, and induced a similar level of peripheral fatigue. It has also been shown that submaximal exercise followed by incremental cycling to exhaustion in 24 °C and 40 °C conditions induced a similar level of force loss in the knee extensors when performing a 5 s maximal voluntary contraction, along with an analogous reduction in voluntary muscle activation [46]. This led to the conclusion that neuromuscular failure could not explain the earlier volitional fatigue observed in the heat during incremental exercise, especially as heart rate was higher under heat stress, the rating of perceived exertion (RPE) was similar between conditions, and peripheral fatigue (i.e. twitch response) was lower in the heat due to terminating exercise sooner. Interestingly, Thomas et al. [33] showed that fatigue following cycling time trials in temperate conditions was task dependent, with a greater degree of peripheral fatigue arising after shorter (6 min) more intense efforts, and a greater level of central fatigue presenting after longer (>30 min) slightly lower-intensity trials.

Importantly, it must be recognised when inferring fatigue during cycling that a maximal isometric contraction recruits and maximally activates all available motor units in an isolated limb. In contrast, motor unit recruitment is submaximal and cyclical during cycling, involving several muscle groups and alternating neuromuscular activation patterns. Performance of a maximal isometric contraction also generates high intramuscular tension, impeding blood flow and the clearance of metabolic by-products. This causes the activation of group III and IV muscle afferents [47–49]. As well as being important modifiers of cortical and motoneurone excitability [50], these afferents play a role in the disproportionately large increases in heart rate and blood pressure relative to metabolic activity noted during isometric exercise [51]. This contrasts with the well-matched cardiovascular, haemodynamic, and metabolic responses during dynamic exercise such as cycling. The culmination of these factors therefore renders it difficult to infer a similar aetiology of neuromuscular fatigue between maximal isometric and submaximal dynamic exercise [52].

13.2.3 Perceptual Responses

Thermal perception, or an elevated level of thermal discomfort, has been suggested to increase RPE in response to an increase in skin temperature and consequently decrease work rate during the early stages of self-paced exercise [53]. Whilst this behavioural thermoregulatory response represents a powerful mechanism via which conscious decisions contribute to preserve thermal homeostasis [54], few studies have demonstrated a lower starting power output during cycling in the heat when the magnitude of hyperthermia was minimal and/or similar to that noted in cool conditions [55, 56]. Indeed, a lower work rate is not typically observed in the early stages of self-paced exercise in the heat, as most time trials are initiated at a similar power output regardless of the environment, whether in a laboratory or field setting [4, 11, 28, 30, 34, 57–59]. Moreover, using a menthol spray to induce the feeling of coolness, Barwood et al. [60] demonstrated that thermal perception did not drive changes in pacing during the early stages of a 40-km time trial in 32 °C conditions, nor did it affect overall performance.

On the other hand, Sawka et al. [61] have highlighted that high skin temperature alone may impair aerobic performance, such as prolonged self-paced cycling. This occurs as hot skin narrows the core-to-skin temperature gradient, which increases peripheral thermoregulatory blood flow requirements and in turn circulatory strain—reducing cardiac filling and elevating heart rate for a given cardiac output [19, 24]. As outlined above, the increased cardiovascular strain associated with high skin temperatures may contribute to modulate performance in advance of a rise in core body temperature (i.e. hyperthermia).

13.3 Influence of Heat Stress on Sprint Cycling Performance

In contrast to endurance exercise, a rise in whole-body temperature, muscle temperature in particular, enhances explosive (e.g. sprinting) exercise performance via improvements in metabolic and contractile function, nerve conduction velocity, and conformational changes associated with muscle contraction [62–64]. These adjustments stem in part from the relationship between tissue temperature and metabolic processes (i.e. Q_{10} temperature coefficient), as the rate of biochemical reactions increases with a rise in temperature. For instance, single sprint performance (i.e. faster time to complete 35 revolutions, ~9.375 kJ) on a cycle ergometer has been shown to be enhanced by 5% for every 1 °C rise in muscle temperature (Fig. 13.3) [67]. This is analogous to the findings of Bergh and Ekblom [65] demonstrating a rise in speed of ~4.5% per 1 °C increase in muscle temperature. Sargeant [66] also showed that maximal power output during a 20 s sprint increased by 11% in conjunction with a rise in muscle temperature of 2.7 °C, whereas a 12% decrease was noted when muscle temperature was lowered by 5.7 °C, relative to resting values (36.6 °C) [66]. The magnitude of improvement in heated muscle was also velocity dependent, with higher cadences resulting in greater peak power outputs. The increased maximal peak power at higher muscle temperatures however was associated with an increased rate of fatigue [66]. Notwithstanding, others have confirmed that peak power output during all-out sprint cycling (30 s) is improved [68, 69]. Given that skeletal muscle temperature at rest hovers around 36 °C and that a rise of 3–4 °C can occur during intensive cycling in the heat [70, 71], explosive sprint performance can be expected to improve in the absence of marked peripheral fatigue.

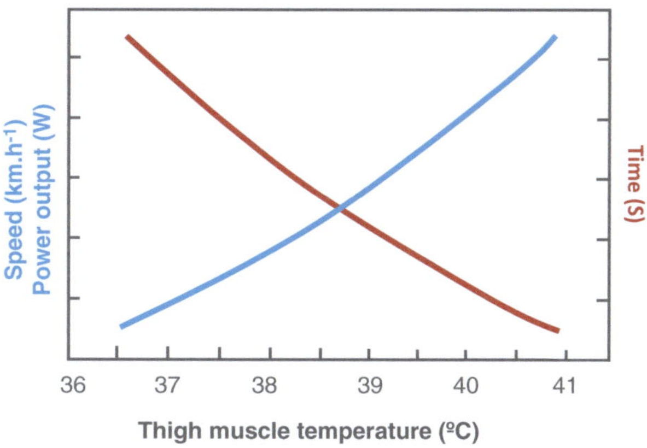

Fig. 13.3 Maximal speed, power output, and total time required to complete a brief (15–20 s) maximal sprint relative to thigh (e.g. *vastus lateralis*) muscle temperature. Based on data from Bergh and Ekblom [65], Sargeant [66], and Asmussen and Boje [67]

It has even been demonstrated that repeated-sprint ability (10 × 6 s, 30 s recovery) is enhanced at a core temperature of ~38 °C when cycling in 35 °C conditions, compared to cycling in 24 °C conditions with a core temperature of ~37.7 °C [7]. This improvement in repeated-sprint ability likely depends on the absence of marked hyperthermia, as intermittent sprint cycling ability (8 × 6 s, 60 s recovery) was also reported to be unaltered in hot (40 °C) versus cool (24 °C) conditions at a core temperature of ~37.7 °C [6].

In contrast, Drust et al. [8] have shown that both peak and mean power output during repeated maximal sprints (5 × 15 s, 15 s recovery) decrease to a larger extent under heat stress (40 °C) relative to cool conditions (20 °C) with core temperatures of 39.5 and 38.2 °C, respectively. The authors suggested that the performance impairment was not related to the accumulation of metabolic fatigue agents (e.g. muscle lactate and plasma potassium), but rather to the influence of a high core temperature on voluntary muscle activation. This premise is attractive and previous research has suggested that central fatigue may be involved in the hyperthermia-induced development of fatigue during prolonged dynamic exercise [41]. However, more recent studies have proposed that the hyperthermia-induced decrement in cycling performance likely relates to cardiovascular and perceptual limitations, rather than a physiological incapacity to generate motor drive [45, 46]. Whether a heat-related lack of motivation can influence repeated-sprint ability however remains to be determined, especially given the nature of the effort and the potential for pacing to occur during 'maximal' repetitions.

As recently summarised by Girard et al. [9], when performing repeated-sprint efforts, a point occurs whereby the muscle-temperature-related benefits are overridden by the exacerbated cardiovascular and metabolic strain, along with a potential reduction in voluntary skeletal muscle activation or motivation. These adjustments are purported to stem from the greater elevation in whole-body temperature occurring when repeatedly sprinting in hot environments.

13.4 Cycling in the Laboratory and the Field

Laboratory studies provide a controlled setting in which performance and the related physiological responses can be accurately and reliably evaluated. In contrast, field studies allow for a more authentic representation of what happens during actual events. As such, several differences can surface when cycling in a laboratory environment compared to an outdoor setting, namely, in relation to changes in terrain, race tactics, race type, and climatic conditions. For example, gross efficiency has been shown to decrease during uphill cycling in conjunction with modifications in pedalling technique [72]. Changes in body orientation during very steep climbing (i.e. 20% gradient) alter lower extremity neuromuscular patterns [73], and a change from sitting to standing decreases cadence and increases power output (~13%) at a given speed [74]. Most experienced cyclists however choose a near optimal cadence based on the power output:cadence relationship to minimise fatigue, independent of the terrain [75, 76].

At high speeds (e.g. 40 km h^{-1}), air resistance is a major obstacle to forward movement. The ability to draft, or follow in the slipstream of other cyclists, allows for reducing energy costs (i.e. VO_2) by 25–40%, depending on whether riding behind one person or a group (e.g. eight cyclists) [77, 78]. Another important aspect to consider is the change in air density associated with different ambient temperatures, which affects the resistance for displacement at different speeds [79, 80]. For example, a 20 °C increase in temperature decreases air density by ~7%, allowing for a ~6% increase in speed for a given power output, assuming that 90% of the resistance is aerodynamic in a solo cyclist. As such, the reduction in air density noted during cycling in hot outdoor conditions may partially offset the impairment in performance afforded by the development of thermal and cardiovascular strain. This phenomenon was highlighted by Racinais et al. [59] when comparing 43 km time trial performances in cool (~8 °C) and hot (37 °C) conditions. The authors calculated that air density was 9.4% lower in the heat (1.131 kg m^{-3}) than in cool (1.249 kg m^{-3}) conditions, representing a power output economy of ~8%. It was reported that despite the slightly lower power output maintained during the time trial in the heat, average speeds did not significantly differ. This was after a 2-week heat acclimatisation regimen.

Junge et al. [81] recently highlighted that power output during prolonged cycling time trials is reduced by ~15% in the heat (>30 °C), relative to cooler conditions (≤20 °C). They also indicated that power output during indoor trials may be maintained at ambient temperatures of up to 27 °C if relative humidity remained under 40% and air velocity matched the movement speed generated during outdoor cycling. This emphasises the importance of air displacement in allowing for effective sweat evaporation [82, 83] and maintaining a compensable environment (see Chap. 2). These factors, in combination with the type of race (e.g. road race, time trial) and individual goals/team tactics, outline the potential differences between laboratory and field settings. Notwithstanding, controlling of environmental factors during laboratory-based experiments, especially wind speed, and understanding the limitations associated with race tactics (e.g. drafting) may yet provide an ecologically valid platform for comparison with the field.

13.5 Countermeasures to Optimises Health and Performance

Along with influencing performance, the development of hyperthermia increases the risk of exertional heat illness (see Chap. 5). In order to minimise the risk of heat illness and enhance performance when cycling in the heat, several countermeasures may be adopted. These include adopting a hydration strategy (Chap. 6), utilising cooling methods (Chap. 7), and the most influential approach, heat acclimation (Chap. 8).

13.5.1 Hydration

Briefly, the effects of dehydration and recommendations about how and when to drink can be a contentious topic, especially in relation to prolonged cycling. A previous meta-analysis suggested that exercise-induced dehydration of up to 4% did not alter cycling performance in an outdoor setting and that relying on thirst sensations to gauge fluid replacement needs was the optimal strategy for time trial performance [84]. However, several comprehensive reviews have indicated that body mass losses beyond 2% are associated with impaired thermoregulatory function and elevated cardiovascular strain, especially during prolonged intensive exercise in the heat, whether in a laboratory or field setting [85–89]. Kenefick [90] recently highlighted that high intensity exercise in which sweat rate is elevated along with activities lasting >90 min in the heat should be accompanied with a planned hydration strategy. In contrast, drinking to thirst may be sufficient to offset fluid losses during low intensity exercise of shorter duration (<90 min) in cool climates.

As with most endurance sports, it is commonly accepted that undertaking a cycling race or training in hot conditions in a rested, well-fed, and well-hydrated state is optimal [91]. It is thus advisable to consume 5–6 mL of water per kg of body mass every 2–3 h prior to training or competing in the heat. This strategy should be tested prior to racing in a major competition to determine if volumes need to be altered. During exercise, cyclists should try to minimise body mass losses induced via sweating, whilst using caution to avoid over drinking. Those that sweat profusely may require supplementing their fluids with additional sodium (e.g. 3.0 g of salt added to 0.5 L of a carbohydrate–electrolyte solution).

13.5.2 Cooling

Various cooling strategies can be applied before (pre-cooling) and during (per-cooling) cycling events to lower core temperature and optimise endurance performance [92–94]. These include external (e.g. application of iced garments, towels, water immersion, or fanning) and internal (e.g. ingestion of cold fluids or ice slurry) methods. Pre-cooling has been proposed to benefit prolonged activities such as cycling in the heat, with the use of a mixed-methods approach being optimal. This includes the use of fans and commercially available ice-cooling vests to provide effective cooling without negatively lowering muscle temperature, a strategy that can be used during a stationary warm-up before time trials. It has also been suggested that vigorous cooling techniques covering a large surface area or techniques that can be frequently applied seem to have the most impact on enhancing performance [93]. It must be emphasised that cooling which is too severe may in fact be detrimental to performance, as locomotor muscle temperature may decrease.

Ingesting ~1 L of crushed ice (4 °C) prior to warming-up may also help reduce core temperature and improve performance [95]; however, the ingestion of ice slurry beverages has been suggested to be an effective strategy more so as per-cooling [92, 93]. Per-cooling via the ingestion of cold fluids or ice slurry in combination with neck and face cooling may further benefit endurance performance in the heat [94]. Ultimately, combining pre- and per-cooling is likely the most effective strategy to improve prolonged cycling performance in the heat [92]. As with any new training or racing strategy, cooling methods should be tested and individualised to minimise any disruption or untoward consequence to the athlete.

13.5.3 Heat Acclimation

The benefits of heat acclimation are well known and achieved through an expansion of plasma volume, enhanced cardiovascular stability, improved sweating and skin blood flow responses, better fluid balance (i.e. hydration), and acquired thermal tolerance [96–100]. The process of adaptation can occur in artificial/laboratory settings (i.e. heat acclimation) and following exposure to natural environments (i.e. heat acclimatisation). In a population of well-trained cyclists it was demonstrated that the reduction in performance (i.e. power output) observed during prolonged (43 km) cycling time trials in hot outdoor conditions (~37 °C) was partly recovered after 1 week of training in the heat and almost fully restored after 2 weeks (Fig. 13.4) [59]. Accordingly, the time course for the development of heat

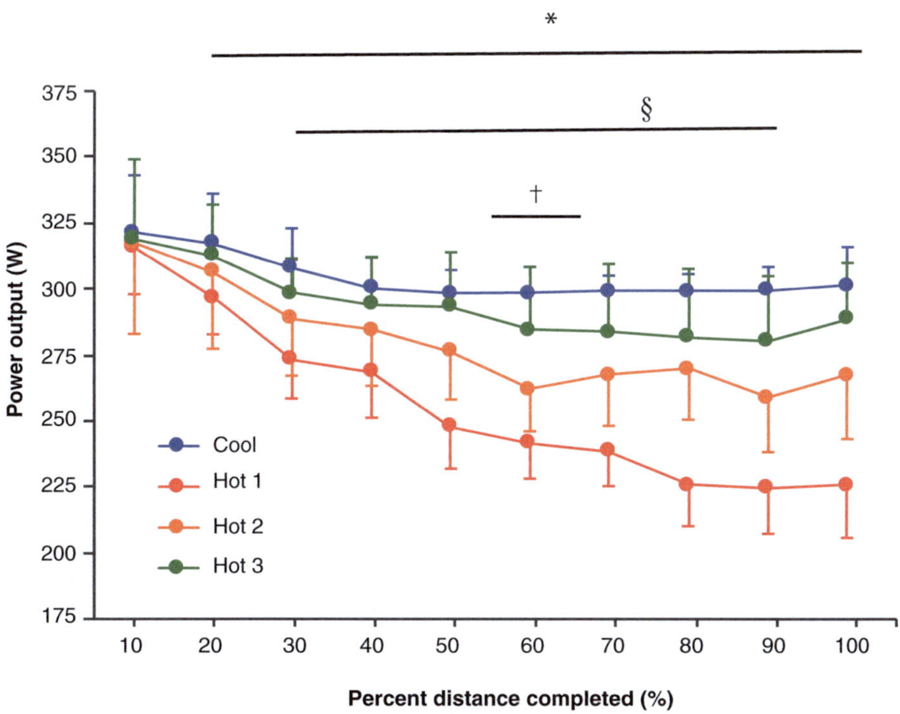

Fig. 13.4 Power output during a 43-km cycling time trial in cool (8 °C) conditions (blue line) and in hot (37 °C) conditions after 1 (red line), 6 (orange line), and 14 (green line) days of heat exposure. Reproduced with permission from [59]

acclimation is relatively quick, with a significant fraction of the adaptations developing during the first week of heat exposure, and 10–14 days being optimal for complete or near-complete acclimation [96, 98, 100, 101]. The rate of decay for the main adaptive benefits (i.e. lowered heart rate and core temperature during exercise) is ~2.5% per day without heat exposure [102]. As such, a cyclist tapering for a major race after a heat acclimation regimen will lose ~35% of their adaptation after 2 weeks without heat exposure. However, these adaptations are re-induced at a faster rate if heat re-acclimation takes place within 1 month of undertaking the first regimen [102].

13.6 Heat Policy in Cycling

As with many sports, cycling's governing body, the Union Cycliste Internationale (UCI), is proactive in trying to enhance the performance and safety of its cyclists. For example, during the 2016 UCI Road World Championships in Qatar, a campaign was launched to 'Beat the Heat' http://www.uci.ch/mm/Document/News/News/17/78/85/UCI_BEATTHEHEAT_GB_2016_English.pdf. This involved disseminating information about how heat impacts on performance in different disciplines, and how to best prepare for competing in the heat with suggestions on heat acclimation, hydration, and cooling strategies. A specific section also addressed some of the differences between males, females, and junior cyclists.

One of the main issues with cycling in the heat is that of health and the development of exertional heat illness, which represents a continuum of medical conditions with potentially severe consequences. Exertional heat illness can affect cyclists in both hot and cool environments and its severity can escalate from heat exhaustion, to heat injury, and on to heat stroke [103]. Currently, the UCI has an Extreme Weather Protocol to subjectively decide on whether to reschedule, shorten, or cancel an event (http://www.uci.ch/mm/Document/News/Rulesandregulation/17/30/54/ProtocoleEWP-E-20160603_English.pdf). This is typically done with an evaluation of the prevailing ambient conditions and using the wet-bulb-globe temperature (WBGT) index. The WBGT is calculated from outdoor measurements of natural wet-bulb, black-globe, and dry-bulb ambient temperatures. Although the WBGT does not directly measure air velocity, the effects of wind speed are reflected somewhat by natural wet-bulb temperature.

The WBGT and the Universal Thermal Climate Index (UTCI), an index of thermal and regulatory behaviour in occupational settings, were recently evaluated in the context of time trial cycling [81]. In conjunction with ambient air temperature being a poor predictor of environmental heat stress, both the WBGT and UTCI failed to predict the impact of environmental heat stress on prolonged self-paced cycling performance. It was purported that these indices do not sufficiently afford wind speed an important enough weighting in their formulas, as trained cyclists travel at speeds greater than 10 m s^{-1} (36 km h^{-1}) when cycling outdoors on flat terrain. This elevated speed can potentially enhance convective and evaporative heat loss, which led Junge et al. [81] to develop an integrated heat stress index with a greater weighing for wind speed. Although this index remains to be experimentally validated in different environmental settings, exercise modalities, and in individuals of varying fitness levels and heat acclimation status, it did correlate well ($R^2 = 0.77$) with performance decrements in the heat. It must also be acknowledged that whilst racing speeds are elevated in professional cycling, the airflow being generated is dependent on position in the peloton during a road race, the direction of the prevailing winds, and isolation from the wind by terrain and flora.

A recent consensus statement on training and competing in the heat highlighted that the WBGT is an environmental heat stress index and not a representation of human heat strain [91]. As such, implementing preventive countermeasures and evaluating the specific demands of the sport when developing extreme heat policies was recommended, as absolute participation cut-off values across different disciplines within a sport are difficult to establish. The decision to cancel or alter an event should be based not only on the ambient conditions, but also on the amount of heat produced by the athlete, along with factors such as time spent in the heat, the athletic population, and the time of year. Additional research is therefore required.

13.7 Summary

Cycling as a sport involves many sub-disciplines. When undertaken in the heat, performance in endurance events is impaired as circulatory and cardiovascular adjustments related to temperature regulation render the maintenance of a given absolute workload (i.e. power output) progressively more difficult. The development of peripheral muscle fatigue that accompanies cyclical pedalling during protracted events along with the potential hyperthermia-induced reduction in voluntary drive further contributes to the impairment in performance. In conjunction with the elevated thermal discomfort and unpleasant environmental sensations originating from exercising in the heat, a rise in perceived exertion may occur in relation to cycling at a particular work rate in cooler conditions. This stems from the progressive decrease in maximal aerobic capacity that occurs during prolonged cycling in the heat. Despite these deleterious consequences, an increase in muscle temperature can improve power output during a single cycling sprint, and in the absence of marked hyperthermia, can enhance repeated-sprint ability. As such, depending on the event, several approaches can be utilised to optimise performance. Whether it be a vigorous warm-up to increase muscle temperature and augment sprint performance or utilising a heat countermeasure (e.g. heat acclimation, pre-cooling) to enhance endurance capacity. It is important to remember that certain responses, especially with regard to thermoregulation, when cycling in a laboratory versus a field setting, may differ. Various responses may also differ between trained and untrained individuals, as well as heat acclimated and non-acclimated cyclists. Along with performance, the health and safety of cyclists is a primary concern and developing appropriate approaches to ensure this safety requires additional research.

References

1. Saltin B, Hermansen L. Esophageal, rectal, and muscle temperature during exercise. J Appl Physiol. 1966;21(6):1757–62.
2. Sawka MN, Leon LR, Montain SJ, Sonna LA. Integrated physiological mechanisms of exercise performance, adaptation, and maladaptation to heat stress. Compr Physiol. 2011;1(4):1883–928.
3. Tatterson AJ, Hahn AG, Martin DT, Febbraio MA. Effects of heat stress on physiological responses and exercise performance in elite cyclists. J Sci Med Sport. 2000;3(2):186–93.
4. Périard JD, Cramer MN, Chapman PG, Caillaud C, Thompson MW. Cardiovascular strain impairs prolonged self-paced exercise in the heat. Exp Physiol. 2011;96(2):134–44.
5. Galloway SD, Maughan RJ. Effects of ambient temperature on the capacity to perform prolonged cycle exercise in man. Med Sci Sports Exerc. 1997;29(9):1240–9.
6. Almudehki F, Girard O, Grantham J, Racinais S. Hot ambient conditions do not alter intermittent cycling sprint performance. J Sci Med Sport. 2012;15(2):148–52.
7. Girard O, Bishop DJ, Racinais S. Hot conditions improve power output during repeated cycling sprints without modifying neuromuscular fatigue characteristics. Eur J Appl Physiol. 2013;113(2):359–69.
8. Drust B, Rasmussen P, Mohr M, Nielsen B, Nybo L. Elevations in core and muscle temperature impairs repeated sprint performance. Acta Physiol Scand. 2005;183(2):181–90.
9. Girard O, Brocherie F, Bishop DJ. Sprint performance under heat stress: a review. Scand J Med Sci Sports. 2015;25:79–89.
10. Nichols D, Travers GJS, Moussay S, Belfekih T, Farooq A, Schumacher YO, et al. Health status, preparedness and medical events when competing in the heat: a cohort study at the 2016 UCI Road World Cycling Championships. Unpublished.
11. Peiffer JJ, Abbiss CR. Influence of environmental temperature on 40 km cycling time-trial performance. Int J Sports Physiol Perform. 2011;6:208–20.
12. Otani H, Kaya M, Tamaki A, Watson P, Maughan RJ. Air velocity influences thermoregulation and endurance exercise capacity in the heat. Appl Physiol Nutr Metab. 2018;43(2):131–8.
13. Otani H, Kaya M, Tamaki A, Watson P, Maughan RJ. Effects of solar radiation on endurance exercise capacity in a hot environment. Eur J Appl Physiol. 2016;116(4):769–79.
14. Levels K, de Koning J, Broekhuijzen I, Zwaan T, Foster C, Daanen H. Effects of radiant heat exposure on pacing pattern during a 15-km cycling time trial. J Sports Sci. 2014;32(9):845–52.
15. Maughan RJ, Otani H, Watson P. Influence of relative humidity on prolonged exercise capacity in a warm environment. Eur J Appl Physiol. 2012;112(6):2313–21.
16. Périard JD, Caillaud C, Thompson MW. The role of aerobic fitness and exercise intensity on endurance performance in uncompensable heat stress conditions. Eur J Appl Physiol. 2012;112(6):1989–99.
17. Rowell LB. Human cardiovascular adjustments to exercise and thermal stress. Physiol Rev. 1974;54(1):75–159.
18. Coyle EF, Gonzalez-Alonso J. Cardiovascular drift during prolonged exercise: new perspectives. Exerc Sport Sci Rev. 2001;29(2):88–92.
19. Cheuvront SN, Kenefick RW, Montain SJ, Sawka MN. Mechanisms of aerobic performance impairment with heat stress and dehydration. J Appl Physiol (1985). 2010;109(6):1989–95.
20. Brengelmann GL, Johnson JM, Hermansen L, Rowell LB. Altered control of skin blood flow during exercise at high internal temperatures. J Appl Physiol. 1977;43(5):790–4.
21. Johnson JM, Minson CT, Kellogg DL. Cutaneous vasodilator and vasoconstrictor mechanisms in temperature regulation. Compr Physiol. 2014;4:33–89.
22. Jose AD, Stitt F, Collison D. The effects of exercise and changes in body temperature on the intrinsic heart rate in man. Am Heart J. 1970;79(4):488–98.
23. Gorman AJ, Proppe DW. Mechanisms producing tachycardia in conscious baboons during environmental heat stress. J Appl Physiol Respir Environ Exerc Physiol. 1984;56(2):441–6.

24. Rowell LB. Circulatory adjustments to dynamic exercise and heat stress: competing controls. Human circulation: regulation during physical stress. New York: Oxford University Press; 1986. p. 363–406.
25. Arngrimsson SA, Stewart DJ, Borrani F, Skinner KA, Cureton KJ. Relation of heart rate to percent VO2 peak during submaximal exercise in the heat. J Appl Physiol (1985). 2003;94(3):1162–8.
26. Nybo L, Jensen T, Nielsen B, Gonzalez-Alonso J. Effects of marked hyperthermia with and without dehydration on VO2 kinetics during intense exercise. J Appl Physiol. 2001;90:1057–64.
27. Pirnay F, Deroanne R, Petit JM. Maximal oxygen consumption in a hot environment. J Appl Physiol. 1970;28(5):642–5.
28. Périard JD, Racinais S. Heat stress exacerbates the reduction in middle cerebral artery blood velocity during prolonged self-paced exercise. Scand J Med Sci Sports. 2015;25(Suppl 1):135–44.
29. Arngrimsson SA, Petitt DS, Borrani F, Skinner KA, Cureton KJ. Hyperthermia and maximal oxygen uptake in men and women. Eur J Appl Physiol. 2004;92(4–5):524–32.
30. Périard JD, Racinais S. Self-paced exercise in hot and cool conditions is associated with the maintenance of %VO2peak within a narrow range. J Appl Physiol. 2015;118:1258–65.
31. Mattern CO, Kenefick RW, Kertzer R, Quinn TJ. Impact of starting strategy on cycling performance. Int J Sports Med. 2001;22(5):350–5.
32. Neary JP, Hall K, Bhambhani YN. Vastus medialis muscle oxygenation trends during a simulated 20-km cycle time trial. Eur J Appl Physiol. 2001;85(5):427–33.
33. Thomas K, Goodall S, Stone M, Howatson G, Gibson AS, Ansley L. Central and peripheral fatigue in male cyclists after 4-, 20-, and 40-km time trials. Med Sci Sports Exerc. 2015;47(3):537–46.
34. Périard JD, Racinais S. Performance and pacing during cycle exercise in hyperthermic and hypoxic conditions. Med Sci Sports Exerc. 2016;48(5):845–53.
35. Mora-Rodriguez R, Del Coso J, Hamouti N, Estevez E, Ortega JF. Aerobically trained individuals have greater increases in rectal temperature than untrained ones during exercise in the heat at similar relative intensities. Eur J Appl Physiol. 2010;109(5):973–81.
36. Sawka MN, Young AJ, Latzka WA, Neufer PD, Quigley MD, Pandolf KB. Human tolerance to heat strain during exercise: influence of hydration. J Appl Physiol (1985). 1992;73(1):368–75.
37. Cheung SS, McLellan TM. Heat acclimation, aerobic fitness, and hydration effects on tolerance during uncompensable heat stress. J Appl Physiol. 1998;84(5):1731–9.
38. Selkirk GA, McLellan TM. Influence of aerobic fitness and body fatness on tolerance to uncompensable heat stress. J Appl Physiol. 2001;91(5):2055–63.
39. Mora-Rodriguez R. Influence of aerobic fitness on thermoregulation during exercise in the heat. Exerc Sport Sci Rev. 2012;40(2):79–87.
40. Morrison S, Sleivert GG, Cheung SS. Passive hyperthermia reduces voluntary activation and isometric force production. Eur J Appl Physiol. 2004;91(5–6):729–36.
41. Nybo L, Nielsen B. Hyperthermia and central fatigue during prolonged exercise in humans. J Appl Physiol. 2001;91(3):1055–60.
42. Périard JD, Caillaud C, Thompson MW. Central and peripheral fatigue during passive and exercise-induced hyperthermia. Med Sci Sports Exerc. 2011;43(9):1657–65.
43. Nielsen B, Savard G, Richter EA, Hargreaves M, Saltin B. Muscle blood flow and muscle metabolism during exercise and heat stress. J Appl Physiol. 1990;69(3):1040–6.
44. Nielsen B, Hales JRS, Strange S, Christensen NJ, Warberg J, Saltin B. Human circulatory and thermoregulatory adaptations with heat acclimation and exercise in a hot, dry environment. J Physiol. 1993;460:467–85.
45. Périard JD, Cramer MN, Chapman PG, Caillaud C, Thompson MW. Neuromuscular function following prolonged intense self-paced exercise in hot climatic conditions. Eur J Appl Physiol. 2011;111(8):1561–9.
46. Racinais S, Girard O. Neuromuscular failure is unlikely to explain the early exercise cessation in hot ambient conditions. Psychophysiology. 2012;49(6):853–65.

47. Adreani CM, Hill JM, Kaufman MP. Responses of group III and IV muscle afferents to dynamic exercise. J Appl Physiol (1985). 1997;82(6):1811–7.
48. Kaufman MP, Hayes SG, Adreani CM, Pickar JG. Discharge properties of group III and IV muscle afferents. Advances in experimental medicine and biology, vol. 508; 2002. p. 25–32.
49. Kaufman MP, Rybicki KJ. Discharge properties of group III and IV muscle afferents: their responses to mechanical and metabolic stimuli. Circ Res. 1987;61(4 Pt 2):I60–5.
50. Sidhu SK, Weavil JC, Mangum TS, Jessop JE, Richardson RS, Morgan DE, et al. Group III/IV locomotor muscle afferents alter motor cortical and corticospinal excitability and promote central fatigue during cycling exercise. Clin Neurophysiol. 2017;128(1):44–55.
51. Rowell LB. Human cardiovascular control. New York: Oxford University Press; 1993.
52. Périard JD. Hyperthermia and supraspinal fatigue. Exp Physiol. 2016;101(11):1323–4.
53. Flouris AD, Schlader ZJ. Human behavioral thermoregulation during exercise in the heat. Scand J Med Sci Sports. 2015;25:52–64.
54. Attia M. Thermal pleasantness and temperature regulation in man. Neurosci Biobehav Rev. 1984;8:335–42.
55. Schlader ZJ, Stannard SR, Mundel T. Is peak oxygen uptake a determinant of moderate-duration self-paced exercise performance in the heat? Appl Physiol Nutr Metab. 2011;36(6):863–72.
56. Schlader ZJ, Simmons SE, Stannard SR, Mundel T. Skin temperature as a thermal controller of exercise intensity. Eur J Appl Physiol. 2011;111(8):1631–9.
57. Abbiss CR, Burnett A, Nosaka K, Green JP, Foster JK, Laursen PB. Effect of hot versus cold climates on power output, muscle activation, and perceived fatigue during a dynamic 100-km cycling trial. J Sports Sci. 2010;28(2):117–25.
58. Périard JD, De Pauw K, Zanow F, Racinais S. Cerebrocortical activity during self-paced exercise in temperate, hot and hypoxic conditions. Acta Physiol (Oxf). 2018;222:1–13.
59. Racinais S, Périard JD, Karlsen A, Nybo L. Effect of heat and heat-acclimatization on cycling time-trial performance and pacing. Med Sci Sports Exerc. 2015;47(3):601–6.
60. Barwood MJ, Corbett J, White D, James J. Early change in thermal perception is not a driver of anticipatory exercise pacing in the heat. Br J Sports Med. 2012;46(13):936–42.
61. Sawka MN, Cheuvront SN, Kenefick RW. High skin temperature and hypohydration impair aerobic performance. Exp Physiol. 2012;97(3):327–32.
62. Fitts RH. Cellular mechanisms of muscle fatigue. Phys Rev. 1994;74(1):49–94.
63. Allen DG, Lamb GD, Westerblad H. Skeletal muscle fatigue: cellular mechanisms. Physiol Rev. 2008;88(1):287–332.
64. Racinais S, Oksa J. Temperature and neuromuscular function. Scand J Med Sci Sports. 2010;20(Suppl 3):1–18.
65. Bergh U, Ekblom B. Influence of muscle temperature on maximal strength and power output in human skeletal muscles. Acta Physiol Scand. 1979;107:33–7.
66. Sargeant AJ. Effect of muscle temperature on leg extension force and short-term power output in humans. Eur J Appl Physiol. 1987;56:693–8.
67. Asmussen E, Boje O. Body temperature and capacity for work. Acta Physiol Scand. 1945;10:1–22.
68. Ball D, Burrows C, Sargeant AJ. Human power output during repeated sprint cycle exercise: the influence of thermal stress. Eur J Appl Physiol Occup Physiol. 1999;79(4):360–6.
69. Linnane DM, Bracken RM, Brooks S, Cox VM, Ball D. Effects of hyperthermia on the metabolic responses to repeated high-intensity exercise. Eur J Appl Physiol. 2004;93(1–2):159–66.
70. Périard JD, Racinais S, Thompson MW. Adjustments in the force-frequency relationship during passive and exercise-induced hyperthermia. Muscle Nerve. 2014;50(5):822–9.
71. Périard JD, Christian RJ, Knez WL, Racinais S. Voluntary muscle and motor cortical activation during progressive exercise and passively induced hyperthermia. Exp Physiol. 2014;99(1):136–48.
72. Arkesteijn M, Jobson SA, Hopker J, Passfield L. Effect of gradient on cycling gross efficiency and technique. Med Sci Sports Exerc. 2013;45(5):920–6.
73. Sarabon N, Fonda B, Markovic G. Change of muscle activation patterns in uphill cycling of varying slope. Eur J Appl Physiol. 2012;112(7):2615–23.

74. Bouillod A, Grappe F. Physiological and biomechanical responses between seated and standing positions during distance based uphill time trials in elite cyclists. J Sports Sci. 2018;36(10):1173–8.
75. Emanuele U, Horn T, Denoth J. The relationship between freely chosen cadence and optimal cadence in cycling. Int J Sports Physiol Perform. 2012;7(4):375–81.
76. Lucia A, Hoyos J, Chicharro JL. Preferred pedalling cadence in professional cycling. Med Sci Sports Exerc. 2001;33(8):1361–6.
77. McCole SD, Claney K, Conte JC, Anderson R, Hagberg JM. Energy expenditure during bicycling. J Appl Physiol (1985). 1990;68(2):748–53.
78. Faria EW, Parker DL, Faria IE. The science of cycling: factors affecting performance - part 2. Sports Med. 2005;35(4):313–37.
79. di Prampero PE. Cycling on earth, in space, on the moon. Eur J Appl Physiol. 2000;82(5–6):345–60.
80. Lazzer S, Plaino L, Antonutto G. The energetics of cycling on Earth, Moon and Mars. Eur J Appl Physiol. 2011;111(3):357–66.
81. Junge N, Jørgensen R, Flouris AD, Nybo L. Prolonged self-paced exercise in the heat – environmental factors affecting performance. Temperature. 2016;3(4):539–48.
82. Brotherhood JR. Heat stress and strain in exercise and sport. J Sci Med Sport. 2008;11(1):6–19.
83. Saunders AG, Dugas JP, Tucker R, Lambert MI, Noakes TD. The effects of different air velocities on heat storage and body temperature in humans cycling in a hot, humid environment. Acta Physiol Scand. 2005;183(3):241–55.
84. Goulet ED. Effect of exercise-induced dehydration on time-trial exercise performance: a meta-analysis. Br J Sports Med. 2011;45(14):1149–56.
85. Cheuvront SN, Kenefick RW. Dehydration: physiology, assessment, and performance effects. Compr Physiol. 2014;4(1):257–85.
86. Kenefick RW, Cheuvront SN. Hydration for recreational sport and physical activity. Nutr Rev. 2012;70(Suppl 2):S137–42.
87. Kenefick RW, Cheuvront SN. Physiological adjustments to hypohydration: impact on thermoregulation. Auton Neurosci. 2016;196:47–51.
88. Sawka MN, Cheuvront SN, Kenefick RW. Hypohydration and human performance: impact of environmental and physiological mechanisms. Sports Med. 2015;45(Suppl 1):51–60.
89. Maughan RJ, Shirreffs SM. Dehydration and rehydration in competitive sport. Scand J Med Sci Sports. 2010;20(Suppl 3):40–7.
90. Kenefick RW. Drinking strategies: planned drinking versus drinking to thirst. Sports Med. 2018;48(Suppl 1):31–7.
91. Racinais S, Alonso J-M, Coutts AJ, Flouris AD, Girard O, Gonzalez-Alonso J, et al. Consensus recommendations on training and competing in the heat. Scand J Med Sci Sports. 2015;25(Suppl 1):6–19.
92. Bongers CC, Thijssen DH, Veltmeijer MT, Hopman MT, Eijsvogels TM. Precooling and percooling (cooling during exercise) both improve performance in the heat: a meta-analytical review. Br J Sports Med. 2015;49(6):377–84.
93. Bongers CCWG, Hopman MTE, Eijsvogel TMH. Cooling interventions for athletes- an overview of effectiveness, physiological mechanisms, and practical considerations. Temperature. 2017;4(1):60–78.
94. Stevens CJ, Taylor L, Dascombe BJ. Cooling during exercise: an overlooked strategy for enhancing endurance performance in the heat. Sports Med. 2017;47(5):829–41.
95. Racinais S, Cocking S, Periard JD. Sports and environmental temperature: from warming-up to heating-up. Temperature (Austin). 2017;4(3):227–57.
96. Périard JD, Racinais S, Sawka MN. Adaptations and mechanisms of human heat acclimation: applications for competitive athletes and sports. Scand J Med Sci Sports. 2015;25(S1):20–38.
97. Taylor NAS. Human heat adaptation. Compr Physiol. 2014;4(1):325–65.
98. Tyler CJ, Reeve T, Hodges GJ, Cheung SS. The effects of heat adaptation on physiology, perception and exercise performance in the heat: a meta-analysis. Sports Med. 2016;46(11):1699–724.

99. Horowitz M. Heat acclimation, epigenetics, and cytoprotection memory. Compr Physiol. 2014;4(1):199–230.
100. Périard JD, Travers GJS, Racinais S, Sawka MN. Cardiovascular adaptations supporting human exercise-heat acclimation. Auton Neurosci. 2016;196:52–62.
101. Guy JH, Deakin GB, Edwards AM, Miller CM, Pyne DB. Adaptation to hot environmental conditions: an exploration of the performance basis, procedures and future directions to optimise opportunities for elite athletes. Sports Med. 2015;45(3):303–11.
102. Daanen HAM, Racinais S, Periard JD. Heat acclimation decay and re-induction: a systematic review and meta-analysis. Sports Med. 2018;48(2):409–30.
103. Leon LR, Bouchama A. Heat stroke. Compr Physiol. 2015;5(2):611–47.

Open-Water Swimming 14

Carl D. Bradford, David F. Gerrard, and James D. Cotter

A Relatively New, Challenging and Unique Sport Open-water swimming (OWS) is defined by its governing body, FINA (Fédération Internationale de Natation), as any competition that takes place in rivers, lakes, oceans or water channels. Major competitions are typically 5, 10 or 25 km, whereas Marathon swimming specifically defines any 10-km event in open-water conditions. FINA first hosted an OWS Marathon at its World Championships in Perth, Western Australia in 1991, and it has been included in the Olympic aquatics programme since Beijing, 2008. A Marathon World Cup is now also well established and overseen by technical and medical officials. An additional FINA Grand Prix series has also emerged, featuring races from 10 to 88 km at various international venues, while other swims typically include crossings of famous waterways. OWS is governed by FINA's OWS Rules and Regulations, and is swum front crawl. Specific By-Laws govern the risk management and safety of competitors in all FINA-sanctioned events. OWS is also represented over shorter distances (typically 0.75–3.8 km) in multisport events such as Triathlon and Ironman (see Chap. 15), with rules set largely by the International Triathlon Union (ITU).

The major differences with OWS relative to pool swimming are the variability of environmental factors (e.g. water temperature; T_w) and the distance (1500 m being the longest pool event). Longer distance and hence longer duration of swimming

C. D. Bradford
School of Physical Education, Sport and Exercise Sciences, University of Otago, Dunedin, New Zealand

D. F. Gerrard
Dunedin School of Medicine, University of Otago, Dunedin, New Zealand

J. D. Cotter (✉)
Exercise and Environmental Physiology, School of Physical Education, Sport and Exercise Sciences, Division of Sciences, University of Otago, Dunedin, New Zealand
e-mail: jim.cotter@otago.ac.nz

J. D. Périard, S. Racinais (eds.), *Heat Stress in Sport and Exercise*,
https://doi.org/10.1007/978-3-319-93515-7_14

introduces the concept of feeding stations (pontoons) from which competitors may receive food and drink from support staff as they swim past (especially for 10-km events). Elite competition, held at a variety of international venues, requires extensive travel with attendant circadian dysrhythmia, climatic variation and other challenges of the local aquatic venue. Great variance may occur in T_w, dry bulb air temperature (T_{db}), humidity, solar radiation, water currents, water type and quality, wind, chop and swells. Warm T_w is mostly accompanied by high T_{db} (e.g. 35–40 °C; dry tropics) or moderate T_{db} with high humidity (e.g. >2 kPa; wet tropics or season). Such environments may seriously challenge the thermoregulatory capacity of swimmers with high rates of metabolic heat production held for durations of 1–2 h in 5- and 10-km events, respectively. Immersion (hydrostatic pressure), the prone posture and the upper-body-dominated nature of exercise add further complexity to the physical and physiological effects of OWS.

Thermodynamics of OWS Metabolic rate can be increased up to 20 times (to as much as 84 kJ min^{-1}) during endurance exercise. Due to the relatively poor efficiency of oxidative metabolism, more than 75% of the energy is lost as heat even in the most efficient modes of locomotion. Swimming has a low efficiency of mechanical work (~5–10%, or ~8–10% at the speeds involved in elite OWS; i.e. ~1.4 m s^{-1}; [1]), and no net storage of kinetic or potential energy. The metabolic rate of swimming at ~1.4 m s^{-1} has been reported to be 3.25 L min^{-1} (~1000 W; [1]) up to 5 L min^{-1} (~1750 W; [2]). Fortunately for swimmers, however, the upper limbs and trunk muscles produce much of this heat and perhaps upward of half is conducted and convected directly to skin overlaying the active muscles [3].

Of particular interest for OWS are the differences in thermoregulation and heat exchange that arise from exercising in water. Evaporation, the body's most powerful heat loss mechanism in air, is nullified for continuously immersed skin. Evaporation can occur particularly from an upper limb during its recovery phase but the surface area of one upper limb is small (~10%) and the time exposed to air is limited (<50%). However, water has a much higher heat capacity than air (4.18 kJ kg^{-1} $°C^{-1}$ vs. 1.30 kJ $(m^3)^{-1}$ $°C^{-1}$, respectively), and a 25-fold greater conductivity (0.600 vs. 0.025 W/(m K)) [4, 5], thus facilitating many-fold faster conductive and convective transfer. Nevertheless, with the major impediment to evaporation, a largely unanswered question is whether individuals who exercise in warm water are also susceptible to larger increases in body temperature. We have preliminary data from ~90 maximal effort performance swims in T_w of 30–33 °C (at similar T_{db}), which showed that such increases are no larger than for running in warm conditions outdoors (unpublished data and [6]).

OWS in warm water normally involves considerable radiant heat load, especially given the prone posture of exercise. This aspect of OWS thermodynamics has been overlooked in almost all laboratory-based studies, yet the findings from such studies are implicitly or explicitly applied to outdoor settings. Irradiance (light energy) from the midday sun on a clear day in tropical locations can reach 1 kW m^{-2}. Nielsen et al. [7] have shown that exercising outdoors under clear sunny skies (solar radiation intensities of 300–700 W m^{-2}) can impose an additional heat load of ~100 W. In terrestrial exercise this additional load appears to incur increased sweating of ~150–180 g h^{-1}, along with an elevated heart rate by 8 b min^{-1} and VO_2 by 0.15 L min^{-1} [7]. Other research has determined that the T_{db} equivalent for solar radiation experienced in a desert environment in summer is an additional ~7–10 °C [8]. We have not found a measureable increase in rectal temperature (T_{re}), heart rate or sweat rate when adding radiant heat load (~400–800 W m^{-2}) in a small pilot study of 20- and 120-min swim performance trials in ambient heat stress (32 °C T_w and T_{db}). Thus, despite the potential for radiant heat load to add to the heat burden of OWS, the few available (laboratory-based) data do not yet support it having a notable impact.

Rate of heat gain/loss when exercising nearly naked in water is linearly dependent on surface area and the skin temperature (T_{sk})—T_w gradient, and non-linearly (diminishingly) on the velocity of skin relative to water. Change in thermal energy and thus mean body temperature depends further on passive and active factors. Passive factors are body mass, mass:area ratio, tissue specific heat and the distribution and extent of subcutaneous fat. Active factors are rate of heat production and blood flow to skin and superficial muscle, which is governed mostly by T_c. Active and passive factors interact in numerous ways, for example, when immersed in cool and cold water the reduction in T_c and increase in VO_2 are proportional to both the subcutaneous fat thickness and the T_w [9–12]. Within a given individual, the rate of forced convective heat transfer therefore depends largely on the difference between T_w and T_c and on the activity [4].

Heat transfer is maximised for swimming for multiple reasons, one of which is that almost the entire skin surface is perpendicular to the water flow (i.e. in contrast to running in water or on land, for example). Swimming also uses muscles throughout the body. Combined upper- and lower-body exercise, and upper-body exercise alone, cause a lower rise in T_c than for lower-body exercise alone, within cool and temperate T_w (i.e. within 20 and 26 but not 33 °C water) [13]. These larger effects for exercise that utilises the arms are likely due to: (1) the smaller muscle mass, which can lead to relatively greater blood flow than the legs; (2) the relative closeness of this arm-muscle blood supply to the skin (coupled with a smaller subcutaneous fat layer compared to the legs) reducing the conductive distance leading to higher convective heat transfer and (3) the greater surface area:mass ratio causing a greater conductive and convective heat loss at the skin–water interface [13]. Therefore, the primary use of arms in swimming, especially over longer distances, may lead to greater heat losses in water. This also indicates that the results obtained from immersion research using cycling and running exercise (especially if the running does not include much arm movement) would be erroneously applied to upper-body-dominated exercise like swimming, due to these differences in heat loss and other issues such as the head and face being largely immersed when swimming. Greater heat loss may be beneficial in warmer water but possibly detrimental in cool and cold water.

In addition to autonomic and metabolic thermoeffector responses, behavioural mechanisms are critically important in helping maintain thermal homeostasis.

In fact, behavioural thermoregulation is more sensitive [14] than autonomic thermoregulation and is of utmost importance for safe and effective performance of endurance exercise (see Chap. 1). In warm water, the loss of evaporative cooling potentially increases one's reliance on behaviour for effective thermoregulation (especially via pacing), although many relevant behavioural responses—such as adjusting posture, donning clothing (wetsuit) or moving to a warmer environment—are constrained or unavailable in many OWS settings. The basic drive for behavioural thermoregulation is thermal displeasure or discomfort. Thermal discomfort appears more dependent on T_{sk} as a thermal input than autonomic thermoregulation does, so the importance of T_{sk} (and thus T_w) should not be underestimated [14]. Yet, data are lacking as to whether swimmers can sense their thermal strain and respond appropriately. This cannot be assumed. The control of thermal comfort during exercise differs from that at rest [15, 16]. Most naturally occurring bodies of warm water will be <33 °C, which clamps the entire skin at a uniform temperature and at a level that would—in air—be associated with thermal comfort (~33 °C) or cool-related discomfort (<30 °C) [17]. Thus, thermo-afferent drive from the skin could be counteractive and thus counterproductive to that from the core when considering that high-intensity exercise will impact T_c in the face of a T_{sk} that is clamped at 'inappropriately' low levels from a sensory perspective. This leads to the question of whether swimmers are able to perceive their thermal strain, or whether the perception and reality are uncoupled. Two studies provide preliminary data relevant to this question. We found that swimmers were in fact able to accurately detect their heat strain (as measured rectally, in controlled T_w and T_{db} conditions), at least as strongly as during terrestrial exercise, despite T_{sk} being clamped during swims in T_w of 30, 32 and 33 °C [6, 18]. Similarly, McKenzie [19] had eight males cycle for ~45 min at ~60% VO_2 max in 29.7 and 31.5 °C water, and heat-related discomfort was rated 8/10 in the 31.5 °C water despite T_{re} plateauing at only 38.06 °C. Therefore, swimmers are not necessarily at increased risk of insidious hyperthermia.

14.1 Influence of Heat on Performance and Health

Few studies have examined performance per se, or performance- or health-related physiological effects of intense or prolonged exercise in water (especially warm water), and even fewer used swimming. Knowledge pertaining to OWS competition is further limited because most studies using swimming exercise have used different strokes, relatively low and controlled exercise intensities (≤50% VO_2 max) and short exercise durations (≤30 min). Also, these studies have mostly been conducted in a swimming pool, thereby restricting the ability to collect physiological data while also impairing ecological validity (e.g. fixed T_w and turns at each end). Effects of radiant heat load in warm T_w have not been examined other than as mentioned above. These points are important if the results are to be transferred to OWS, especially for longer swims (>20 min or 1.5 km).

Only five studies, to our knowledge, have examined effects of warm T_w over distances and intensities relevant to OWS. Robinson and Somers [20] undertook the first of these and also used elite swimmers. Six male Olympic-level swimmers swam as far as possible in 60 min in each of three T_w, averaging 33.5, 29.0 and 21.0 °C. The authors also included a comparison to a single runner on a treadmill (at approximately the same metabolic rate; 585 W m^{-2}) for 60 min in two T_{db}, averaging 25.0 and 9.3 °C. The T_{re} of the two fastest swimmers increased to 38.4 °C after 60 min in the 33.5 °C water (metabolic rate 580–605 W m^{-2}). This was a similar peak T_{re} to that seen in the runner after 60 min (38.6 °C) exercising in the cool air (9.3 °C), but the runner appears to have started with a T_{re} ~0.5 °C lower, therefore his rise was even more than the swimmers. Further, the T_{re} of the runner in 25.0 °C air ended much higher (39.7 °C) than in the fastest swimmers in 33.5 °C water (i.e. 38.4 °C). In the cool water (21.0 °C), the slightly slower swimmers (metabolic rate of 390–465 W m^{-2}) showed a small drop in T_{re}, from 37.5 to 37.1 °C. However, while also struggling in the cool conditions, the two fastest swimmers (metabolic rate of 560–595 W m^{-2}) showed a small increase in T_{re}, from 37.3 to 37.9 °C. Thus, it appears that after 60 min of intense exercise at a similar metabolic rate, the T_{re} of a runner in temperate air (25 °C) can rise to a potentially concerning level of 39.7 °C, while a swimmer in warm water (33.5 °C) appears to remain a modest T_{re} of 38.4 °C. This finding cannot be considered to apply to all swimmers due to likely differences in swimming efficiency and body composition that may affect heat balance.

Holmér and Berg [10] had five males (incl. two swimmers) complete incremental swimming tests to exhaustion lasting 5–8 min after a 20-min preload of swimming at 50% VO_2 max in each of 18, 26 and 34 °C T_w, in a swimming flume. Each participant undertook an identical running protocol in 20–22 °C air on a treadmill. The authors found no significant difference in performance between T_w (detail not reported). Other relevant findings included (1) highly variable oesophageal temperature (T_{oes}) in cool water but homogeneous T_{oes} in warm water, (2) vastus lateralis temperature averaged ~1 °C above T_{oes} and (3) both of these were higher after running than swimming, regardless of T_w.

Macaluso et al. [21] had competitive Masters swimmers complete a 5-km race simulation in three T_w: 23, 27 and 32 °C. This OWS simulation was completed in an indoor 25-m pool, and the average split times (94–95 s 100 m^{-1}) indicate that the participants were at least moderately trained. Performance was impaired by both cool and warm water; mean times were 5.3% slower in 23 °C and 3.7% slower in 32 °C than in 27 °C. Yet, the T_{re} (from mercury thermometers before and after each swim) showed minimal change in 23 °C water and mean rises of 0.9 °C in 27 °C water and 1.1 °C in 32 °C water. The peak T_{re} recorded after the 5-km swims in 27 and 32 °C water (which took 75–80 min) was only ~38 °C, and thus supports the earlier findings [20] in showing only modest heat strain effects of swimming in overtly warm T_w. Thus, performance was optimised—*on average*—when swimming in normal pool temperatures, and impairments in cooler or warmer water were (again, on average) not attributable to swimmers reaching the limits of tolerable T_c.

Two studies examining the effect of cold-water ingestion in international-level OWS during pool training (T_w: ~30 °C) and competition (T_w: 28–29 °C) showed T_c increases of <1.0 °C during swims with neutral-water ingestion [22, 23]. Swim pace was similar, although no maximal performance trials were completed to examine the effects of fluid temperature.

Unpublished data from our laboratory indicate performance swim distance was 4–7% less in 32 °C compared to 27 °C across 20-, 60- and 120-min distance trials in well-trained swimmers. Effects were clear for 20 and 60 min only. The magnitude of impairment was also more homogeneous than the performance effects of swimming in the coolest water used for non-wetsuit swimming in ITU events, i.e. 20 °C, wherein some swimmers were faster and some slower than in 27 °C. These performance effects are logical because 32 °C is heat stressful for all swimmers and thus incurs multiple forms of physiological and psychophysical strain that could contribute to fatigue, whereas cool water will be a net cold stress to some swimmers (especially lean or slow) [10, 24], but thermally optimal for others (especially a combination of large, adipose and fit, for reasons described above and observed previously [25]).

We also had eight trained swimmers undertake self-paced performance swims in 33 °C before and after 6 days of heat acclimation (1 h of interval and continuous swimming) in a swimming flume at this same T_w and T_{db} [6]. Swims included a 10-min warm-up, a 20-min distance trial and 30 min at an intensity aimed at maximising training load and heat strain. Thermoregulatory responses during warm-water swimming were compared against those of terrestrial exercise (cycle ergometry) under conditions intended to incur equivalent T_{sk} (~33 °C; T_{db} 29 °C, 60% RH, 3 m/s air velocity; v_a). Swimmers' heat strain at completion of the 20-min distance trial was modest, i.e. peak T_{re} was on average 38.3 °C (range: 37.5–38.7). Notably, this did not exceed the peak T_{re} of terrestrial exercise (38.4 °C; range: 37.7–39.0), was comparable with the peak obtained during the daily 60-min heat acclimation swims (~38.0 °C) and was minimally higher than the peak recorded from an overlapping cohort doing 20-min distance trials in temperate (27 °C) water (38.0 °C; range: 37.2–38.8).

Safety monitoring of athletes' T_c (gastrointestinal pill) has been undertaken in some international OWS competitions held in Asia (T_w 25.0–32.0 °C). As shown in Fig. 14.1, T_c from five such races is typically higher than in flume-based equivalent T_w, and while peak T_c is ~39 °C, on average, T_c up to 40.0 °C was evident. The higher T_c seems unlikely to be a methodological issue of the site of T_c measurement [26], and thus highlights the importance of gaining further such race data during OWS competition. It is equally important to contextualise these against the at-least-as-high T_c that occur during terrestrial endurance competition of similar durations in these geographic regions and to understand the importance of intact behavioural thermoregulation (discussed below).

The Physiology of OWS Brief discussion of the physiology of OWS training and competition is warranted here because of its importance to performance, health and potential countermeasures to heat stress. A far more extensive and contemporary review of the physiology of immersion and swimming is available elsewhere [27]. Table 14.1 is a summary of the causes of strain and potential countermeasures. It is critical for the reader to appreciate that the multiple impacts of OWS in warm water

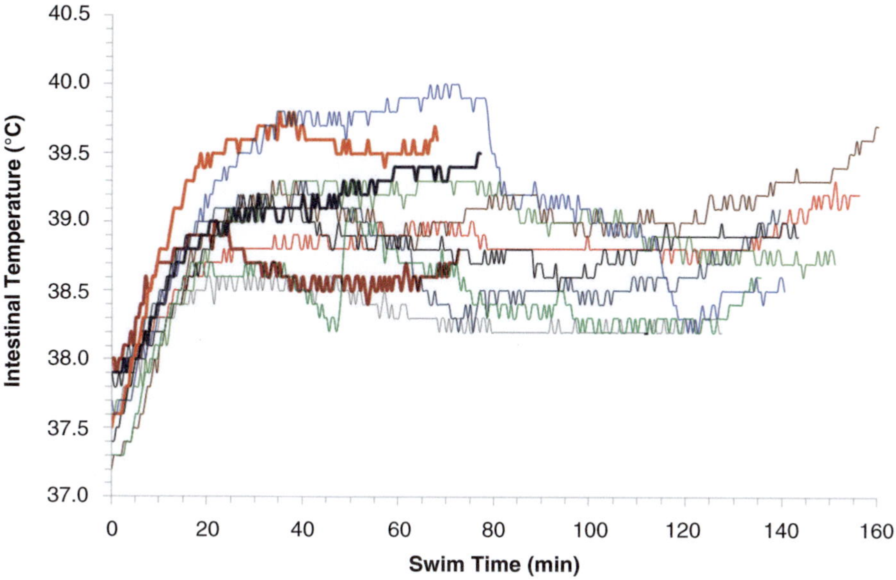

Fig. 14.1 Gastrointestinal pill thermometer-recorded temperatures from five swimmers during OWS competitions in Asian ocean and lake swims of 5 km (thick lines) and 10 km (thin lines). T_w ranged from 25 to 32 °C with similar T_{db}. Data reproduced with the permission of the Singapore Sport Institute. Any omission or error remains with the authors

(shown in Table 14.1) will have interactive effects that could acutely impair health or performance [28].

Cardiovascular and Metabolic Another consequence of predominantly arm-based exercise is that cardiorespiratory, autonomic and metabolic strain of exercise is greater than would be incurred from equivalent leg-based exercise [29–31]. This has numerous implications for health, performance and potential countermeasures. For example, the higher rate pressure product will increase the risk of myocardial infarction in susceptible swimmers, particularly in cool water. Conversely, rates of local and whole-body glycogenolysis are higher ([32]; and presumably more so in warmer T_w [33]), which would make OWS advantageous as a form of exercise for facilitating glucose removal from the blood, but disadvantageous for endurance swimming performance. Hypoglycaemia can develop despite regular intake of carbohydrate (~0.7 g/kg/h) during arm-based exercise (e.g. kayaking) within other ultra-endurance settings, and seems to be specific to the arm exercise [34], so warrants consideration as a potential problem especially during longer OWS events.

Inflammatory Considerations OWS racing in warm water will increase competition for blood flow by the skin and muscles, further reducing gut blood flow for extended durations, a response that can contribute to exercise-induced endotoxaemia in terrestrial settings ([35]; Chap. 5). Indeed, some of the increased T_c during

Table 14.1 Causes of strain relevant to open-water swimming and potential countermeasures

Physical or physiological effect	Predisposing factors (**Bold = relatively specific to OWS**)			Potential countermeasures			
	Person	Exercise	Environment	Conditioning and tapering	On the day (pre-race)	During race	Effective?
Cold strain (especially ↓ T_c)	↓ mass, ↓ mass:area, ↓ adiposity, slow, fatigue	Whole-body but arm dominant, immersed, prone	↑ **Specific heat**, ↓ T_w	Fitness, adiposity, CHO	Passive warming? CHO	Mildly hot rehydration	Minimal
Heat strain (especially ↑ T_c)	↑ mass:area, ↑ adiposity		**Negligible evaporation**	Intense heat acclimation?	Precool, minimise warm-up, hydration	Pacing, hydration, no cap	Minimal
↑ then ↓ blood volume	Unfit	Prone, sweating	Immersed, ↓ T_w, often sea water	Upright exercise in heat?	Sodium and water loading	Hydration	Minimal
↓ Glycogen, hypoglycaemia?	Unfit, diet	Arm dominant	↓ or ↑ T_w	Training and dietary	CHO	CHO	Yes

Some effects are largely unavoidable and thus minimally conducive to amelioration via countermeasures
CHO carbohydrate, T_c core temperature, T_w water temperature

prolonged exercise in hot ambient conditions appears to be due to inflammatory mediators [36], but the extent to which this occurs in endurance-trained swimmers is unknown. It is possible that sustained, intense and hence energy-depleting exercise that relies mostly on upper-limb and -trunk muscles may also lead to exercise-induced muscle inflammation despite the absence of eccentric contraction and ground reaction forces. Importantly, these may be two key pathways (heat-induced endotoxaemia and heat-induced tissue damage) that can contribute to the development of heat stroke, both of which appear to be associated with systemic inflammation ([37, 38]; discussed in Chap. 5). Thus, exercise may increase local inflammation and compromise heat tolerance by increasing the pro-inflammatory cytokine response through endotoxaemia. Swimmers with high training loads may be at increased risk of heat stroke due to a combination of factors such as immunosuppression, musculoskeletal inflammation, subclinical infections (including from ingestion of water while swimming) and gastrointestinal disturbances contributing to increased pro-inflammatory and decreased anti-LPS (lipopolysaccharide) activity during heat stress and exercise [37]. Such effects seem worthy of studying, even if via observational studies within elite and recreational swimmers.

Blood Volume Dehydration (i.e. net loss of body water) is common with endurance exercise and, while not apparent in a study of elite swimmers training in a pool [39], may be exacerbated for OWS in warm water, for a few reasons. Reasons include the (1) prolonged duration in warm water, (2) limited opportunities to rehydrate during races and (3) unique neuroendocrine and hydrostatic effects that combine to drive a greater diuresis than in terrestrial activity. Specifically, an initial increase in blood volume can occur as the external hydrostatic pressure may reverse the normal trans-capillary pressure gradient (Starling forces), facilitating movement of interstitial fluid, particularly in the legs, into the intravascular space [40, 41]. However, the hydrostatic squeeze and a prone posture both act to increase venous return and the resultant increase in mean arterial pressure and central blood volume and pressure will ultimately favour diuresis (after ~30–60 min) by way of increased secretion of atrial natriuretic peptide and suppression of antidiuretic hormone [41–43]. Modest and cool T_{sk} will exacerbate this effect by limiting venous pooling and further increasing central blood volume. Speculatively, swimmers may benefit from the initial immersion and posture-related increases in blood volume, and this may even allow increased skin blood flow [44], which in warm water would assist heat loss. However, the combined effect of renal- and sweat-induced dehydration from more prolonged immersion may increase the competition for blood flow, thus reducing the capacity to store and remove heat and circulate oxygen and substrates [45]. This is complicated by exercising in water, which may have an opposite effect on plasma volume (i.e. decrease) and neuroendocrine responses [46–49].

Sweating Sweat rates during terrestrial endurance events in temperate and warm conditions can vary between individuals but are typically ~0.5–2.5 L h^{-1} in the OWS-comparable sport of Marathon running [50]. These sweat rates are similar to those reported during 5-km performance swims in 27 and 32 °C T_w [21] and also our unpublished observations from 60- and 120-min swims in similar T_w (range: ~0.1–2.2 L h^{-1}), despite hidromeiosis presumably acting to constrain sweating [51]. Even so, these rates are unhelpful to swimmers because they contribute nothing to the (very small) evaporation that will occur from exposed skin, and exacerbate renal losses. Presumably, most elite swimmers would already be aware of their body mass changes in training and racing, thus these magnitudes may not be surprising. However, since there is a strong likelihood many swimmers ingest water (during the stroke, and more than they are aware of), these similar rates of mass loss compared to terrestrial athletes might underestimate sweat loss. This has implications for hydration and feeding strategies of 10-km swimmers and highlights a possibly problematic situation for swimmers in sea water events where the swallowing of salty sea water may exacerbate dehydration (transiently) through intestinal absorption of extracellular fluid. This is an area where further research would certainly be warranted.

The unique thermodynamics of OWS would appear to produce very a challenging environment for at least three settings. One is elite swimmers competing in 5- and 10-km events in cool (~20 °C) water, by virtue of swimmers' low adiposity, high area:mass ratio and high velocity. Multiple laboratory-based studies demonstrate the inability of such swimmers to maintain T_c in cool water regardless of their velocity, so in the absence of a wetsuit or any other effective and realistic countermeasure (Table 14.1), their safety depends on race duration, intact behavioural thermoregulation and an adequate opportunity to exit the water. Another group is small and slow swimmers in prolonged mass participation events in cool water, by virtue of low metabolic rate, low thermal mass and high area:mass ratio and often a limited opportunity to remove themselves from the setting. Their risk is evident in field studies showing high casualty rates (e.g. [24]), especially compared with specialist long-distance events (e.g. channel swims), which are undertaken by experienced, fit and morphologically suited swimmers. For both of the above-mentioned settings, a wetsuit is of obvious potential value, but also provides buoyancy and lessens drag and was therefore illegal in FINA-sanctioned events and in ITU events above category-specific T_w. Another concern is whether the performance advantage afforded by the buoyancy makes its use ubiquitous and hence exposes cold-tolerant swimmers to heat injury. These issues are addressed below under countermeasures.

The third thermally challenging setting is elite swimmers in warm water, by virtue of their high metabolic rate. This potential is highlighted by the death of an elite and experienced, American swimmer, Fran Crippen, during a FINA 10-km World Cup event in Fujairah, UAE in 2010. Official reports indicate that T_w at this event was 29 °C, but anecdotal athlete reports suggest it was more likely 31–33 °C [52]. The ambient conditions were reported as T_{db} 35 °C, and while no humidity was officially recorded, meteorological records show the daily humidity to have ranged from 45 to 75% (i.e. ~3 kPa). Being late morning, radiant heat load will also have been present. The consequences of that tragedy are further addressed in the final section below.

Health and performance effects of OWS in warm water must be kept in perspective relative to those of cool water. Hypothermia appears to be relatively prevalent in cool water, and the prevalence of mortality also appears to be far higher in cool [53] than warm water, but of course so is the exposure, so exposure-normalised risk would be valuable to establish. Competitive swimming in cool water carries additional and unique risk factors, especially to cardiac safety. The cold shock response and cold-induced incapacitation are well-recognised risks, but 'autonomic conflict' (dual sympathetic and parasympathetic activation; [54]) may constitute a special hazard warranting attention (see: [53, 55]). Notably, the above-mentioned OWS simulations undertaken at the University of Otago showed more problems with cool (20 °C) water than 32 °C water in regard to athletes' tolerance.

Given that both cool and warm/hot water exert so many physiological effects, it is unsurprising (but also somewhat ironic) that both passive and active cool and warm water immersion are also of rapidly growing interest for their potential to stimulate health-related adaptations [56–59], including aerobic fitness adaptations [60].

14.2 Countermeasures to Optimise Performance and Health

Acute Thermoregulation Behavioural and autonomic thermoeffector responses are both vital in helping maintain thermal homeostasis, and are vital prerequisites to optimising performance and health. Behavioural thermoregulation is the more sensitive and more powerful of these, as detailed in Chap. 1 and discussed earlier in this chapter. Thermoregulatory behaviour during heat-stressful competition is most easily achieved by reducing pace (i.e. heat production), whereas in cold-stressful competition it would seem prudent not to enter the event in the first place *if* the swimmer knows themselves to be cold intolerant. Increasing (or decreasing) adiposity and thus one's passive thermoregulatory capacity (body mass, mass:area ratio and especially insulation) over the months preceding cold (or heat) stressful OWS competition is theoretically possible but may not be practicable, not least because subcutaneous fat on the arms is most important in both circumstances but is minimally affected.

Adaptive Thermoregulation Heat acclimation, as reviewed thoroughly in Chap. 8, is a common strategy used in terrestrial sports to improve performance and tolerance in hot environments. Heat acclimation modifies several elements of active thermoregulation that collectively lessen heat content and increase the capacity for its storage. The adaptations most relevant for OWS include reduced resting T_c, increased blood volume, and increased skin blood flow, whereas any attendant increase in sweating power would be counterproductive (i.e. hasten dehydration without any evaporative benefit). The reduction in resting T_c and thus exercising T_c develops rapidly (mostly within seven exposures), as does hypervolaemia [61, 62], whereas peripheral aspects of the increased skin blood flow develop more slowly [63].

A commonality to most heat acclimation and acclimatisation protocols used to date is that they are terrestrial, whereas OWS is obviously aquatic. Resting in hot water or swimming in warm water may be more specific for swimmers, especially if undertaking heat acclimatisation. However, the main adaptations that arise from repeated bouts of exercise in air may be limited especially for swimming in water. For example, the increased conductivity of water and primarily arm-based exercise may reduce thermal strain, while the prone posture and hydrostatic pressure will attenuate the acute reductions in central venous pressure and renal blood flow. Accordingly, we [6] found that short-term (1 h/day for 6–7 days) heat acclimation using swimming in warm (33 °C) water did not confer measurable heat adaptations for trained swimmers, nor did it improve performance in warm or temperate T_w, or warm terrestrial exercise. There is one report of heat acclimatisation leading to improved performance (by 10 s, or ~4%) in temperate water 30 days after the acclimatisation [64] but this requires verification. It remains unknown whether resting in hot water and/or exercise in hot air would provide physiological, psychophysical or functional benefits for swimmers, but such research would be valuable. Matching time of day of heat stress bouts to the impending competition has been considered important for gaining the thermal advantage of a lower T_c [65] although recent research using post-exercise hot water immersion indicates this is possibly unnecessary [66]. It would also seem prudent to exercise upright (to target cardiovascular rather than sudomotor adaptations), incorporate some upper-limb exercise and provide intensity rather than volume of heat stress in each session.

Thermal Protection As mentioned above, one problem facing the organisations overseeing OWS (i.e. FINA, ITU and IOC) is whether wearing a wetsuit helps protect against hypothermia in susceptible swimmers without imposing undue heat strain in tolerant swimmers. We therefore undertook, and report here, a small study of the effect of wetsuit usage in 22 °C T_w (21.4 °C T_{db}), i.e. the ITU threshold at and above which wetsuits were not permitted to be worn. Swimmers undertook 1500-m performance trials with and without a wetsuit, in crossover fashion on separate days. Participants were eight well-trained surf swimmers and triathletes (4 males, 4 females); body fat averaged 14% (SD 5; range: 6–22%) as determined using 8-electrode bioimpedance analysis. Body mass averaged 67.3 kg (9.9; range 49.1–77.5), and body mass index (BMI) averaged 22.3 kg m^{-2} (1.8; range 18.7–25.0). Without a wetsuit, oesophageal temperature (T_{oes}) fell in two of the eight swimmers—reaching 35.3 °C in one (Fig. 14.2, top panels)—without any notable cold discomfort (Fig. 14.2, middle panels), whereas the wetsuit prevented a decline in T_{oes} in both swimmers. When wearing the wetsuit, T_{oes} rose 0.6 ± 0.6 °C (mean ± 95% CI) more than when not wearing it (0.8 vs 0.2 °C; P = 0.03; ES = 0.94). Adiposity predicted one third (32%) of the variability in T_{oes} response when not wearing a wetsuit, but none (1%) when wearing a wetsuit. The wetsuit effect on T_{oes} was largest in those who were coolest without it ($r = 0.77$; Fig. 14.2, bottom panel), and had negligible effect (<0.3 °C) for the two swimmers whose T_{oes} rose >1.0 °C without it. Perhaps most importantly, thermal discomfort was closely coupled with rising T_{oes}

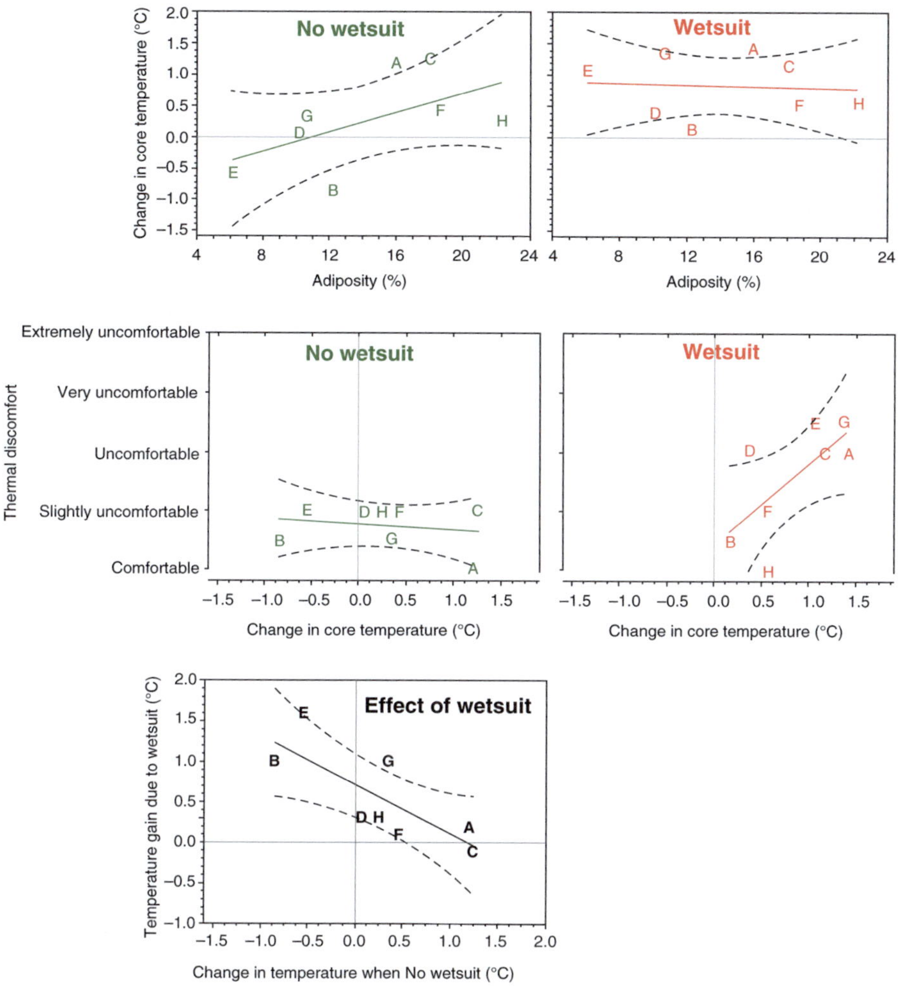

Fig. 14.2 Oesophageal temperature (T_{oes}; top panels) and thermal discomfort (middle panels) in response to swimming 1500 m with or without a wetsuit in 22 °C water. The top panels show T_{oes} responses in relation to adiposity, while the middle ones show thermal discomfort in relation to T_{oes}. The bottom panel shows the effect of wearing a wetsuit, relative to the response in the no wetsuit swim. These data show that a wetsuit prevented core cooling in cold susceptible swimmers, without adding notably to the rise in T_{oes} in other swimmers of this cohort, and any swimmer whose T_{oes} rose while wearing the wetsuit also became uncomfortably hot

with a wetsuit, but not when swimming without it (Fig. 14.2, middle panels). These data preliminarily indicate that wetsuit usage can provide thermal protection in cold-intolerant swimmers without imposing excessive heat stress or insidious heating of the core in others, whereas core cooling was evident without a wetsuit and failed to elicit discomfort.

The use of a swim cap may also influence the thermal status of swimmers in warm and cool water. One study showed that wearing a silicon cap during warm-up and an 800-m time trial led to 0.3 °C higher T_c, compared with no cap, and was associated with ~2% faster performance despite similar physiological and subjective measures in ~33 °C water [67].

Thermal Manipulation Swimmers' heat content can be manipulated up or down during the hour before competition using an active warm-up or a precooling strategy, respectively. These are often used in combination, but both require scrutiny. The rise in T_c incurred by a 20-min warm-up will easily wipe out all of the calorimetric (thermal reserve) benefit gained from a 7- to 10-day heat acclimation. Warm-ups are also typically far longer than is physiologically necessary even for severe-intensity dynamic exercise performed at VO_2 max [68]. Performance gains are modest for such exercise so may be either non-existent or counterproductive for the prolonged, concentric-contraction and body-weight-supported nature of OWS. Therefore, an obvious countermeasure that warrants an open mind and careful investigation by sport scientists and swimmers is limiting the warm-up (e.g. to 3–4 min of upper-body exercise) or potentially removing it before competitions in warm water. This would aid heat storage capacity while also minimising loss of substrates (water and energy) and time, along with psychological dependence.

Pre-race cooling strategies (as discussed in Chap. 7) may help mitigate some of the heat stress encountered by OWS in warm water, but seem likely to provide only modest physiological and performance benefit for 5-km races and negligible benefit for 10-km events. Specifically, the overwhelming dominance of convective heat transfer in OWS will reduce or reverse the normal heat transfer gradients between skin and water, and thereby further reduce the already-modest benefits of precooling that exist in ecologically valid terrestrial exercise settings. Ingesting an ice slurry before competition seems more thermodynamically appropriate for OWS.

OWS is conspicuous among endurance sports in affording athletes no meaningful opportunity for supplemental cooling during competition, other than by reducing their pace. Ingestion of cool fluids has been shown to have little effect [69], although there is some evidence it may be thermally beneficial during endurance swimming in warm water [22, 23]. Ingesting an ice slurry *during* a race seems unwise for at least two reasons, however. First, its rate of ingestion is too slow due to the intense cold, whereas the time available for ingesting fluid and energy is so heavily constrained. Second, the gut has recently been shown to be thermosensitive and thus participates actively in thermoregulatory control. This will potentially reduce athletes' thermal safety reserve for heat injury and might explain findings of athletes achieving higher performance and peak T_c during heat-stressful exercise after consuming an ice slurry within laboratory-based experiments.

Energy and Hydration Beginning a race with high muscle and liver glycogen content seems essential because of the prolonged, mostly arm-based exercise and limited opportunity for ingestion from the pontoons while racing. Optimal rates of intake of up to >90 g carbohydrate per hour are realistic for elite terrestrial athletes ingesting a composite of simple sugars [70], but we are unaware whether this applies also to OWS. Gastric emptying and intestinal absorption might be enhanced by the mechanical action, posture or hydrostatic effects of swimming, but could alternatively be impaired if sympathetic activation is higher than for running or cycling. Swimmers may also swallow water frequently (unpublished observations from our studies), which could impair gut function if they were ingesting sea water or non-potable water. Considerations for hydration are similar to those for energy replenishment, except that swimmers' ability to remain euhydrated will be constrained for reasons described above.

14.3 Heat Policy and Implementation

Following the tragic events in a FINA-sanctioned event in October 2010, there was an urgent call for greater control and scrutiny of the open-water competitive environment. Until then, the well-intended rules governing OWS were less informed by science and more by experience and anecdotal evidence gleaned from swimmers, coaches and officials. In early 2011, FINA joined with the IOC and ITU to invite expressions of interest to undertake specific investigations surrounding the safety of different T_w for OWS, especially warmer water. Research from The University of Otago was reported to these organisations in January 2013 outlining the findings from ~200 swims from a cohort of 24 experienced and elite swimmers. Participants completed 20-, 60- and 120-min distance-trial swims in a flume (to simulate 1.5-, 5- and 10-km OWS) in T_w of 20, 27, 30 and 32 °C with matching T_{db}. Swims in 32 °C T_w also included simulated radiant load (400–800 W m^2). Several physiological and psychophysical variables were measured, including those summarised above. The outcome for FINA and ITU has been the implementation of specific rules that now stipulate the maximum upper T_w of 31 and 32 °C, respectively, for OWS events. Additionally, all ITU OWS distances are reduced to 750 m in T_w over 31 °C [71, 72]. For FINA and ITU, respectively, the rules demand that the venue T_w is taken 2 and 1 h prior to the start of the event at an agreed mid-course site and at a depth of 40 and 60 cm. Thereafter, FINA regulates that T_w is monitored by race officials hourly, with the authority to halt the event if subsequent readings are outside the FINA-approved range for water temperature (16–31 °C).

A more recent development has been the revision of FINA Rules governing the use of wetsuits. From 1 January 2017 a new By-Law (BL 8.5) declared that wetsuits of an appropriate design are optional in T_w of 18–20 °C and compulsory below 18 °C. ITU maintains its wetsuit policies, that vary slightly between elite and age group athletes, but make their use mandatory below 16 °C and forbidden above 22 or 25 °C for events over 1500 m.

Local bodies can enforce their own variations of these rules for non-FINA or ITU sanctioned events. For example, after 2010, USA Swimming set an upper T_w limit for cancellation of local events of 29.45 °C or the sum of T_{db} and T_w exceeding 63 °C. These criteria remain in the current 2018 rule handbook. Also, USA Triathlon allow wetsuits use for $T_w < 25.6$ °C.

As noted above, thermal perceptions appear to remain intact for swimmers in warm water, and feeling hot is important. Therefore, while these rules are evidence-based, they cannot be considered definitive. An onus remains on all athletes to heed their internal cues and for coaches to be vigilant in monitoring their athletes for uncharacteristic signs and demeanour. This is especially important in the face of competitive behaviour and illness and is no different to what is expected and acted upon by terrestrial athletes. A marked difference for open-water swimmers, however, is that their environment is much less forgiving if they get into trouble.

Acknowledgements We are grateful to FINA, ITU and the IOC Medical Commission for funding support for research referred to herein. We are grateful also to our colleagues who collaborated on the swimming research: Drs Monique Francois and Samuel Lucas, Associate Professors Chris Button and Zhifa Sun, Mr. Terry Hill and Professor Eugene Wissler.

Disclaimer The material and opinions presented do not represent the views of FINA, ITU, the IOC, the University of Otago or the Singapore Sport Institute.

References

1. Toussaint HM, Knops W, De Groot G, Hollander AP. The mechanical efficiency of front crawl swimming. Med Sci Sports Exerc. 1990;22(3):402–8.
2. Holmér I. Oxygen uptake during swimming in man. J Appl Physiol. 1972;33(4):502–9.
3. González-Alonso J, Quistorff B, Krustrup P, Bangsbo J, Saltin B. Heat production in human skeletal muscle at the onset of intense dynamic exercise. J Physiol. 2000;524(Pt 2):603–15.
4. Nadel ER, Holmér I, Bergh U, Astrand PO, Stolwijk JA. Energy exchanges of swimming man. J Appl Physiol. 1974;36(4):465–71.
5. Ramires MLV, de Castro CAN, Nagasaka Y, Nagashima A, Assael MJ, Wakeham WA. Standard reference data for the thermal conductivity of water. J Phys Chem Ref Data. 1995;24(3):1377–81.
6. Bradford CD, Lucas SJ, Gerrard DF, Cotter JD. Swimming in warm water is ineffective in heat acclimation and is non-ergogenic for swimmers. Scand J Med Sci Sports. 2015;25(Suppl 1):277–86.
7. Nielsen B, Kassow K, Aschengreen FE. Heat balance during exercise in the sun. Eur J Appl Physiol Occup Physiol. 1988;58(1–2):189–96.
8. Lee DH, Vaughan JA. Temperature equivalent of solar radiation on man. Int J Biometeorol. 1964;8(1):61–9.
9. Choi JS, Ahn DW, Choi JK, Kim KR, Park YS. Thermal balance of man in water: prediction of deep body temperature change. Appl Hum Sci. 1996;15(4):161–7.
10. Holmér I, Bergh U. Metabolic and thermal response to swimming in water at varying temperatures. J Appl Physiol. 1974;37(5):702–5.
11. Šrámek P, Šimečková M, Janský L, Šavlíková J, Vybíral S. Human physiological responses to immersion into water of different temperatures. Eur J Appl Physiol. 2000;81(5):436–42.
12. Lounsbury DS, DuCharme MB. Arm insulation and swimming in cold water. Eur J Appl Physiol. 2008;104(2):159–74.
13. Toner MM, Sawka MN, Pandolf KB. Thermal responses during arm and leg and combined arm-leg exercise in water. J Appl Physiol. 1984;56(5):1355–60.
14. Schlader ZJ, Perry BG, Jusoh MR, Hodges LD, Stannard SR, Mundel T. Human temperature regulation when given the opportunity to behave. Eur J Appl Physiol. 2013;113(5):1291–301.
15. Chatonnet J, Cabanac M. The perception of thermal comfort. Int J Biometeorol. 1965;9(2):183–93.

16. Thompson RL, Hayward JS. Wet-cold exposure and hypothermia: thermal and metabolic responses to prolonged exercise in rain. J Appl Physiol. 1996;81(3):1128–37.
17. Craig AB, Dvorak M. Thermal regulation during water immersion. J Appl Physiol. 1966;21(5):1577–85.
18. Bradford CD, Gerrard DF, Lucas SJE, Sun ZF, Cotter JD (eds.) Is swimming in warm water actually putting swimmers in hot water? In: Proceedings of the 15th international conference on environmental ergonomics (ICEE); Queenstown, New Zealand; 2013.
19. McKenzie JM. Challenges to exercise thermoregulation during warm-water immersion [Ph.D.]. Ann Arbor: Indiana University; 2015.
20. Robinson S, Somers A. Temperature regulation in swimming. J Physiol. 1971;63(3):406–9.
21. Macaluso F, Di Felice V, Boscaino G, Bonsignore G, Stampone T, Farina F, et al. Effects of three different water temperatures on dehydration in competitive swimmers. Sci Sports. 2011;26(5):265–71.
22. Hue O, Monjo R, Lazzaro M, Baillot M, Hellard P, Marlin L, et al. The effect of time of day on cold water ingestion by high-level swimmers in a tropical climate. Int J Sports Physiol Perform. 2013;8(4):442–51.
23. Hue O, Monjo R, Riera F. Imposed cold-water ingestion during open water swimming in internationally ranked swimmers. Int J Sports Med. 2015;36(11):941–6.
24. Brannigan D, Rogers IR, Jacobs I, Montgomery A, Williams A, Khangure N. Hypothermia is a significant medical risk of mass participation long-distance open water swimming. Wild Environ Med. 2009;20(1):14–8.
25. Costill DL, Cahill PJ, Eddy D. Metabolic responses to submaximal exercise in three water temperatures. J Appl Physiol. 1967;22(4):628–32.
26. O'Brien C, Hoyt RW, Buller MJ, Castellani JW, Young AJ. Telemetry pill measurement of core temperature in humans during active heating and cooling. Med Sci Sports Exerc. 1998;30(3):468–72.
27. Pendergast DR, Moon RE, Krasney JJ, Held HE, Zamparo P. Human physiology in an aquatic environment. Compr Physiol. 2015;5(4):1705–50.
28. Akerman AP, Lucas SJE, Katare R, Cotter JD. Heat and dehydration additively enhance cardiovascular outcomes following orthostatically-stressful calisthenics exercise. Front Physiol. 2017;8:756.
29. Ahlborg G, Jensen-Urstad M. Metabolism in exercising arm vs. leg muscle. Clin Physiol. 1991;11(5):459–68.
30. Davies CT, Sargeant AJ. Physiological responses to standardised arm work. Ergonomics. 1974;17(1):41–9.
31. Vokac Z, Bell H, Bautz-Holter E, Rodahl K. Oxygen uptake/heart rate relationship in leg and arm exercise, sitting and standing. J Appl Physiol. 1975;39(1):54–9.
32. Harvey C. The effect of exercise mode and muscle mass on plasma glucose oxidation during exercise in the fasted state. Dunedin: University of Otago; 2011.
33. Febbraio MA. Temperature, muscle metabolism and performance. In: Lamb DL, Murray R, editors. The metabolic bases of performance in sport and exercise. Perspectives in exercise science and sports medicine. Carmel: Cooper Publishing Group; 1999. p. 315–53.
34. Helge JW, Rehrer NJ, Pilegaard H, Manning P, Lucas SJ, Gerrard DF, et al. Increased fat oxidation and regulation of metabolic genes with ultraendurance exercise. Acta Physiol (Oxf). 2007;191(1):77–86.
35. Selkirk GA, McLellan TM, Wright HE, Rhind SG. Mild endotoxemia, NF-κB translocation, and cytokine increase during exertional heat stress in trained and untrained individuals. Am J Physiol Regul Integr Comp Physiol. 2008;295(2):R611–R23.
36. Bradford CD, Cotter JD, Thorburn MS, Walker RJ, Gerrard DF. Exercise can be pyrogenic in humans. Am J Physiol Regul Integr Comp Physiol. 2007;292(1):R143–9.
37. Lim CL, Mackinnon LT. The roles of exercise-induced immune system disturbances in the pathology of heat stroke: the dual pathway model of heat stroke. Sports Med. 2006;36(1):39–64.
38. Lim CL, Wilson G, Brown L, Coombes JS, Mackinnon LT. Pre-existing inflammatory state compromises heat tolerance in rats exposed to heat stress. Am J Physiol Regul Integr Comp Physiol. 2007;292(1):R186–R94.

39. Maughan RJ, Dargavel LA, Hares R, Shirreffs SM. Water and salt balance of well-trained swimmers in training. Int J Sport Nutr Exerc Metab. 2009;19(6):598–606.
40. Johansen LB, Jensen T, Ulrik S, Pump B, Norsk P. Contribution of abdomen and legs to central blood volume expansion in humans during immersion. J Appl Physiol. 1997;83(3):695–9.
41. Stocks JM, Patterson MJ, Hyde DE, Jenkins AB, Mittleman KD, Taylor NAS. Effects of immersion water temperature on whole-body fluid distribution in humans. Acta Physiol Scand. 2004;182(1):3–10.
42. Wolf JP, Nguyen NU, Dumoulin G, Baulay A, Berthelay S. Relative effects of the supine posture and of immersion on the renin aldosterone system at rest and during exercise. Eur J Appl Physiol Occup Physiol. 1987;56(3):345–9.
43. Wolf JP, Nguyen NU, Baulay A, Dumoulin G, Berthelay S. The role of posture on the changes in plasma atrial natriuretic factor and arginine vasopressin levels during immersion. Eur J Appl Physiol Occup Physiol. 1990;61(3–4):284–8.
44. Nielsen B, Rowell LB, Bonde-Petersen F. Cardiovascular responses to heat stress and blood volume displacements during exercise in man. Eur J Appl Physiol Occup Physiol. 1984;52(4):370–4.
45. González-Alonso J, Teller C, Andersen SL, Jensen FB, Hyldig T, Nielsen B. Influence of body temperature on the development of fatigue during prolonged exercise in the heat. J Appl Physiol. 1999;86(3):1032–9.
46. McMurray RG. Plasma volume changes during submaximal swimming. Eur J Appl Physiol Occup Physiol. 1983;51(3):347–56.
47. Ertl AC, Bernauer EM, Hom CA. Plasma volume shifts with immersion at rest and two exercise intensities. Med Sci Sports Exerc. 1991;23(4):450-7.
48. Sheldahl LM, Tristani FE, Connelly TP, Levandoski SG, Skelton MM, Cowley AW. Fluid-regulating hormones during exercise when central blood volume is increased by water immersion. Am J Physiol Regul Integr Comp Physiol. 1992;262(5):R779–R85.
49. Bonifazi M, Bela E, Carli G, Lodi L, Lupo C, Maioli E, et al. Responses of atrial natriuretic peptide and other fluid regulating hormones to long distance swimming in the sea. Eur J Appl Physiol Occup Physiol. 1994;68(6):504–7.
50. Noakes TD, Myburgh KH, du Plessis J, Lang L, Lambert M, van der Riet C, et al. Metabolic rate, not percent dehydration, predicts rectal temperature in marathon runners. Med Sci Sports Exerc. 1991;23(4):443–9.
51. Candas V, Libert JP, Vogt JJ. Effect of hidromeiosis on sweat drippage during acclimation to humid heat. Eur J Appl Physiol Occup Physiol. 1980;44(2):123–33.
52. Werner G, Vervaecke H, Pelliccia A, Smith T-L, Towle G. FINA appointed task force report—part 4 climatic conditions. Lausanne: FINA; 2011.
53. Tipton MJ. Sudden cardiac death during open water swimming. Br J Sports Med. 2014;48(15):1134–5.
54. Shattock MJ, Tipton MJ. 'Autonomic conflict': a different way to die during cold water immersion? J Physiol. 2012;590(14):3219–30.
55. Tipton M, Bradford C. Moving in extreme environments: open water swimming in cold and warm water. Extreme Physiol Med. 2014;3:12.
56. Thomas KN, van Rij AM, Lucas SJ, Cotter JD. Lower-limb hot-water immersion acutely induces beneficial hemodynamic and cardiovascular responses in peripheral arterial disease and healthy, elderly controls. Am J Physiol Regul Integr Comp Physiol. 2017;312(3):R281–r91.
57. Hooper PL. Hot-tub therapy for type 2 diabetes mellitus. N Engl J Med. 1999;341(12):924–5.
58. Brunt VE, Howard MJ, Francisco MA, Ely BR, Minson CT. Passive heat therapy improves endothelial function, arterial stiffness and blood pressure in sedentary humans. J Physiol. 2016;594(18):5329–42.
59. Sosner P, Gayda M, Dupuy O, Garzon M, Lemasson C, Gremeaux V, et al. Ambulatory blood pressure reduction following high-intensity interval exercise performed in water or dryland condition. J Am Soc Hypertens. 2016;10(5):420–8.
60. Bailey TG, Cable NT, Miller GD, Sprung VS, Low DA, Jones H. Repeated warm water immersion induces similar cerebrovascular adaptations to 8 weeks of moderate-intensity exercise training in females. Int J Sports Med. 2016;37(10):757–65.

61. Shido O. Can our thermoregulatory system anticipate temperature exposure? Med Hypotheses. 1998;50(3):199–202.
62. Racinais S, Periard JD, Karlsen A, Nybo L. Effect of heat and heat acclimatization on cycling time trial performance and pacing. Med Sci Sports Exerc. 2015;47(3):601–6.
63. Lorenzo S, Minson CT. Heat acclimation improves cutaneous vascular function and sweating in trained cyclists. J Appl Physiol. 2010;109(6):1736–43.
64. Hue O, Antoine-Jonville S, Sara F. The effect of 8 days of training in tropical environment on performance in neutral climate in swimmers. Int J Sports Med. 2007;28(1):48–52.
65. Shido O, Sugimoto N, Tanabe M, Sakurada S. Core temperature and sweating onset in humans acclimated to heat given at a fixed daily time. Am J Physiol Regul Integr Comp Physiol. 1999;276(4):R1095–R101.
66. Zurawlew MJ, Mee JA, Walsh NP. Heat acclimation by post-exercise hot water immersion in the morning reduces thermal strain during morning and afternoon exercise-heat-stress. Int J Sports Physiol Perform. 2018;13(10):1281–86.
67. Hue O, Galy O. The effect of a silicone swim cap on swimming performance in tropical conditions in pre-adolescents. J Sports Sci Med. 2012;11(1):156–61.
68. Palmer CD, Jones AM, Kennedy GJ, Cotter JD. Effects of prior heavy exercise on energy supply and 4000-m cycling performance. Med Sci Sports Exerc. 2009;41(1):221–9.
69. Tan P, Lee J. The role of fluid temperature and form on endurance performance in the heat. Scand J Med Sci Sports. 2015;25(S1):39–51.
70. Stellingwerff T, Cox GR. Systematic review: carbohydrate supplementation on exercise performance or capacity of varying durations. Appl Physiol Nutr Metab. 2014;39(9):998–1011.
71. Federation Internationale de Natation. FINA open water swimming rules. 2016.
72. International Triathlon Union. ITU competition rules. Competition rules. 2018. 10/12/2017.

Triathlon and Ultra-Endurance Events in Tropical Environments

15

Clovis Chabert, Eric Hermand, and Olivier Hue

15.1 Influence of Heat on Performance and Health

15.1.1 Specificity of the Triathlon

The Ironman and Challenge triathlon series are trademark brands for long-distance triathlons consisting of a 3.8-km swim, 180-km cycling leg, and full marathon (42.195 km). The best professional triathletes complete the distance in under 7 h and 50 min, whereas age-group athletes vary from 9 to 13 h and more, making this one of the longest endurance races. The duration depends on several factors, including terrain (i.e., course route and geography) and environmental conditions. For example, hot and humid climates impact on performance and the rate of withdrawal, as observed in the Kona (Hawaii) Ironman World Championships every October [1, 2]; top age groupers are slower in Kona than at their qualifying races [2], and withdrawal rate can reach 10%, a high proportion considering the fitness level of athletes participating in this particular event [3]. Similarly, more than 25% of the athletes did not reach the finish line at the inaugural Ironman Vichy in 2015, which was held in unusually hot conditions (31–35 °C), in contrast to North American races with a minimum 95% finish rate [4].

The following sections detail how thermal stress can specifically impact each of the triathlon legs: swimming, cycling, and running.

Author contributed equally with all other contributors.C. Chabert and E. Hermand

C. Chabert · E. Hermand · O. Hue (✉)
Laboratory “Adaptations au Climat Tropical, Exercice et Santé” (ACTES; EA 3596), French West Indies University, Pointe-à-Pitre, Guadeloupe, France
e-mail: olivier.hue@univ-antilles.fr

J. D. Périard, S. Racinais (eds.), *Heat Stress in Sport and Exercise*,
https://doi.org/10.1007/978-3-319-93515-7_15

15.1.1.1 Swimming

Most long-distance triathlons allow neoprene wetsuits to be worn for the swim leg as long as water temperature does not exceed 24.5 °C. Between 24.5 °C and 28.8 °C, athletes who choose to wear a wetsuit are allowed to participate, but are not eligible for age-group awards or for qualifying slots for the Ironman World Championships, and over 28.8 °C wetsuits are prohibited [3]. With its enhanced buoyancy, the neoprene wetsuit reduces ventilation and O_2 consumption ($\dot{V}O_2$) at a given swimming speed, and therefore metabolic heat production [5]. However, wetsuits reduce heat dissipation capacity and can lead to high core body temperatures and dehydration [6, 7]. Hence, although banning wetsuits in warm water might impair swimming performance, it might also decrease the risk of hyperthermia.

Interestingly, Kerr et al. [6] showed in a simulated Olympic triathlon that core temperature did not change by wearing a wetsuit, as excess heat was still transferred at the periphery, leading to a higher skin temperature. A subsequent field study confirmed that core temperature remained around 38 °C in well-trained athletes wearing a wetsuit in 20.5 °C water during a 3.8-km swim [7]. This was again confirmed by a case report in which an athlete's core temperature was continuously monitored during an Ironman [8]. In 29.5 °C water, another study showed that the core temperature of moderately trained athletes increased by an average 0.7 °C and remained under 38 °C during a non-wetsuit half-Ironman (Fig. 15.1; [9]). Thus, no study appears to have shown a dangerous elevation in core temperature during the swim

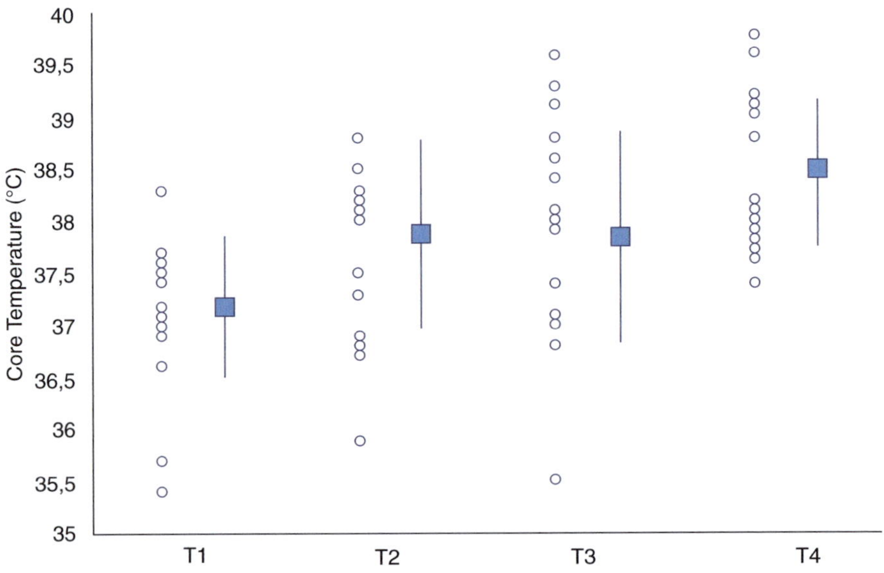

Fig. 15.1 Individual and mean values (±SD) of core temperature at each stage of the Guadeloupe half-Ironman held in tropical climate. T1: just before the race; T2: after the swim phase; T3: after the cycle phase; T4: at the end of the run phase [9]. Temperatures were obtained from telemetric intestinal temperature devices ingested at least 6 h before the race

leg of long-distance triathlon events. Moreover, while 13 of the 14 deaths in triathlon races from 2006 to 2008 occurred during the swim leg, pre-existing cardiovascular abnormalities seem to have been the major factor, ruling out hyperthermia as a potential cause [10].

15.1.1.2 Cycling

When cycling, some of the heat produced by the muscles is dissipated via convection. This however depends on a sufficient temperature gradient between the skin and ambient air. Convection is therefore limited in hot conditions when air temperature exceeds 35 °C. Heat is also dissipated during cycling via the evaporation of sweat, which depends on the water vapor partial pressure gradient between air and skin. Like convection, this mechanism is reduced in humid environments, making performance in a tropical environment challenging as thermoregulatory efficiency is reduced [11]. In a tropical climate (31–33 °C, 70–75% RH), 1 h of pedaling at a submaximal intensity increases core temperature, heart rate, sweat rate, and water loss [12–14]. These results can be extrapolated to describe the physiological impact of 4–6-h events, especially if the cycling course includes hills, where the diminished air speed reduces heat dissipation [11]. Laboratory-controlled studies showed that increasing ambient temperature impaired cycling capacity when relative humidity was clamped at 70% [15, 16]. During a 9-day (2.5 h/day) cycling race held in Guadeloupe (31.1 °C, 75.6% RH), tympanic temperatures measured immediately after the stages never exceeded 2 °C of resting pre-race values [17]. In contrast to the variability associated with stage racing in a peloton, triathletes will adopt a much steadier pace during racing in some cases to "save their legs" for the following marathon [18]. Interestingly, Baillot and Hue [9] showed a negative correlation between body mass and the evolution of core temperature throughout the cycling leg, implying greater heat storage and inertia. A bigger body thus presents a larger surface to preserve heat exchange at high velocity, even in a hot and humid environment [19].

In summary, the high velocities and individual nature of cycling in the triathlon support heat dissipation by convection on the condition that air temperature is lower than skin temperature. In a thermally challenging environment, sweat evaporation is the main avenue for heat dissipation but it is limited by the high humidity of a tropical environment. Nevertheless, triathletes appear to show adequate thermoregulatory capability to prevent excessive hyperthermia, mainly through intensity management and pacing strategies [9].

15.1.1.3 Running

Of the three sports in the triathlon, running may carry the greatest risk of hyperthermia for triathletes competing in a tropical climate. Several factors come into play. Firstly, although running is faster than swimming, air possesses neither the heat capacity nor transfer celerity of water, hence limiting the quantity of heat energy transferred from the body. Secondly, unlike in cycling, the lower speed associated with running does not allow for excess heat dissipation via convection [19]. These factors can have a strong negative impact on running performance [20], particularly the marathon [21]. This negative impact is amplified (1) by a heavier body mass, which, at a given pace, requires more energy for running propulsion and therefore produces more heat, and (2) in heavier athletes running in a hot/humid environment by limiting heat exchange by convection and evaporation, who display greater imbalance between heat production and dissipation, even at slower paces [22].

While the core temperature data of professional triathletes in the Hawaii Ironman or other tropical locations are currently not available, a few studies of moderately to very well-trained age-group athletes have shown that none of them suffered from heat-related illness and excessive increases in core temperature [7] (Fig. 15.1; [9, 23]). In the report from Laursen et al. [7] 7 out of 10 well-trained triathletes performed a high-level sub-10-h performance at Ironman Western Australia in temperate but humid conditions (23.3 °C, 60% RH), and the average core body temperature remained close to 38 °C. Despite a much hotter and more humid environment in Guadeloupe (27.2 °C, 80% RH), the average body temperature of moderately trained athletes during a half-Ironman run was 38.2 °C and none of them reached 40 °C [9]. These values are not surprising given the relative intensity and absolute workload at which most long-distance triathlon races are performed, which is in contrast to shorter events performed at a higher intensity and thus higher heat production [9, 24]. The importance of exercise intensity was confirmed in dryer conditions in a study showing that the fastest athletes exhibited the highest core temperatures [23].

15.1.2 Specificity of Ultra-Endurance Events

15.1.2.1 Ultra-Endurance in the Heat

Ultramarathons last more than 6 h [25]. In the past decade, trail running races have become a major ultra-endurance sport. Since the first edition of the Ultra Trail du Mont Blanc (160 km and +10,000 m of elevation), the number of finishers has grown from 67 in 2003 to 1685 in 2017. This kind of effort involves several physiological and psychological parameters that are modulated by environmental factors. Ultramarathon races are performed with a variety of off-road terrains, elevation profiles (positive: D+ and negative: D− changes), and distances, all of which greatly complicate the attempt to understand how environmental conditions affect performance. Marathon races have more standardized characteristics, facilitating the comparison of performances in order to study the role played by heat (see Chap. 12). Based on the analysis of winners' times in several marathons, Maughan [26] determined that the best running temperature is between 10 °C and 12 °C for elite athletes. Beyond this temperature, the analysis of 136 marathons performed between a wet-bulb globe temperature (WBGT) of 5 and 25 °C showed a decrease in performance with an increase in WBGT (Fig. 15.2; [21]). Interestingly, these authors also showed that slower runners, who are exposed to heat for a longer time than faster runners, are more affected by temperature. For ultramarathon distances, the same effect of hot environments on performance was observed in a study that examined running times in the Western States Endurance Run (161 km and 6000 m D+) in

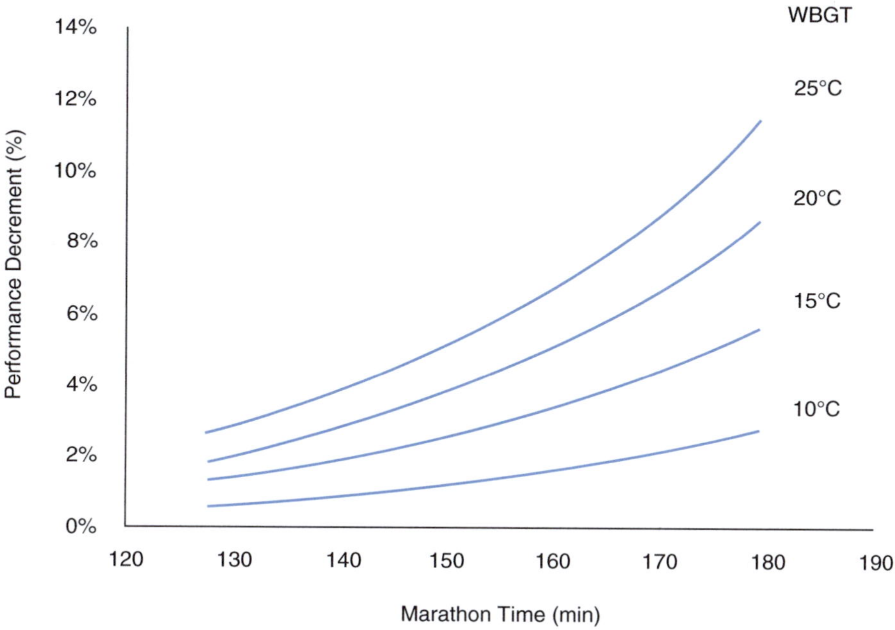

Fig. 15.2 Nomogram representing the impact of WBGT on the relation between marathon finishing times according to quartiles and relative performance decrement in comparison with WBGT at 5 °C [21]

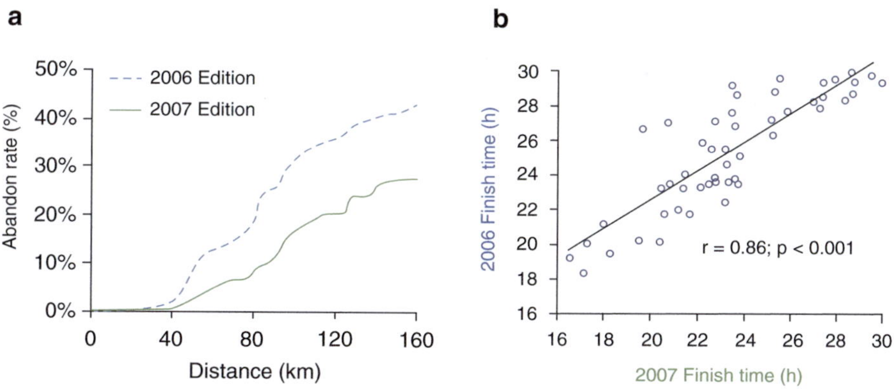

Fig. 15.3 Proportion of runners abandoning the Western States Endurance Run (161 km; 6000 m D+) at each checkpoint along the course (**a**), and relationship between finishing times for the 2006 (7–38 °C) and 2007 (2–30 °C) editions (**b**) [27]

2006 and 2007 [27]. In these two editions (Fig. 15.3), the authors followed 50 runners who had finished the race in both years to investigate the effect of temperature changes from 7–38 °C in 2006 to 2–30 °C in 2007. Independently of their level, the athletes were 8% slower in the hot condition (2006) than in the cooler environment

(2007), and the withdrawal rate was 14% higher in 2006. In this study, slower runners did not appear to be more impacted by the hot environment, but this observation was probably due to the longer time running at a comfortable temperature (i.e., night).

15.1.2.2 Hydration Requirements During Ultra-Endurance Exercise in the Heat

During muscle contraction, only ~25% of the substrate energy is converted to mechanical work, with the other ~75% released as heat. Effective thermoregulation during exercise depends on the balance between the absolute mechanical work generated by the athlete and the heat loss process (Chap. 2). Ultramarathons, characterized by the maintenance of a low relative exercise intensity over very long periods, do not significantly challenge the heat balance in temperate environments. However, maintaining mechanical load over a longer time may challenge heat balance in athletes, especially when competing in hot and/or humid environments. Furthermore, environmental heat stress (i.e., temperature) reduces dry heat loss, due to the small (or even negative) temperature gradient between the skin and air, which progressively places sweat evaporation as the only mechanism available for the organism to dissipate endogenous heat [28, 29]. Indeed, at in 30 °C conditions a dry heat loss of 75 W is observed, whereas at 40 °C, it induces a dry heat gain of 75 W, increasing the sweating rate required to maintain core body temperature [30]. Thus, ultramarathons performed in a hot environment require bigger volumes of water intake to compensate for increased water losses. During the Badwater Ultramarathon of 2012 (217 km and 4000 m D+), the temperature oscillated from 10.1 °C at night to 46.6 °C during the day. The follow-up of four runners who completed the race in 36 ± 3 h showed a mean water intake of 34 ± 13 L [31]. On average, each athlete drank 0.93 L h^{-1}, whereas on completion of the Biel Ultramarathon in Switzerland (100 km and 645 m D+), in a temperate environment varying from 8 to 28 °C, the runners consumed 0.65 L h^{-1} [32]. When the distances of these two races were linked to elevation changes (100 m D+ corresponds to 1 km of distance), the corrected distance for the Biel Ultramarathon was 106.45 km and 257 km for Badwater. According to these corrected distances, the Badwater runners drank almost twice as much water as the Biel runners (0.13 L km^{-1} and 0.069 L km^{-1}, respectively), despite the higher mechanical workload developed during the Biel Ultramarathon due to its shorter distance and lower elevation change (6.6 min km^{-1} vs. 8.5 min km^{-1} for Badwater). The characteristics of the 2007 Peninsula Ultra Fun Run (PUFfeR) of South Africa (80 km and 1000 m D+) were close to those of the Biel Ultramarathon, but the participants were subjected to lower environmental temperatures (8–20 °C). The runners drank less during this more temperate race than during the Biel Ultramarathon (0.028 L km^{-1} and 0.069 L km^{-1}, respectively) [33]. Analysis of the hydration status of 16 ultramarathon runners (161 km, 7000 m D+) in a hot environment (4.8–37.8 °C) showed that finishers ($n = 6$) had drunk significantly more water at the 48th kilometer than non-finishers ($n = 10$) [34]. However, the authors were unable to determine whether beverage intake was directly linked to finishing capacity or if it was due to other factors, such as the experience or endurance capacity of the runners. A second study performed during a 160-km foot race (positive elevation not known) in temperatures that peaked at 38 °C followed only those runners who had completed at least one previous ultramarathon among the top 50% [35]. The authors showed that, despite the homogenous level and experience of these runners, non-finishers also drank significantly less than finishers (−35%, $p < 0.01$). All these data confirmed the relationship between event temperature and water intake that is crucial to health and performance. A study during the Gwada Run of 2011, a 6-day multi-stage race for a total of 142 km under tropical conditions (30 ± 2.4 °C and 82 ± 4% RH) showed a sweat loss of 0.19 L km^{-1} [36]. Yet water intake (1.5 ± 0.3 L per stage) was probably distorted by the shortness of the stages (from 16 to 21 km), which enabled athletes to tolerate transient dehydration during the races (−4.2 ± 0.9 L per stage).

15.2 Countermeasures to Optimize Performance and Health

15.2.1 Hydration

Limiting dehydration during triathlons and ultramarathons in hot environments seems to be essential to maintain the exercise workload over many hours. The role of hydration in a tropical climate (see Chap. 6) is critical since it has been shown that hyperthermia and dehydration are worse than hyperthermia alone [37, 38]. Despite contradictory observations on influencing thermoregulation and core temperature [39, 40], it seems that hyperhydration can delay the development of dehydration [39, 41, 42]. However, hyperhydration does not seem to lead to better performances during a laboratory-based 46 km cycling time trial by elite cyclists [40] or a 60 min run in endurance trained runners [43]. As hyperhydration results in non-negligible added body weight, it may not be an advantage for running performance, which is greatly affected by extra weight [44]. In addition, it does not enhance thermoregulation in a hot and humid climate and augments the risk of hyponatremia [45].

In the triathlon, although hydration depends very much on the availability of adequate means set up by the race organization, the descriptive papers to date report that no triathlete has suffered from dehydration symptoms in hot and humid events, despite occasional significant water loses [7, 9, 23, 46]. This suggests that the hydration strategies employed by the athletes meet the body's water requirements [47].

Maintaining optimal hydration is more complicated than it would seem in the ultramarathon. Indeed, analysis of the hydration status of ultramarathon runners in two different races showed that finishers had drunk significantly more water than non-finishers, even if the runners' experience was taken into account [34, 35]. In contrast, slower runners became over-hydrated because they feared dehydration, which may lead to hyponatremia [48, 49]. Hyponatremia, however, is not only due to over-hydration, but may be the consequence of insufficient sodium intake during the race [34, 35]. In a hot environment, hyponatremia affected 30–50% of ultramarathon finishers, reflecting the inadequate or incorrect hydration strategies of many runners [35, 50, 51]. Thus, better sodium intake during ultramarathons could contribute to better performances by stimulating thirst, increasing voluntary fluid intake, enhancing intestinal glucose and water absorption, optimizing extracellular and intracellular fluid balance, and potentially mitigating the occurrence of clinically significant episodes of hyponatremia [52–55].

15.2.2 Hydration Policy and Implementation

In a triathlon field study, oral salt supplementation improved half-Ironman performance through faster cycling ($p < 0.05$) and showed a similar trend in the running leg ($p = 0.06$), with reduced sweat rate and limited electrolyte deficit [56]. The consensus recommendation is thus to ingest 0.5–0.7 g L^{-1} h^{-1} for long endurance races [57] and up to 1.5 g L^{-1} h^{-1} for athletes prone to develop muscle cramping [58]. However, adding salt to the consumed water is not sufficient to avoid hyponatremia if athletes overdrink during a race (73% of severe symptomatic hyponatremia found after an Ironman [54]).

Hoffman and Stuempfle [59] observed no advantage to sodium-enriched beverages during a 161-km ultramarathon performed by all levels of athletes in heat (38 °C). This result may have been due to a variation in sodium intake from solid food and/or to the tolerance of faster runners to hypohydration and hyperthermia [36, 60].

15.2.3 Cooling

Cooling strategies such as cold drink ingestion or cold-water immersion to reduce the thermal load may enhance performance in the heat (see Chap. 7). This effect has been manifested in cyclists (Fig. 15.4; [61–63]) and runners as a longer time to exhaustion [24, 64, 65]. Wearing cooling garments prior to or during exercise has proven to be performance-effective in hot and humid climates [66–68], but they are difficult, impractical, and potentially not allowed to be used during official long-distance events.

Other studies have focused on cooling strategies, but the exercise duration has been relatively short compared with ultra-endurance events. Nonetheless, they show interesting results that may be extrapolated for long to ultra-long events. Spraying or pouring water over the face and/or body can improve performance in tropical conditions. For example, pouring cold water over the skin will reduce skin temperature before dripping off the body, and transiently improve thermal comfort. Cooling the head in this manner resulted in a 51% increase in cycling time to exhaustion at 75% VO_{2max} [69], with similar effects recently observed in running [70]. Neck cooling during a 90-min running trial in a hot environment (30.4 °C and 53% RH) increased the distance covered by 7.4% with no change in rectal temperature [71]. A similar effect was found with menthol ingestion, as it activates cold dermal sensors [72]. Mixed into a cold beverage, menthol did not

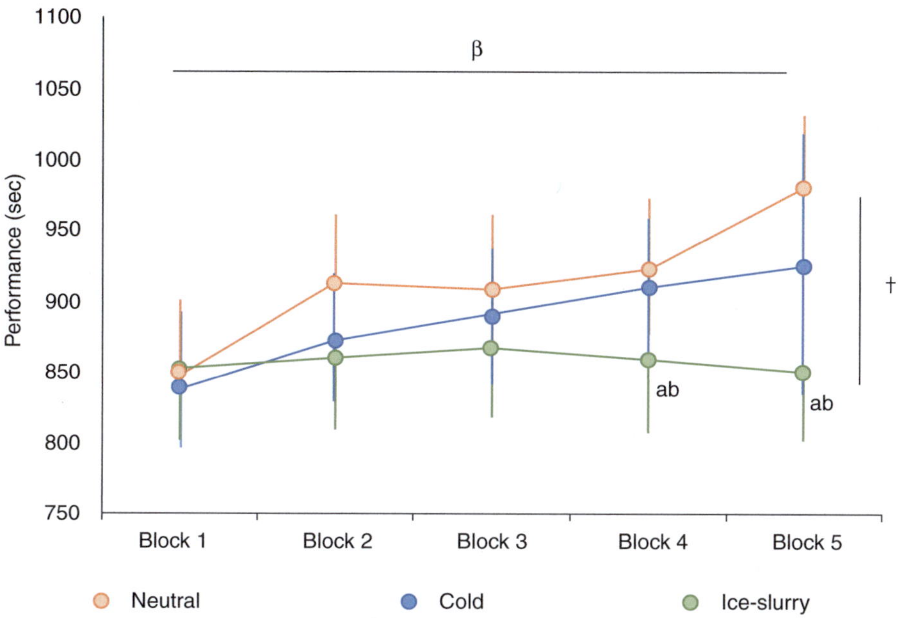

Fig. 15.4 Performance trial times for 5 successive blocks (4 km cycling + 1.5 km running) with the ingestion of neutral water (orange), cold water (blue), and ice slurry (green) in a natural outdoor setting (28 °C and 32% RH and 26 km/h wind speed). [a] Significantly different from neutral water cycling and running performance using ice slurry/menthol ($P < 0.05$). [b] Significantly different from cold ($P < 0.05$). Mean values and SD are shown. β,† denote that block performance was affected by time period ($P < 0.007$) and the time period × drink temperature interaction ($P < 0.004$), respectively [61]

lower core body temperature but had a positive effect on thermal sensation and running/cycling performances over various distances in tropical climate [61, 73]. Indeed, Stevens et al. [73] showed that a menthol mouth rinse every kilometer (25 mL at a concentration of 0.01%) during a running time trial in the heat significantly improved 5-km performance time by 3%. A cumulative effect of menthol and ice slurry or cold water was observed on performances during a 20-km cycling trial in hot environment (30.7 ± 0.8 °C and 78 ± 0.03% RH) [61]. Ultimately, menthol mouth rinsing and ingesting influence thermal perception and thermal comfort, which in turn might contribute to enhance performance in hot climates [73, 74].

15.2.4 Cooling Policy and Implementation

According to the literature, pre-cooling by ice or cold water has been successfully studied during short duration exercise (<60 min) that is too brief to be a key factor of performance during ultra-endurance trials [64, 70]. Conversely, per-cooling by cold water or ice slurry ingestion during exercise seems to be a very interesting intervention to improve performances in a hot environment and possibly increase fluid intake, particularly for well-trained athletes (see Chap. 7). Indeed, most high-performing runners show the biggest increase in internal temperature, probably due to higher thermogenesis consecutive to higher workloads [36, 60]. Faster trail runners (27 km; [60]) and multi-stages trail runners (127 km on 6 days; [36]) also present greater post-race dehydration, which could be reduced by cold water due to its effect on voluntarily increasing water absorption during exercise. The higher temperatures and greater dehydration observed in faster trail runners also seem to indicate that performance in heat is associated with a better tolerance to hyperthermia in elite athletes [75].

15.2.5 Heat Acclimation

Heat acclimation (HA) before a race is an efficient way to increase performance in a hot environment [76, 77] (see Chap. 8 for further details). The physiological adaptations of HA lead to better cardiovascular function (output, stroke volume, and heart rate) and a decreased core temperature at rest and during exercise, in part due to an enhanced sweat rate and expanded plasma volume [76]. HA is also known to reduce sodium loss by sweating [78]. However, HA does not fully restore long to ultra-long endurance performance to the level reached by athletes in temperate conditions [79].

If early arrival at the race location is not possible, HA in an environmental chamber is a strategy that may help preserve ultramarathon performance. Costa et al. [80] examined the effects of six 2-h sessions of running at 60% of VO_{2max} on a motorized treadmill in an environmental heat chamber at 30 °C or 35 °C. From the third session onward, mean heart rate was lower at 30 °C, whereas mean heart rate and thermal comfort were lower at 35 °C. The authors concluded that two bouts of running at 60% of VO_{2max} in a 30 °C air temperature conditions were sufficient to induce heat acclimation in ultra-runners, which may enhance their performance in such environments.

15.3 Conclusions

The maintenance of triathlon or ultramarathon performance in a hot environment is a considerable challenge which requires meticulous preparation and management during the race. To limit the performance decrements induced by these harsh environmental conditions, several countermeasures have been proposed, such as cold–water or ice slurry ingestion, external pre- and per- cooling, and/or menthol use. However, the limited duration of exercise performed in these studies does not allow for direct application of the results to an ultra-endurance or ultra-triathlon context. Thus, further studies are required to validate these approaches in long to triathlon and ultra-endurance events.

Currently, one of the most relevant strategies to employ when performing in the heat is to consume adequate fluids. Maintaining water availability along different race courses is thus crucial due to the detrimental effects dehydration has on fluid balance and consequently heat loss via sweating. Individual sweat rates are highly dependent on the athlete's morphology, the intensity of exercise, and the environment temperature and humidity. Sweat losses are accompanied by electrolyte losses, particularly sodium, which may contribute to hyponatremia if not compensated. To limit this risk, exogenous intake by enriched sodium beverages is the most effective strategy. In addition to the traditional compounds found in most sport drinks (e.g., glucose, magnesium), it seems that ultramarathon beverages should contain about 0.7–1.2 g L^{-1} of salt when conducted in the heat. However, sodium-enriched beverages are insufficient if athletes overdrink by fear of dehydration during the race, placing management of water intake as a key determinant of performance.

Heat acclimation is also a relevant strategy that may enhance ultra-endurance and triathlon performance in hot environments. Most of the acclimation benefits occur in the first 14 days of exposure, but a short-term protocol of 5 days will induce significant early adaptations.

References

1. Giambelluca T et al. Evapotranspiration of Hawai'i: final report to the US Army Corps of Engineers—Honolulu District and the Commission on Water Resource Management, State of Hawai'i. 2014: 178 p.
2. Stiefel M, Knechtle B, Rüst CA, Rosemann T. Analysis of performances at the 'Ironman Hawaii triathlon' and its qualifier events with respect to nationality. J Sci Cycl. 2013;2:27.
3. Ironman (2018) Rules and regulations. http://m.eu.ironman.com/triathlon/events/emea/ironman/tallinn/athletes/rules-and-regulations.aspx.
4. Britt R (2011) North American Ironman DNF rates: finishers and DNF by race. http://www.runtri.com/2010/09/north-american-ironman-dnf-rates.html.
5. Trappe T, Pease D, Trappe S, Troup J, Burke E. Physiological responses to swimming while wearing a wet suit. Int J Sports Med. 1996;17:111–4.
6. Kerr CG, Trappe TA, Starling RD, Trappe SW. Hyperthermia during Olympic triathlon: influence of body heat storage during the swimming stage. Med Sci Sports Exerc. 1998;30:99–104.
7. Laursen PB, et al. Core temperature and hydration status during an Ironman triathlon. Br J Sports Med. 2006;40:320–5; discussion 325
8. Laursen PB, Watson G, Abbiss CR, Wall BA, Nosaka K. Hyperthermic fatigue precedes a rapid reduction in serum sodium in an ironman triathlete: a case report. Int J Sports Physiol Perform. 2009;4:533–7.
9. Baillot M, Hue O. Hydration and thermoregulation during a half-ironman performed in tropical climate. J Sports Sci Med. 2015;14:263–8.
10. Harris KM, Henry JT, Rohman E, Haas TS, Maron BJ. Sudden death during the triathlon. JAMA. 2010;303:1255–7.
11. Nybo L. Cycling in the heat: performance perspectives and cerebral challenges. Scand J Med Sci Sports. 2010;20(Suppl 3):71–9.
12. Saat M, Sirisinghe RG, Singh R, Tochihara Y. Effects of short-term exercise in the heat on thermoregulation, blood parameters, sweat secretion and sweat composition of tropic-dwelling subjects. J Physiol Anthropol Appl Hum Sci. 2005a;24:541–9.

13. Saat M, Tochihara Y, Hashiguchi N, Sirisinghe RG, Fujita M, Chou CM. Effects of exercise in the heat on thermoregulation of Japanese and Malaysian males. J Physiol Anthropol Appl Hum Sci. 2005b;24:267–75.
14. Voltaire B, Berthouze-Aranda S, Hue O. Influence of a hot/wet environment on exercise performance in natives to tropical climate. J Sports Med Phys Fitness. 2003;43:306–11.
15. Galloway SD, Maughan RJ. Effects of ambient temperature on the capacity to perform prolonged cycle exercise in man. Med Sci Sports Exerc. 1997;29:1240–9.
16. Maughan RJ, Otani H, Watson P. Influence of relative humidity on prolonged exercise capacity in a warm environment. Eur J Appl Physiol. 2012;112:2313–21.
17. Hue O, Voltaire B, Hertogh C, Blonc S. Heart rate, thermoregulatory and humoral responses during a 9-day cycle race in a hot and humid climate. Int J Sports Med. 2006;27:690–6.
18. Wu SS, Peiffer JJ, Brisswalter J, Nosaka K, Lau WY, Abbiss CR. Pacing strategies during the swim, cycle and run disciplines of sprint, Olympic and half-Ironman triathlons. Eur J Appl Physiol. 2015;115:1147–54.
19. Nielsen B. Olympics in Atlanta: a fight against physics. Med Sci Sports Exerc. 1996;28:665–8.
20. Kenefick RW, Cheuvront SN, Sawka MN. Thermoregulatory function during the marathon. Sports Med. 2007;37:312–5.
21. Ely MR, Cheuvront SN, Roberts WO, Montain SJ. Impact of weather on marathon-running performance. Med Sci Sports Exerc. 2007;39:487–93.
22. Marino FE, Mbambo Z, Kortekaas E, Wilson G, Lambert MI, Noakes TD, Dennis SC. Advantages of smaller body mass during distance running in warm, humid environments. Pflugers Arch. 2000;441:359–67.
23. Del Coso J, et al. Relationship between physiological parameters and performance during a half-ironman triathlon in the heat. J Sports Sci. 2014;32:1680–7.
24. Siegel R, Mate J, Brearley MB, Watson G, Nosaka K, Laursen PB. Ice slurry ingestion increases core temperature capacity and running time in the heat. Med Sci Sports Exerc. 2010;42:717–25.
25. Zaryski C, Smith DJ. Training principles and issues for ultra-endurance athletes. Curr Sports Med Rep. 2005;4:165–70.
26. Maughan RJ. Distance running in hot environments: a thermal challenge to the elite runner. Scand J Med Sci Sports. 2010;20(Suppl 3):95–102.
27. Parise CA, Hoffman MD. Influence of temperature and performance level on pacing a 161 km trail ultramarathon. Int J Sports Physiol Perform. 2011;6:243–51.
28. Brotherhood JR. Heat stress and strain in exercise and sport. J Sci Med Sport. 2008;11:6–19.
29. Taylor NA. Challenges to temperature regulation when working in hot environments. Ind Health. 2006;44:331–44.
30. Gagnon D, Jay O, Kenny GP. The evaporative requirement for heat balance determines whole-body sweat rate during exercise under conditions permitting full evaporation. J Physiol. 2013;591:2925–35.
31. Brown JS, Connolly D. Food and fluid intake during extreme heat: experiences from the Badwater Ultramarathon. Wilderness Environ Med. 2015;26:e4.
32. Knechtle B, Knechtle P, Rosemann T. Low prevalence of exercise-associated hyponatremia in male 100 km ultra-marathon runners in Switzerland. Eur J Appl Physiol. 2011;111:1007–16.
33. Tam N, Hew-Butler T, Papadopoulou E, Nolte H, Noakes TD. Fluid intake and changes in blood biochemistry, running speed and body mass during an 80 km mountain trail race. Med Sport. 2009;13:108–15.
34. Stuempfle KJ, Hoffman MD, Weschler LB, Rogers IR, Hew-Butler T. Race diet of finishers and non-finishers in a 100 mile (161 km) mountain footrace. J Am Coll Nutr. 2011;30:529–35.
35. Glace BW, Murphy CA, McHugh MP. Food intake and electrolyte status of ultramarathoners competing in extreme heat. J Am Coll Nutr. 2002;21:553–9.
36. Hue O, Henri S, Baillot M, Sinnapah S, Uzel AP. Thermoregulation, hydration and performance over 6 days of trail running in the tropics. Int J Sports Med. 2014;35:906–11.
37. Sawka MN, Montain SJ, Latzka WA. Hydration effects on thermoregulation and performance in the heat. Comp Biochem Physiol A Mol Integr Physiol. 2001;128:679–90.

38. Sawka MN, Noakes TD. Does dehydration impair exercise performance? Med Sci Sports Exerc. 2007;39:1209.
39. Latzka WA, Sawka MN, Montain SJ, Skrinar GS, Fielding RA, Matott RP, Pandolf KB. Hyperhydration: thermoregulatory effects during compensable exercise-heat stress. J Appl Physiol (1985). 1997;83:860–6.
40. Ross ML, Jeacocke NA, Laursen PB, Martin DT, Abbiss CR, Burke LM. Effects of lowering body temperature via hyperhydration, with and without glycerol ingestion and practical precooling on cycling time trial performance in hot and humid conditions. J Int Soc Sports Nutr. 2012;9:55.
41. Latzka WA, Sawka MN. Hyperhydration and glycerol: thermoregulatory effects during exercise in hot climates. Can J Appl Physiol. 2000;25:536–45.
42. Sawka MN, Francesconi RP, Young AJ, Pandolf KB. Influence of hydration level and body fluids on exercise performance in the heat. JAMA. 1984;252:1165–9.
43. Scheadler CM. Glycerol hyperhydration and endurance running performance in the heat. Ohio: The Ohio State University; 2009.
44. Teunissen LP, Grabowski A, Kram R. Effects of independently altering body weight and body mass on the metabolic cost of running. J Exp Biol. 2007;210:4418–27.
45. Noakes TD, Goodwin N, Rayner BL, Branken T, Taylor RK. Water intoxication: a possible complication during endurance exercise, 1985. Wilderness Environ Med. 2005;16:221–7.
46. Speedy DB, et al. Fluid balance during and after an ironman triathlon. Clin J Sport Med. 2001;11:44–50.
47. Kenefick RW. Drinking strategies: planned drinking versus drinking to thirst. Sports Med. 2018;48:31–7.
48. Hew-Butler T, et al. Statement of the second international exercise-associated hyponatremia consensus development conference, New Zealand, 2007. Clin J Sport Med. 2008;18:111–21.
49. Noakes TD. Drinking guidelines for exercise: what evidence is there that athletes should drink "as much as tolerable", "to replace the weight lost during exercise" or "ad libitum"? J Sports Sci. 2007;25:781–96.
50. Hoffman MD, Stuempfle KJ, Rogers IR, Weschler LB, Hew-Butler T. Hyponatremia in the 2009 161-km western states endurance run. Int J Sports Physiol Perform. 2012;7:6–10.
51. Lebus DK, Casazza GA, Hoffman MD, Van Loan MD. Can changes in body mass and total body water accurately predict hyponatremia after a 161-km running race? Clin J Sport Med. 2010;20:193–9.
52. Maughan RJ. Fluid and electrolyte loss and replacement in exercise. J Sports Sci. 1991;9(S1):117–42.
53. Shirreffs SM, Sawka MN. Fluid and electrolyte needs for training, competition, and recovery. J Sports Sci. 2011;29(Suppl 1):S39–46.
54. Speedy DB, et al. Hyponatremia in ultradistance triathletes. Med Sci Sports Exerc. 1999;31:809–15.
55. Vrijens DM, Rehrer NJ. Sodium-free fluid ingestion decreases plasma sodium during exercise in the heat. J Appl Physiol (1985). 1999;86:1847–51.
56. Del Coso J, et al. Effects of oral salt supplementation on physical performance during a half-ironman: a randomized controlled trial. Scand J Med Sci Sports. 2016;26:156–64.
57. Von Duvillard SP, Braun WA, Markofski M, Beneke R, Leithauser R. Fluids and hydration in prolonged endurance performance. Nutrition. 2004;20:651–6.
58. Bergeron MF. Heat cramps: fluid and electrolyte challenges during tennis in the heat. J Sci Med Sport. 2003;6:19–27.
59. Hoffman MD, Stuempfle KJ. Hydration strategies, weight change and performance in a 161 km ultramarathon. Res Sports Med. 2014;22:213–25.
60. Baillot M, Le Bris S, Hue O. Fluid replacement strategy during a 27-km trail run in hot and humid conditions. Int J Sports Med. 2014;35:147–52.
61. Tran Trong T, Riera F, Rinaldi K, Briki W, Hue O. Ingestion of a cold temperature/menthol beverage increases outdoor exercise performance in a hot, humid environment. PLoS One. 2015;10:e0123815.

62. Burdon CA, Hoon MW, Johnson NA, Chapman PG, O'Connor HT. The effect of ice slushy ingestion and mouthwash on thermoregulation and endurance performance in the heat. Int J Sport Nutr Exerc Metab. 2013;23:458–69.
63. Riera F, Trong T, Rinaldi K, Hue O. Precooling does not enhance the effect on performance of midcooling with ice-slush/menthol. Int J Sports Med. 2016;37:1025–31.
64. Siegel R, Mate J, Watson G, Nosaka K, Laursen PB. Pre-cooling with ice slurry ingestion leads to similar run times to exhaustion in the heat as cold water immersion. J Sports Sci. 2012;30:155–65.
65. Yeo ZW, Fan PW, Nio AQ, Byrne C, Lee JK. Ice slurry on outdoor running performance in heat. Int J Sports Med. 2012;33:859–66.
66. Arngrimsson SA, Petitt DS, Stueck MG, Jorgensen DK, Cureton KJ. Cooling vest worn during active warm-up improves 5-km run performance in the heat. J Appl Physiol (1985). 2004;96:1867–74.
67. Hasegawa H, Takatori T, Komura T, Yamasaki M. Wearing a cooling jacket during exercise reduces thermal strain and improves endurance exercise performance in a warm environment. J Strength Cond Res. 2005;19:122–8.
68. Luomala MJ, Oksa J, Salmi JA, Linnamo V, Holmér I, Smolander J, Dugue B. Adding a cooling vest during cycling improves performance in warm and humid conditions. J Therm Biol. 2012;37:47–55.
69. Ansley L, Marvin G, Sharma A, Kendall MJ, Jones DA, Bridge MW. The effects of head cooling on endurance and neuroendocrine responses to exercise in warm conditions. Physiol Res. 2008;57:863–72.
70. Stevens CJ, Kittel A, Sculley DV, Callister R, Taylor L, Dascombe BJ. Running performance in the heat is improved by similar magnitude with pre-exercise cold-water immersion and mid-exercise facial water spray. J Sports Sci. 2017;35:798–805.
71. Tyler CJ, Sunderland C. Neck cooling and running performance in the heat: single versus repeated application. Med Sci Sports Exerc. 2011;43:2388–95.
72. Stevens CJ, Best R. Menthol: a fresh ergogenic aid for Athletic performance. Sports Med. 2017;47:1035–42.
73. Stevens CJ, Thoseby B, Sculley DV, Callister R, Taylor L, Dascombe BJ. Running performance and thermal sensation in the heat are improved with menthol mouth rinse but not ice slurry ingestion. Scand J Med Sci Sports. 2016;26:1209–16.
74. Stevens CJ, Bennett KJ, Sculley DV, Callister R, Taylor L, Dascombe BJ. A comparison of mixed-method cooling interventions on preloaded running performance in the heat. J Strength Cond Res. 2017;31:620–9.
75. McKeag D, Moeller JL. ACSM's primary care sports medicine. Philadelphia: Wolters Kluwer/Lippincott Williams & Wilkins; 2007.
76. Armstrong LE, Maresh CM. The induction and decay of heat acclimatisation in trained athletes. Sports Med. 1991;12:302–12.
77. Nielsen B. Heat stress and acclimation. Ergonomics. 1994;37:49–58.
78. Kirby CR, Convertino VA. Plasma aldosterone and sweat sodium concentrations after exercise and heat acclimation. J Appl Physiol (1985). 1986;61:967–70.
79. Voltaire B, et al. Effect of fourteen days of acclimatization on athletic performance in tropical climate. Can J Appl Physiol. 2002;27:551–62.
80. Costa RJ, Crockford MJ, Moore JP, Walsh NP. Heat acclimation responses of an ultra-endurance running group preparing for hot desert-based competition. Eur J Sport Sci. 2014;14(Suppl 1):S131–41.

MIX
Papier aus verantwortungsvollen Quellen
Paper from responsible sources
FSC® C105338

If you have any concerns about our products,
you can contact us on
ProductSafety@springernature.com

In case Publisher is established outside the EU,
the EU authorized representative is:
Springer Nature Customer Service Center GmbH
Europaplatz 3, 69115 Heidelberg, Germany

Printed by Libri Plureos GmbH
in Hamburg, Germany